Springer Tracts in Mechanical Engineering

Springer Tracts in Mechanical Engineering (STME) publishes the latest developments in Mechanical Engineering - quickly, informally and with high quality. The intent is to cover all the main branches of mechanical engineering, both theoretical and applied, including:

- Engineering Design
- Machinery and Machine Elements
- Mechanical Structures and Stress Analysis
- Automotive Engineering
- Engine Technology
- Aerospace Technology and Astronautics
- Nanotechnology and Microengineering
- Control, Robotics, Mechatronics
- MEMS
- Theoretical and Applied Mechanics
- Dynamical Systems, Control
- Fluids Mechanics
- Engineering Thermodynamics, Heat and Mass Transfer
- Manufacturing
- Precision Engineering, Instrumentation, Measurement
- Materials Engineering
- Tribology and Surface Technology

Within the scope of the series are monographs, professional books or graduate textbooks, edited volumes as well as outstanding PhD theses and books purposely devoted to support education in mechanical engineering at graduate and post-graduate levels.

Indexed by SCOPUS, zbMATH, SCImago.

Please check our Lecture Notes in Mechanical Engineering at https://link.springer.com/bookseries/11236 if you are interested in conference proceedings.

To submit a proposal or for further inquiries, please contact the Springer Editor **in your region** :

Ms. Ella Zhang (China)
Email: ella.zhang@springernature.com
Swati Meherishi (India)
Email: swati.meherishi@springer.com
Dr. Leontina Di Cecco (All other countries)
Email: leontina.dicecco@springer.com

All books published in the series are submitted for consideration in Web of Science.

Vitalii Ivanov
Editor

Emerging Technologies in Manufacturing and Materials Engineering

Springer

Editor
Vitalii Ivanov
Sumy State University
Sumy, Ukraine

ISSN 2195-9862 ISSN 2195-9870 (electronic)
Springer Tracts in Mechanical Engineering
ISBN 978-3-032-14913-8 ISBN 978-3-032-14914-5 (eBook)
https://doi.org/10.1007/978-3-032-14914-5

This Springer imprint is published by the registered company Springer Nature Switzerland AG
The registered company address is: Gewerbestrasse 11, 6330 Cham, Switzerland

Preface

This book provides a comprehensive overview of emerging technologies in manufacturing and materials engineering, highlighting fundamental principles and practical solutions to contemporary industrial challenges. It integrates studies on process design, quality management, materials innovation, and computational methods.

This book is organized into four thematic parts: Part I, "Process Design and Technological Methodologies," presents innovative approaches to machining and restoration, focusing on process planning, machinability, and function-oriented technological processes. Part II, "Quality Assurance and Measurement Systems," addresses monitoring methods, evaluating, and improving quality, covering agile management practices, system diagnostics, and advanced modeling techniques to ensure consistent performance in complex industrial environments. Part III, "Advanced Materials and Surface Engineering," explores surface modification, residual stress evaluation, and alloy treatments that enhance product durability, performance, and sustainability. These contributions demonstrate how material innovations directly support industrial competitiveness and product reliability. Part IV, "Computational Approaches in Manufacturing," highlights comprehensive methods, ranging from cutting dynamics and system stability to predictive modeling and stochastic approaches within Industry 4.0 frameworks.

By combining insights of 17 invited contributions written by 30 authors from 14 leading research institutions, higher education organizations, and companies across 5 countries (Germany, Latvia, the Slovak Republic, Turkey, and Ukraine), this book addresses the industrial challenges by integrating fundamental research with applied solutions to meet the growing demands for efficiency, customization, and digitalization.

A rigorous peer-review process was implemented to ensure high scholarly quality and coherence across 17 invited chapters. The review followed a single-blind format and consisted of two rounds. In the first round, External Review, each chapter was reviewed by independent external reviewers (the list of reviewers is presented below) who are experts in the relevant subject area. A total of 17 reviewers representing 13 institutions and organizations from 8 countries were involved. Avoiding conflicts of interest was considered: reviewers were selected from the same professional backgrounds and different countries than the authors, ensuring impartial evaluation. The reviewers assessed the chapters based on originality, scientific rigor, clarity, coherence, and relevance. Reviewers provided constructive feedback and recommendations for improvements, which the authors incorporated into their revised manuscripts. In the second round, Editorial Review, after revisions, the editor conducted a thorough evaluation to ensure consistency and integration across all chapters. This stage focused on: content integration, logical flow, and quality control, ensuring that each chapter contributes effectively and aligns with the overall goals of the monograph. Moreover, the editor provided harmonization of terminology, style, and formatting across chapters and offered additional suggestions for clarity, depth, and comprehensiveness where needed. Through this two-round peer-review process, this book combines the rigor of expert evaluation with the editorial oversight necessary to create a comprehensible and high-quality scholarly monograph.

This book is aimed at researchers and academics in manufacturing and materials engineering, as well as engineers and practitioners seeking innovative approaches. It will also serve as a valuable reference for graduate students and educators, offering theoretical fundamentals and practical perspectives on the technologies shaping the future of production.

Sumy, Ukraine
September 2025

Vitalii Ivanov

List of Invited External Reviewers (in Alphabetical Order)

Dmytro Adamenko, WAGO GmbH & Co. KG, Germany
Katarzyna Antosz, Rzeszow University of Technology, Poland
Liliana-Laura Badita-Voicu, National Institute of Research and Development in Mechatronics and Measurement Technique, Romania
Yevheniia Basova, National Technical University "Kharkiv Polytechnic Institute," Ukraine
Andre Batako, Liverpool John Moores University, United Kingdom
Emilia Campean, Technical University of Cluj-Napoca, Romania
Olaf Ciszak, Poznan University of Technology, Poland
Predrag Dašić, Engineering Academy of Serbia, Serbia
Ivan Dehtiarov, Sumy State University, Ukraine
Jakub Grabski, Poznan University of Technology, Poland
Dmytro Lesyk, University of the Basque Country, Spain
Mykola Melnychuk, Lutsk National Technical University, Ukraine
Grigore Marian Pop, Technical University of Cluj-Napoca, Romania
Oleksandr Povstianoi, Lutsk National Technical University, Ukraine
Anatolii Tkachuk, Lutsk National Technical University, Ukraine
Jerzy Winczek, Czestochowa University of Technology, Poland
Jinyang Xu, Shanghai Jiao Tong University, China

Acknowledgment

The preparation and publication of this book were supported by the International Association for Technological Development and Innovations and the International Innovation Foundation. Additional support was provided by the EU NextGenerationEU through the Recovery and Resilience Plan for Slovakia under the project no. 09I03-03-V01-00094.

Contents

Process Design and Technological Methodologies

Methodology of Function-Oriented Technological Process of Machining Products

Vadym Stupnytskyy

Abstract It is essential to consider and guarantee effective operational conditions for the most stressed components of mechanical engineering products at both the design and production stages as part of a Product Lifecycle Management (PLM) system. The conventional methodology for establishing the technological phase of an integrated complex automated system for technological preparation of automated production, founded upon a sequential non-iterative synthesis of operations, necessitates structural and methodological enhancement. The planning of functionally oriented engineering technologies is a distinctive aspect of the novel methodology for technological production preparation, predicated upon the tenets of parallelism and iteration of interrelated functional processes, and incorporates the PLM concept. A distinctive aspect of the functionally oriented technological process is the phase of simulation and analysis of the stress-strain state of the product during the machining process, which is integrated into the Computer Aided Forming system. This book's section presents an example of optimizing the structure of a functionally oriented machining operation of an injection mold ejector pin. It has been demonstrated that modifying the structure of the technological process, from grinding to turning with a tungsten carbide-insert tool, influences the predicted wear resistance of the machined functional surface of the ejector. The relative indicator of the operational wear resistance of the machined surface, which is characterized by the utilization of diverse structural variants and processing parameters for this surface, is contingent upon the microtopographic state of the surface layer and the presence of cutting-induced residual stresses. Subsequent research endeavors will enhance the precision of simulation modeling outcomes and facilitate information exchange among designated subsystems to operationalize the Smart Manufacturing concept.

Keywords Function-oriented technological process · Cutting-induced residual stress · Product lifecycle management

V. Stupnytskyy (✉)
Lviv Polytechnic National University, Lviv, Ukraine
e-mail: vadym.v.stupnytskyi@lpnu.ua

V. Ivanov (ed.), *Emerging Technologies in Manufacturing and Materials Engineering*, Springer Tracts in Mechanical Engineering,
https://doi.org/10.1007/978-3-032-14914-5_1

1 Introduction

In modern integrated systems for the preparation of mechanical engineering production, a generally accepted algorithm of automated technological design is used, which involves implementing successive interrelated stages of non-iterative descending structural and parametric synthesis based on the prototyping algorithm. At the same time, the technologist (or automated technological design system) completely abstracts from the problems of the functional (operational) nature of the production object, rightly believing that the appointment of initial data and boundary conditions for technological design is the prerogative of the designer (or automated CAD (Computer Aided Design)/CAE (Computer Aided Engineering) system) (Stupnytskyy et al. 2020). The technological process formed according to this methodology in the classification context of the goal can be considered object-oriented.

However, ensuring the basic operational properties of the product (wear resistance, fatigue strength, lubrication, and corrosion resistance of functional surfaces) is primarily determined not only by the accuracy and quality standards of individual product surfaces set by the designer but also by such essential factors as surface microtopology, residual stresses, and deformations of the surface layer. However, the formation of these product characteristics depends on the structure and parameters of the manufacturing process and cannot be a priori normalized by the designer. The systemic principle of implementing the product life cycle requires a direct connection between the technological and operational stages, not only their indirect connection through the provision of conditions and restrictions formulated at the design stage. This is the methodology of parallel engineering in a PLM system (Product Life Management) (Davim 2010). Therefore, copying the generally accepted algorithm of technological design significantly limits the potential of an integrated automated design and technological system to prepare machine-building production.

An essential research stage is harmonizing simulation modeling results with the experimental studies. It is advisable to use the following research algorithm: formulation of initial data and creation of a model of the machining process; conducting simulation studies; conducting selective experiments and comparing theoretical and actual test results; correction of the studied simulation model by selecting fracture criteria, calculation methods, dependencies of thermodynamic influence on the mechanical properties of the processed material, etc.

2 Fundamentals and Challenges

Therefore, the integrated system of automated preparation of machine-building production utilizes two alternative concepts of technological design: object-oriented and function-oriented. The primary criterion for the formation of the optimal structure and parameters of the object-oriented technological process is to ensure the minimum cost of manufacturing an object of production (product) while observing formalized quality indicators (accuracy, roughness, physical and mechanical properties of specific surfaces) and the production volumes specified by the production schedule, taking into account the technological continuity of the existing production (Eder 2016).

The goal function in making decisions about the optimal structure and parameters of a functionally oriented technological process is an integral qualimetric indicator that systematically characterizes the wear resistance, fatigue strength, corrosion resistance, etc., of the most loaded surfaces of the product (Rangan et al. 2005). This indicator is obtained due to the formation of the surface layer microtopology, residual stresses, and deformations during the forming process. To facilitate rapid and comprehensive analysis of these indicators, state-of-the-art cutting process modeling software systems such as ABAQUS, LS-Dyna, AdvantEdge, and DEFORM are employed. These engineering analysis systems implement the finite element method (FEM) and enable the determination of stress and strain fields in both the chip formation zone and the surface layer of the machined workpiece. This facilitates calculating the tool's and workpiece's stress-strain and thermodynamic state during machining. A common disadvantage of systems employed for simulation modeling of forming processes is that their adequacy to actual processes is mainly contingent on the experience and qualifications of the designer, the correct choice of fracture criteria, constraints, and features of the formation of a finite element mesh, its gradient for determining the acting and residual stresses and strains, etc. A recurrent and iterative connection between the design and technological design stages and the study of the processes of forming the product's functionally important surfaces is lacking. Consequently, existing systems for simulating machining processes can only serve as a basis for creating an automated system for product shaping. A significant challenge in developing such a system is implementing a comprehensive set of operational methodologies for analyzing the structure and parameters of technological operations and transitions. These methodologies must link the outcomes of the stress-strain state of the product surfaces during their processing with the stage of predicting the quality of the product (Demoly et al. 2013).

Implementing function-oriented technologies in the context of PLM methodology in mechanical engineering, predicated on the principle of concurrent design, entails integrating the CAF (Computer Aided Forming) system (Stupnytskyy 2013). This system is predicated on analyzing a simulation rheological model of specific technological steps in conjunction with a set of functional modules and analytical and applied programs to form precision, thermal, deformation, microgeometric, and structural-phase parameters of machined surfaces. The CAF system is notable for its integration of well-established software products, including DEFORM, Abaqus, AdvantEdge, and LS-DYNA, which are widely recognized in rheological modeling for cutting processes. These systems are multipurpose Finite Element Analysis (FEA) programs designed to address two- and three-dimensional dynamic nonlinear problems of plastic deformation and fracture mechanics and related processes. These software products implement explicit and implicit finite element methods, allowing for the construction of various mesh types, including Eulerian and hybrid meshes and multi-component hydrodynamics.

Additionally, they utilize a smoothed lattice method based on the Galerkin method, contributing to their advanced capabilities. The CAF system seamlessly integrates with the existing CAD/CAE/CAPP (Computer Aided Planning Production)/CAM (Computer Aided Manufacturing) systems, offering a comprehensive and predictive simulation modeling capability. It establishes the dependence of the product's primary qualitative indicators on the configuration and parameters of the technological process of mechanical machining of the part (Fig. 1).

The goal function in choosing the optimal structure and parameters of technological operations and transitions of a functionally oriented technological process is an integral qualimetric indicator that systematically characterizes the complex operational properties of the most loaded surfaces of the product (Dmitriev et al. 2009; Thomas 1999; Straffelini 2015; Aramaki et al. 1993; Karpenko and Akay 2001; Kucharski and Starzynski 2014). The most essential characteristics of the functional efficiency of a specific *j*-th surface of the *i*-th workpiece $\{F_i\}$ are evaluated by local criteria: wear resistance (q_{ij1}); tribological quality of mating (q_{ij2}); fatigue strength (q_{ij3}); corrosion resistance (q_{ij4}); lubricant retention capacity (q_{ij5}); optical reflection properties (q_{ij6}); thermal energy absorption properties (q_{ij7}); properties of ensuring the tightness of joints (q_{ij8}), etc.

As illustrated in Fig. 2, the generalized flowchart of the algorithm for structural and parametric optimization of a functionally oriented technological process comprises the following stages:

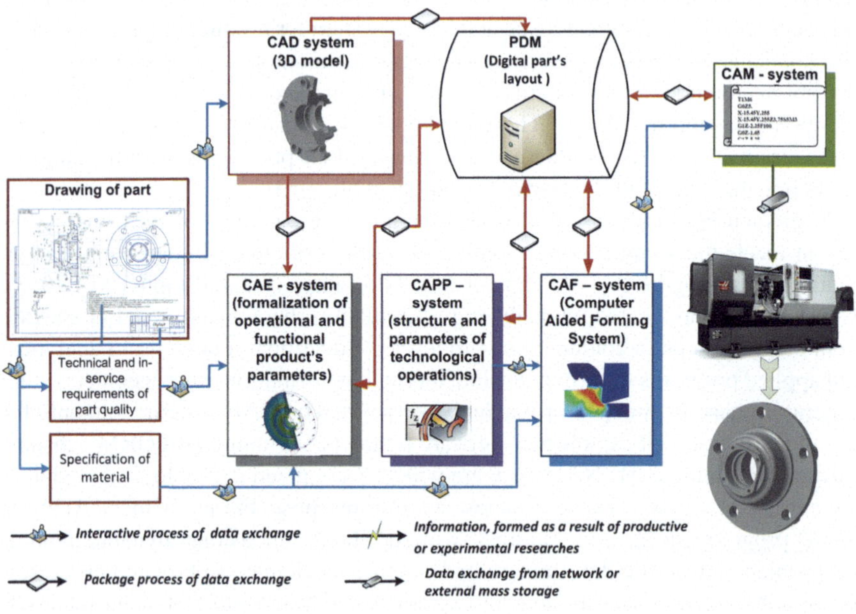

Fig. 1 Flowchart of the modernized integrated system of design and technological preparation of mechanical engineering production

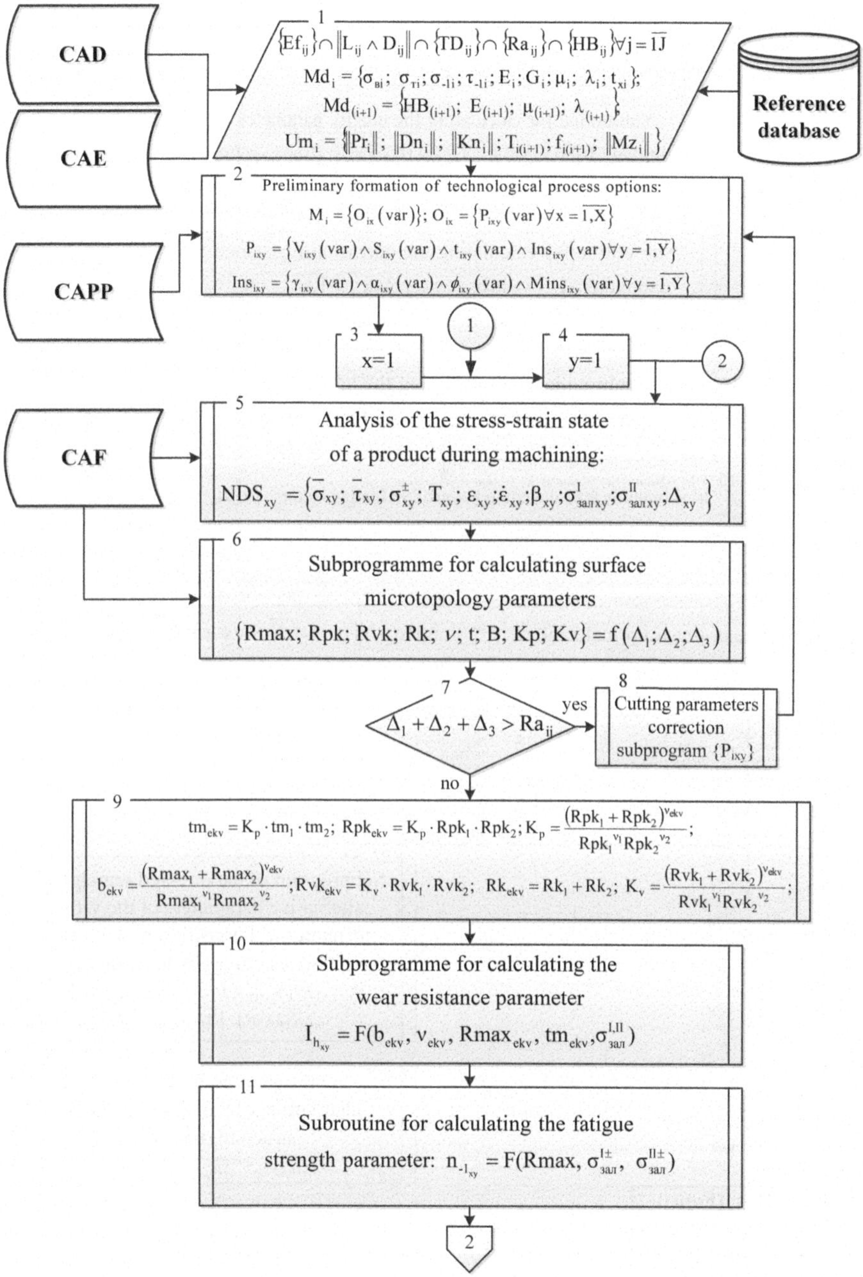

Fig. 2 Flowchart of the algorithm for structural and parametric optimization of a functionally oriented technological process

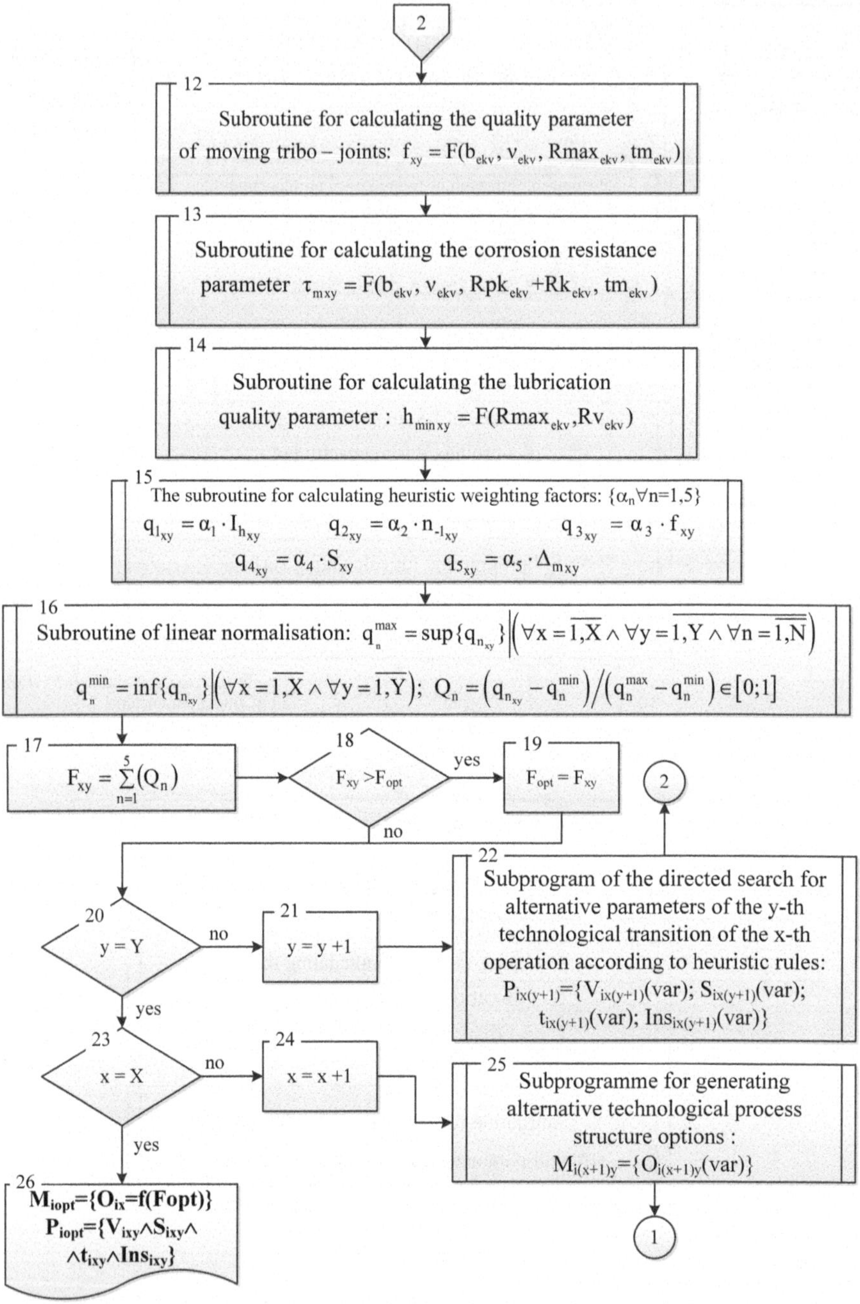

Fig. 2 (continued)

1. The stage of interactive input of information (input data) into the system: product modeling in a CAD system and preliminary engineering analysis in a CAE system. For each *i*-th part, which includes $j = \overline{1J}$ surfaces, the system interactively enters design information $\{Ef_{ij}\} \cap \|L_{ij} \wedge D_{ij}\| \cap \{TD_{ij}\} \cap \{Ra_{ij}\} \cap \{HB_{ij}\}$ about such parameters as the geometric shape $\{Ef_{ij}\}$ of all $\forall j = \overline{1J}$ surfaces of the part, their dimensions $\|L_{ij} \wedge D_{ij}\|$, accuracy, and roughness parameters $\{TD_{ij}\} \cap \{Ra_{ij}\}$, and specified strength parameters $\{HB_{ij}\}$ (Block 1 - Fig. 2). In addition, the physical and mechanical properties of the material of a given part are subject to formalization. It is also important to formally interpret the special conditions Um_i of force loading $\|Pr_i\|$, dynamic $\|Dn_i\|$, kinematic $\|Kn_i\|$, temperature $T_{i(i+1)}$ interaction, and parameters of tribological contact value $f_{i(i+1)}$ with mating parts in the product, as well as kinematic constraints $\|Mz_i\|$ set by the designer in the process of functioning of the part in the product (machine or mechanism).
2. Pre-synthesis of the structure and parameters of the technological process in the CAPP system (block 2). During the implementation of this stage, a basic version of the processing route for the *i*-th part is formed as a set of technological operations of the *x*-th structure: $M_i = \{O_{ix}(var)\}$. Each such operation is characterized by a tuple of parameters (technological steps): $O_{ix} = \left\{P_{ixy}(var)\forall x = \overline{1,X}\right\}$, which are determined by the following indicators: cutting speed $V_{ixy}(var)$, feed rate $S_{ixy}(var)$, depth of cut $t_{ixy}(var)$, and tool parameters $Ins_{ixy}(var)$: $P_{ixy} = \left\{V_{ixy}(var) \wedge S_{ixy}(var) \wedge t_{ixy}(var) \wedge Ins_{ixy}(var)\forall y = \overline{1,Y}\right\}$. Moreover, the formal characterisation of the tool is not limited to geometrical parameters $\gamma_{ixy}(var)$, $\alpha_{ixy}(var)$, $\varphi_{ixy}(var)$, but also includes the choice of tool material and coating type $Mins_{ixy}(var)$: $Ins_{ixy} = \left\{\gamma_{ixy}(var) \wedge \alpha_{ixy}(var) \wedge \varphi_{ixy}(var) \wedge Mins_{ixy}(var)\forall y = \overline{1,Y}\right\}$. The preliminary formation of alternative variants of the structure and parameters of the technological process is based on the prototyping methodology. It includes several formal procedures for recognizing the geometric shape of the production object $\{Ef_{ij}\}$ of all $\forall j = \overline{1J}$ surfaces of the part, analysing their dimensions $\|L_{ij} \wedge D_{ij}\|$, accuracy and roughness parameters $\{TD_{ij}\} \cap \{Ra_{ij}\}$, and specified strength parameters $\{HB_{ij}\}$ (see block 1 of this algorithm). The most relevant analog available in the database of the CAPP system is coordinated and adapted to the input data, taking into account the technological inheritance of the existing production based on logical formalized procedures. At the same time, the machining process is designed considering the technological inheritance of the existing output: its type and organizational form, group of machines, and technological equipment using consistency of methods and their sequence when processing individual surfaces. These initial data are pre-formulated by the CAD system and coordinated by the CAE system following operational requirements: $\{Ef_{ij}\} \cap \|L_{ij} \wedge D_{ij}\| \cap \{TD_{ij}\} \cap \{Ra_{ij}\} \cap \{HB_{ij}\}$.
3. The stage of simulation and analysis of the stress-strain state of the product in the process of its shape formation, implemented in the CAF-system (block 5). For each *x*-th variant of the structure of the technological operation and *y*-th variant of the technological step, the CAF system simulates the pattern of the stress-

strain state NDS_{xy} of the appropriate surface (including residual parameters of stresses of the first and second kind and strain) in the process of its cutting: $NDS_{xy} = \left\{\overline{\sigma}_{xy};\overline{\tau}_{xy};\sigma^{\pm}_{xy};T_{xy};\varepsilon_{xy};\dot{\varepsilon}_{xy};\beta_{xy};\sigma_{resxy};\sigma_{resxy};\Delta_{xy}\right\}$. Based on these measurements, the parameters of the geometric and kinematic components of the roughness, the vibration component, and the deformation component Δ_3 are determined. The solution of this problem formalizes the process of forming micro-irregularities of the profile as a set of parameters of the surface microtopology: $\{Rmax; Rpk; Rvk; Rk; \nu; t; B; Kp; Kv\} = f(\Delta_1; \Delta_2; \Delta_3)$.

4. When modeling the tribological contact, the surface microtopology parameters must be modified to take account of the need to formalize the equivalent contact surface - block 9 (Gao et al. 2016). In this case, the equivalent bearing surface of the contact depends on the tribological load varying from 0.2 to 1.0, with the dominant influence being exerted by the shape of the roughness rather than its height and pitch parameters (Aramaki et al. 1993).
5. The stage of calculating local criteria - the intensity of wear of functional surfaces in simulated conditions of potential operation of the product (blocks 6–10); the margin factor for the fatigue strength of functional surfaces (Karpenko and Akay 2001; Ludema 1996) of the part in the conditions of their future operation (block 11); the indicator of dependence of the dynamic quality of tribo-conjugations (Kucharski and Starzynski 2014; Wei et al. 2009; Hirani 2016) of functionally significant surfaces of the part (block 12); the parameter of dependence of corrosion resistance (fretting corrosion) of the surfaces of the part (Greenwood and Williamson 1966) (block 13); the coefficient of the reserve of liquid tribological contact of the functional surfaces of the part (Nepershin 2015; Hidirov 2022) (block 14) depending on the variable structure and parameters of the functionally oriented technological process of forming these surfaces.
6. The stage of interactive linear normalization of qualimetric criteria Q_n (block 16) and the current calculation of the objective function of the selected structural and parametric variant of the functionally oriented technological process (block 17) based on the calculation of the priorities $\alpha_n \left(\forall n = \overline{1,8}\right)$ of the weighting coefficients of each q_{ijn}-th local criterion of a certain j-th surface ($j = \overline{1,J}$) of the i-th part (block 15). Taking into account the diverse nature of the impact of each q_{ijn} local criteria on the operational properties of the production facility, for each local criterion, the value of the supremum (the operand for selecting the most effective (from a functional point of view) value of the local criterion) and the infimum (the operand for selecting the least effective value of the criterion) from the entire set of its permissible values is determined (block 16):

$$\begin{aligned} q_n^{\max} &= \sup\left\{q_{ijn} \middle| \forall n = \overline{1,N}\right\}; \\ q_n^{\min} &= \inf\left\{q_{ijn} \middle| \forall n = \overline{1,N}\right\} \end{aligned} \tag{1}$$

$$Q_n = \left(q_{n_{xy}} - q_n^{\min}\right) / \left(q_n^{\max} - q_n^{\min}\right) \in [0;1] \tag{2}$$

7. The stage of the directed search for alternative options for the structure and parameters of the functionally oriented technological process (blocks 20–25) of the y-th technological step of the x-th operation $P_{ix(y+1)} = \{V_{ix(y+1)}(var); S_{ix(y+1)}(var); t_{ix(y+1)}(var); Ins_{ix(y+1)}(var)\}$ and the generation of alternative options for the structure of the technological process $M_{i(x+1)y} = \{O_{i(x+1)y}(var)\}$ occurs according to heuristic rules.
8. The stage of formalizing the structural and parametric optimization results in the functionally oriented technological process according to the complex qualimetric criterion (block 26). The most effective toolkit is the method of converting a set of local normalized criteria into a single indicator equivalent in terms of commutativity $\{F_i\}$ and continuous differentiation in the space of indicators for all indicators of functional efficiency $\left\{q_{ijn} \middle| \forall n = \overline{1,N}\right\}$ for the j-th surface $(j = \overline{1,J})$ of the i-th part. In this case, the parametric form of the inverse operator is an additive normalization relation:

$$K(F_i) = \sum_{n=1}^{N} \alpha_n \cdot Q_n \neq 0 \forall n = \overline{1,N} \tag{3}$$

where $\alpha_n \left(\forall n = \overline{1,N}\right)$ is the column matrix of the weighting coefficients of each q_{ijn}-th local criterion of a specific j-th surface $(j = \overline{1,J})$ of the i-th machined part.

3 Ideas and Solutions

For implementing the methodology for planning a functionally oriented technological process described in the previous section, a generalized example of the effectiveness of applying theoretical provisions can be given. The problem of choosing the optimal option for machining mold ejectors' most accurate cylindrical surfaces during the finishing operation is being solved. Option 1: finishing grinding with an abrasive wheel; Option 2: fine turning with a blade tool with a composite insert based on polycrystalline cubic boron nitride (the use of a Sandvik Coromant insert with Wiper (WH) geometry of CB7025 grade (60% CNB) with bimodal grain distribution in a ceramic bond is proposed) (Askeland and Phule 1996).

To determine the main criteria for the effectiveness of the functionally oriented technological process in the CAF system, a simulation rheological model of the formation of the main functional surfaces of the ejector was established and the stress-strain state, parameter values of the surface microtopology and its structural-phase state were defined (Stupnytskyy 2014).

For cutting with a tool with a CB7025 composite insert, the geometric and kinematic components of the microroughness Δ_1 is:

$$\Delta_1 = \frac{S^2}{8r} = 0{,}000625mm = 0{,}625\mu m \tag{4}$$

The value of the vibration component of the microroughness (Δ_2) takes into account the regenerative mechanism of oscillation excitation when the tool moves along the surface of the workpiece formed at the previous step (Ra_{i-1} = 5 μm). The surface microrelief in the radial direction is modeled by superimposing on the array of tool displacements the profile of its top in plan with the amplitude of tool oscillations, taking into account the trace of the tool top on a given technological step H_B. In this case:

$$\Delta_2 = 2A_{\max} - H_B = 0{,}00039mm = 0{,}4\mu m \quad (5)$$

The component of the roughness profile caused by plastic deformations in the contact zone of the tool and workpiece Δ_3 is determined by [] under the conditions $\varphi > \arcsin\left(S/_{2r}\right)$ and $\varphi_1 \geq \arcsin\left(S/_{2r}\right)$ is equal to:

$$\Delta_3 = \frac{\left(1 - \frac{\tau_{xy}}{\sqrt{\tau_{xy}^2 + \bar{\sigma}^2}}\right) \cdot \left(2S_i + \frac{r}{2} \cdot \left(1 - \frac{\tau_{xy}}{\sqrt{\tau_{xy}^2 + \bar{\sigma}^2}}\right)\right)}{64} = 0{,}0001mm \quad (6)$$

where τ_{xy} is the dynamic shear stress, MPa; $\bar{\sigma}$ is the effective mean stress in the chip formation zone, MPa.

Thus, for this example, the average values of the height of microroughness resulting from copying the cutting edge of the tool concerning the kinematic characteristics of its movement (Δ_1) are 55.5%, the height of microroughness determined by the oscillations of the cutting tool (Δ_2) is 34.7%, and the component of the roughness profile caused by plastic deformations in the contact zone of the tool and workpiece (Δ_3) is 9.8%, respectively.

In forming parts, plastic deformation occurs, accompanied by crushing and elongating crystal grains in the direction of deformation, the curvature of sliding planes (Persson 2000), and residual stresses. In this case, two opposite processes simultaneously occur in the surface layer of the workpiece: strain hardening and thermodynamic softening (Davim 2010). The ratio of the intensity and speed of these processes determines the physical state of the surface layer of the workpiece. The analytical formalization of this dynamic process is extremely difficult. The curve of residual stresses of the first kind reflects the interference picture of the manifestation of fluctuating tensile (temperature and frictional) and compressive (force) loads (Balyts'kyi et al. 2020; Stupnytskyy and Hrytsay 2020). The average value of residual stresses in the thermal stabilization zone (at a temperature of about 100 °C) will be approximately 160 MPa when machining an ejector blank made of H21T20821 steel with a tool based on CNB grade CB7025 with a ceramic bond (Fig. 3).

Having modeled the stress-strain and thermodynamic state of the cutting zone in the Deform HT2 software, it is possible to calculate residual stresses of the second kind. According to the calculations of the Johnson-Mehl-Avrami equation (Xuanwei Lei et al. 2016) for steel H21T20821, the martensitic bulk phase on the surface of

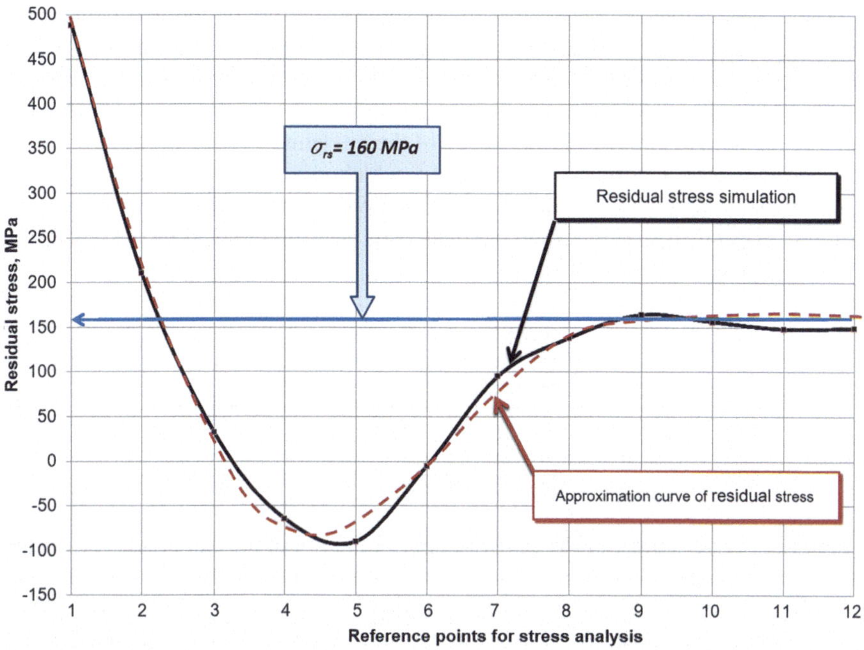

Fig. 3 Results of simulating residual stresses for the technological operation of finishing turning the hardened surface of a part made of steel H21T20821 with a tool with a CNB-based insert CB7025 (Stupnytskyy and Hrytsay 2020)

the workpiece is ζ_M= 0.395 (39.5%), which denies the appearance of a "white layer" on the surface:

$$\zeta_M = 1 - \exp\{0{,}025T - 0{,}002315\sigma_m - 0{,}003314\bar{\sigma} - 5{,}35\} = 0{,}395 \tag{7}$$

The average density value ρ_Σ of the structural phase of the material formed as a result of the influence of thermodynamic and power factors of cutting a machined surface with a polyphase structure is equivalent to the densities of different phases ρi following their percentage content ζ_i. Then, the value of the residual stress of the second kind $\sigma_{res}(l)$ at a depth l (μm) of their occurrence on the machined surface is taken as the value of the residual stress of the second kind:

$$\sigma_{res}(l) = E_M\left(1 - \frac{\sum_i \rho_i \cdot \zeta_i(l)}{\rho_P}\right) \tag{8}$$

where E_M = 224 is Young's modulus for steel H21T20821, GPa; ρ_P = 7760 kg/m^3 is the density of the dominant (pearlite) phase of the material of the part before machining.

The calculated value of the second kind's residual stress on the part's surface is only $\sigma_{res\_2} \approx 5,5$ MPa. This is a consequence of the high temperature's localized and short-term effect on the part's surface due to its blade processing. The error of the simulated and calculated values does not exceed 12.5% in this case, which is quite acceptable to confirm the adequacy of the modeling.

The relative wear resistance index $K(I_h)$, which characterizes the change in the wear intensity depending on the state of functional surfaces formed as a result of different variants of structures and parameters of the technological process, is determined by the equation:

$$K(I_h) = \frac{I_{h1}}{I_{h2}} = \frac{\left[\dfrac{0,6(1-\mu^2)\cdot P_a \cdot \lambda_1}{\sqrt{\nu_1}(\nu_1^2-1)\cdot K_2 \cdot E \cdot n}\right]}{\left[\dfrac{0,6(1-\mu^2)\cdot P_a \cdot \lambda_2}{\sqrt{\nu_2}(\nu_2^2-1)\cdot K_2 \cdot E \cdot n}\right]} = \frac{\sqrt{\nu_2}(\nu_2^2-1)\cdot\lambda_1}{\sqrt{\nu_1}(\nu_1^2-1)\cdot\lambda_2} \tag{9}$$

where ν_i is the parameter of the stepwise approximation of the Abbott curve; λ is the coefficient that takes into account the effect of surface residual stresses on the wear intensity.

Comparing the finishing turning of the mold ejector with a CNB-based tool CB7025 with a ceramic bond and its grinding with a diamond wheel ASV125/100 M5–2, we can obtain a relative wear resistance index $K(I_h)$:

$$K(I_h) = \frac{I_{h1}}{I_{h2}} == \frac{\sqrt{1,9}(1,9^2-1)\cdot 0,75}{\sqrt{1,7}(1,7^2-1)\cdot 1,27} = 0,86 \tag{10}$$

That is, as a result of changing the structure of the technological process of finishing the mold ejector, the wear resistance of the functional surface increases by about 14% due to a change in the formed surface microtopology and a change in the type and nominal values of residual stresses.

The relative fatigue strength index $K(n_{-1})$, which characterizes the change in the fatigue strength factor n_{-1} depending on the state of functional surfaces formed as a result of various structural variants and process parameters, is determined by the equation:

$$K(n_{-1}) = \frac{n_{-1\_1}}{n_{-1\_2}} = \frac{n_{\mathrm{BR}\_1}\cdot n_{ez\_1}}{n_{\mathrm{BR}\_1}\cdot n_{ez\_2}} = \frac{0,94\cdot 1,18}{0,93\cdot 0,83} = 1,43 \tag{11}$$

The equation determines the relative rate of fretting-corrosion fracture:

$$K(\tau) = \frac{\tau_{m\_1}}{\tau_{m\_2}} = \frac{\Delta m_1}{\Delta m_2} = \frac{(Rpk_1 + Rk_1) \cdot l \cdot k_o \cdot \ln\left(\frac{t}{\tau} + 1\right)}{(Rpk_2 + Rk_2) \cdot l \cdot k_o \cdot \ln\left(\frac{t}{\tau} + 1\right)} = \frac{(Rpk_1 + Rk_1)}{(Rpk_2 + Rk_2)} = \frac{0,28 + 0,84}{0,43 + 0,61} = 1,07 \tag{12}$$

The relative indicator of the reliability factor of the mechanical system in the mode of ensuring liquid tribological contact *K*(*S*) can be calculated as:

$$K(S) = \frac{h_{min\_1}}{h_{min\_2}} = \frac{k_S \cdot Rvk_1}{k_S \cdot Rvk_2} = \frac{Rvk_1}{Rvk_2} == \frac{0,17}{0,18} = 0,94 \tag{13}$$

The relative index *K*(*f*), characterizing the change in the friction coefficient in a conjugated tribocouple depending on the state of the functional surfaces formed as a result of various variants of structures and process parameters, is determined by the equation:

$$K(f) = \frac{f_1}{f_2} = \frac{\sqrt{\frac{R\max_1}{rb_1^{1/v_1}}}}{\sqrt{\frac{R\max_2}{rb_2^{1/v_1}}}} = \sqrt{\frac{0,024}{0,096}} = 0,5 \tag{14}$$

The calculated values of the normalized local optimization criteria by qualimetric indicators for finishing turning the functional surface of the mold ejector with a tool based on CNB of the CB7025 brand with a ceramic bond are as follows for this example: relative wear resistance $Q_1 = 0.28$; fatigue strength factor $Q_2 = 0.30$; tribological quality of mating $Q_3 = 0.10$; corrosion resistance $Q_4 = -0.14$; and liquid tribological contact factor $Q_5 = -0.17$. Finally, the additive value of the integral qualimetric indicator, determined by the method of multicriteria optimization of a functionally oriented technological process (Fig. 2), is equal to $F_1 = 0, 383$. Thus, the variant of finishing turning the functional surface of the mold ejector with a tool based on CNB of the CB7025 brand with a ceramic bond, compared to grinding with a diamond wheel ASV 125/100 M5-2, is more optimal in terms of the integral qualimetric indicator, which systematically characterizes wear resistance, fatigue strength, tribological quality of bonds, corrosion resistance and the parameter of ensuring the bearing capacity of the lubricating layer and is obtained as a result of the formation of the microtopology of the surface layer, residual stresses and strains during the forming process.

4 Conclusions

An analysis of the current state of technological maintenance of integrated systems for automated preparation of machine-manufacturing production has shown that the traditional methodology of the technological stage of planning, which is based on a sequential non-iterative synthesis of operations and transitions, requires structural and methodological improvement. Implementing an integrated product Life Cycle Management (PLM system) requires considering the operating conditions of the most loaded parts in the product, not only at the design stage and assignment of product accuracy standards, but also at the stage of designing the technology for their manufacture. For this purpose, an automated system that implements the relationship between the product-forming technology and the conditions of their potential operation (Computer Aided Forming - CAF-system) must be added to the integrated CAD/CAE/CAPP/CAM system. This system is based on the analysis of the rheological simulation model of individual technological operations and transitions, as well as a set of functional modules and analytical applications for the formation of precise, thermo-deformation, microgeometric, and structural-phase parameters of the processed surfaces.

The presented algorithm of structural and parametric optimization of the functionally oriented technological process involves the stage of simulation and analysis of the stress-strain state of the product in the process of its forming, implemented in the CAF-system and the stages of calculation of the technological component of local criteria - the intensity of wear of functional surfaces, the safety factor in terms of fatigue strength, the indicator of dynamic quality of tribo-conjugations, the safety factor of liquid tribo-contact of functional surfaces of the part, etc. The expediency of using relative indicators of the operational quality of machine parts, which are determined by the formation of indicators of the microtopology of the machined surfaces, residual stresses, and strains depending on the structure and parameters of the technological process, is proved.

Based on the analysis of the results of research in the CAF system and the average statistical results of experimental and production studies, the comparative characteristics of alternative options for finishing the mold ejector are presented: turning with a CNB-based tool with a ceramic bond and finishing grinding with a diamond wheel with a metal bond. It is shown that forming a functionally oriented technological process creates the prerequisites for ensuring a more efficient operational quality of mechanical engineering products.

Acknowledgments The research was partially supported by the International Association for Technological Development and Innovations.

References

Aramaki H, Cheng HS, Chung. Y-W (1993) The contact between rough surfaces with longitudinal texture - part I: average contact pressure and real contact area. J Tribol 115(3):1–6. https://doi.org/10.1115/1.2921653

Askeland DR, Phule PP (1996) The science and engineering of materials, 4th edn. Springer Science & Business Media, Oxford. https://doi.org/10.1007/978-1-4899-2895-5

Balyts'kyi OI, Kolesnikov VO, Havrylyuk MR (2020) Influence of modification of 38khN3MFa steel on the structural-phase state and cutting products under variable technological conditions. Mater Sci 55(6):915–920. https://doi.org/10.1007/s11003-020-00387-0

Davim JP (2010) Surface integrity in machining. Springer, London. https://doi.org/10.1007/978-1-84882-874-2

Demoly F, Dutartre O, Yan X-T, Eynard B, Kiritsis D, Gomes S (2013) Product relationships management enabler for concurrent engineering and product lifecycle management. Comput Ind 64(7):833–848. https://doi.org/10.1016/j.compind.2013.05.004

Dmitriev AI, Smolin AY, Popov VL, Psakhie SG (2009) A multilevel computer simulation of Friction and Wear by numerical methods of discrete mechanics and a phenomenological theory. Phys Mesomech 12(1–2):11–19. https://doi.org/10.1016/j.physme.2009.03.002

Eder WE (2016) Theory of technical systems - educational tool for engineering. Univer J Educ Res 4(6):1395–1405. https://doi.org/10.13189/ujer.2016.040617

Gao W, Tan Y, Jiang S, Zhang G, Zang M (2016) A virtual-surface contact algorithm for the interaction between FE and spherical DE. Finite Elem Anal Des 108:32–40. https://doi.org/10.1016/j.finel.2015.09.001

Greenwood JA, Williamson G (1966) Contact of nominally flat surfaces. Proc Royal Society Math Phys Eng Sci. https://doi.org/10.1098/rspa.1966.0242

Hidirov S (2022) Fundamentals of tribology. https://doi.org/10.1080/15732479.2019.1647914

Hirani H (2016) Fundamentals of engineering tribology with applications. University Press, Cambridge. https://doi.org/10.1017/CBO9781107479975

Karpenko YA, Akay A (2001) A numerical model of Friction between rough surfaces. Tribol Int 34:531–545. https://doi.org/10.1016/S0301-679X(01)00044-5

Kucharski S, Starzynski G (2014) Study of contact of rough surfaces: modeling and experiment. Wear 311(1–2):167–179. https://doi.org/10.1016/j.wear.2014.01.009

Ludema KC (1996) Friction, wear, lubrication: a textbook in tribology. CRC-Press, New York. https://doi.org/10.1201/9780429444715

Nepershin RI (2015) Plastic deformation of rough surface by a sliding tool. J Frict Wear 36:293–300. https://doi.org/10.3103/S1068366615040108

Persson B (2000) Sliding Friction: physical principles and applications. Springer Science & Business Media, Berlin. https://doi.org/10.1007/978-3-662-04283-0

Rangan RM, Rohde SM, Peak R, Chadha B, Bliznakov P (2005) Streamlining product lifecycle processes: a survey of product lifecycle management implementations, directions, and challenges. J Comput Inf Sci Eng 5(3):227–237. https://doi.org/10.1115/1.2031270

Straffelini G (2015) Friction and wear: methodologies for design and control. Springer, New York. https://doi.org/10.1007/978-3-319-05894-8

Stupnytskyy V (2013) Computer aided machine-building technological process planning by the methods of concurrent engineering. Europaische Fachhochschule: Wissenschaftliche Zeitschrift 2(3):50–53

Stupnytskyy V (2014) A generalized example of structural and parametric optimization of functionally-oriented process. Bull Nat Tech Univ "KhPI". Series: Techniques in a machine industry 42(1085):116–130

Stupnytskyy V, Hrytsay I (2020) Simulation study of cutting-induced residual stress. In: Advances in design, simulation and manufacturing II. DSMIE 2019, Lecture notes in mechanical engineering. Springer, Cham, pp 341–350. https://doi.org/10.1007/978-3-030-22365-6_34

Stupnytskyy, V., Hrytsay, I.: Computer-aided conception for planning and researching of the functional-oriented manufacturing process. In: Tonkonogyi, V., et al. Advanced manufacturing processes. InterPartner 2019. Lecture notes in mechanical engineering. Springer, Cham, pp. 309–320 (2020). https://doi.org/10.1007/978-3-030-40724-7_32

Thomas TR (1999) Rough surfaces. Imperial College Press, London. https://doi.org/10.1142/p086

Wei L, Gu BQ, Feng X, Sun JJ (2009) Research on friction characteristic of end faces of mechanical seals. In: Luo J, Meng Y, Shao T, Zhao Q (eds) Advanced tribology. Springer, Berlin, Heidelberg. https://doi.org/10.1007/978-3-642-03653-8_95

Xuanwei Lei JH, Xing Jin SC, Xingke Zhao (2016) Application of Johnson-Mehl-Avrami-Kolmogorov type equation in non-isothermal phase process. Mater Lett 181. https://doi.org/10.1016/j.matlet.2016.06.039

Technological Aspects of Gear Spline Joints

Olena Naboka and Viktor Kalinichenko

Abstract The work is devoted to researching the technological features of spline joint machining of gear wheels. The first section analyses the types of spline joints, their applications, and the main types of damage. The conclusions of this section give a general idea of the wide range of uses and possible problems of machining spline joints of gear wheels. The second section is devoted to machining quality assurance. It includes an analysis of the range of materials used to manufacture spline joints, the complex methods of their manufacture, and the technology of manufacturing surfaces. Various machining methods, including milling, grinding, and slotting, are considered in detail, considering their advantages and limitations. The given conclusions allow for development recommendations for planning optimum technology for processing spline joints of gear wheels. In the third section, computer studies of the machining of elements of spline joints are carried out. The overall goal is to develop and offer optimal technologies for machining spline joints, which have practical applications in various industries. The conclusions and recommendations provide a basis for further practical applications in the production environment.

Keywords Spline joint · Machining · Precision · Damage · Manufacturing methods · Milling · Grinding · Slotting · Modelling

1 Introduction

Enhancing the operational reliability of machines requires special attention to the durability of spline joints, which are widely applied in power transmission systems. Their production constitutes a considerable part of modern machine-building, with machining of gear and spline surfaces often accounting for more than half of the

---

O. Naboka (✉) · V. Kalinichenko
National Technical University "Kharkiv Polytechnic Institute", Kharkiv, Ukraine

V. Ivanov (ed.), *Emerging Technologies in Manufacturing and Materials Engineering*, Springer Tracts in Mechanical Engineering,
https://doi.org/10.1007/978-3-032-14914-5_2

total manufacturing effort. Consequently, the search for advanced technological solutions to improve spline wear resistance and service life remains a priority. Recent studies emphasize the importance of novel manufacturing strategies for splines within the framework of digitalized (Basova et al. 2023; Ivanov et al. 2025) and sustainable production (Jasiulewicz-Kaczmarek et al. 2023). This chapter focuses on the technological aspects of spline joint machining and discusses potential approaches to increase their performance.

The most widespread in the machine-building industry today are right-angle spline connections, which account for 80–90% (Malashchenko et al. 2020). Connections with an involute tooth profile have become widely used in recent years. Their load capacity is higher than that of straight splines (Skuratovskyi et al. 2021).

2 Fundamentals and Challenges

The choice of the type of spline joints is related to the design and technological aspects of straight and involute spline joints. The analysis of spline joint machining is an essential aspect in mechanical engineering, as the efficiency of this process directly affects the quality and reliability of the final product (Pavlyshche 2003; Gasanov et al. 2019). Here are some of the key aspects of spline machining analysis:

- dimensional accuracy. Profile geometry changes directly affect connection tightness and may lead to premature wear or failure;
- machining equipment. The precision and productivity of conventional milling or turning machines and specialized spline-machining tools determine the attainable quality of the joint;
- applied methods. Milling, grinding, turning, and electrical discharge machining each provide specific advantages; the selection depends on tolerance requirements and economic feasibility;
- material choice. The mechanical properties of steels or alloys used for spline components influence load-bearing capacity and wear resistance;
- quality assurance. Post-process inspection using micrometers, gauges, or coordinate measuring systems (CMMs) verifies compliance with design specifications and provides feedback for process adjustments;
- process optimisation. Continuous improvement initiatives reduce production time and costs while ensuring stable spline performance.

Analysing these aspects makes it possible to optimise the process of machining spline joints, improve the quality and reliability of the final product, and reduce production costs. Today, spline joints with spur, involute, and triangular tooth profiles are predominantly used in mechanical engineering (Electronic resource n.d.-a).

Spline joints with different tooth profiles are currently used without sufficient technical and economic justification. This leads to spline connections with different tooth profiles used in many components of the same design and loading conditions. As a rule, when selecting certain types of spline joints in production, they are not based on an analysis of loading conditions, design features, and manufacturing complexity, but rather on the technological process adopted at the plant. Often, there is no differentiated approach to selecting a particular type of spline connection, considering the entire range of determining factors, such as the durability and cost-effectiveness of manufacturing spline connection parts.

3 Ideas and Solutions

Analysis and justification of the manufacturing technology of the joint surfaces, with the provision of quality indicators for their production.

The type of spline alignment is selected based on design and process considerations. Outside or inside diameter alignment is used in cases where there are increased requirements for the accuracy of the relative position of the mating parts, for example, in the connection of a gear to a shaft (Electronic resource n.d.-b).

Using keyed connections does not solve the problem of heavily loaded parts and high torque transmission. At first, specialists in manufacturing try to use two 180° keyways to solve the problem of power tension and torque. If this fails, they prevent using spline joints until they need to. A spline shaft is united to a spline bore with rabbets (Fig. 1). Many splines are located along the perimeter of the shaft, with the help of which the high rotational load is evenly distributed to the mating areas. The number of splines on the surface can be $z = 3, 4, 6$, or 8.

A spline solution can ensure axial movement of the gear on the shaft.

In the case of design engineering, the side surfaces must be developed for surface voltage (Fig. 2) (Chen 2013).

The surface pressure on the tooth surface is calculated by the formula:

$$p = \frac{F_{per}}{A \cdot \psi_i} = \frac{M}{r \cdot a \cdot l \cdot z \cdot \psi_i} \leq p_{\max} \tag{1}$$

where M is the torque (N*m);

r—radius of the dividing circle (mm);
a—height of the right-sided tooth (mm);
l—slot length (mm);
z—number of splines,
Ψ_i—error rate (Ψ_i =0,79 ÷ 0,9).

The accuracy of positioning and the quality of the spline fit are maintained by the:

- concentricity and accuracy of the centre diameters;
- lateral dimensions of the slot width, the division accuracy, and the parallelism of the side surfaces to the centre line;
- surface roughness and hardness;
- precision of landings.

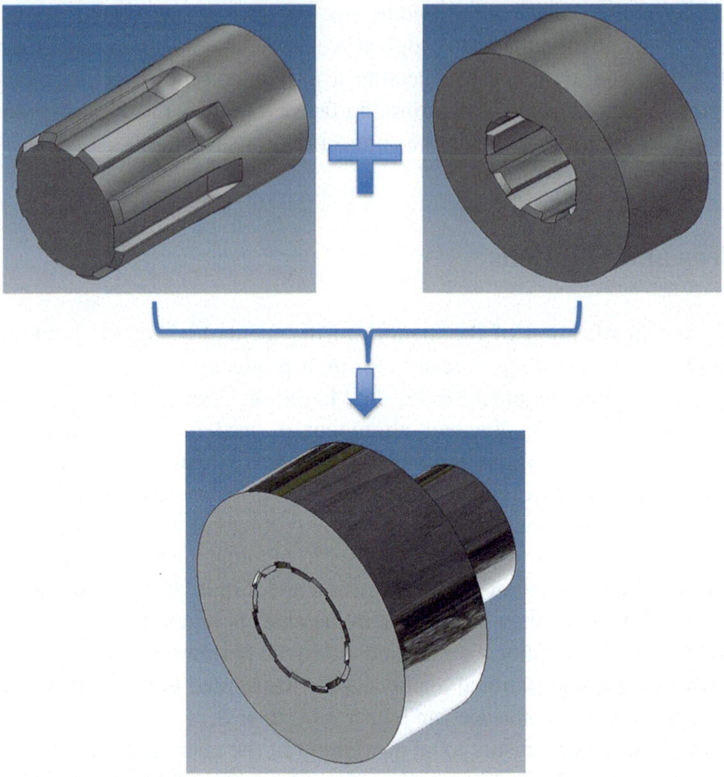

Fig. 1 Spline connections with spur teeth

Fig. 2 Surface pressure distribution on the surface of a spur tooth

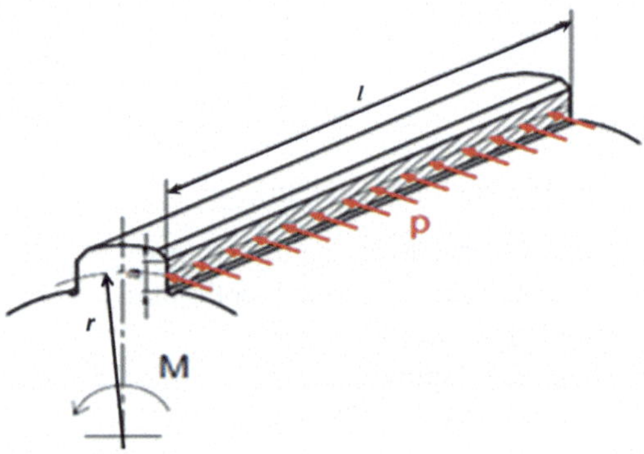

The geometrical dimensions and tolerances of the shaft and bore are standardised as specified in paragraph 1 (DIN 5463, DIN 5480). These are the main traditional production methods that will be analyzed. We will thoroughly analyze the technologies that can be used on conventional and CNC machines. Mathematical analysis can help with CNC process design and subsequent quality checks using a 3D measuring machine (Chen 2013).

A flowchart of the prospective process of designing and manufacturing a spline connection is shown in Fig. 3.

Note that, according to the centralised surface, the spline fit can be as follows (Naboka et al. n.d.): small diameter fit; large diameter fit; Side bearing fit. Let's compare the alignment methods for fitting splines (Fig. 4) (Naboka et al. n.d.; Ben et al. 2018).

Method 1: Small diameter fit (∅d): The connection is made along the diameter of the spline shaft root circle.

Advantages: the manufacture of the spline circumference diameter (D) is simple, the eccentricity of the spline shaft root circumference diameter can be ensured by grinding technology, and the shaft and hole can be tempered.

Disadvantages: Grinding the spline shaft root circumference (∅d) is complex and challenging.

Method 2: Fit in the main diameter (∅d): The connection is made along the circumference of the spline shaft shank.

Advantages: The manufacture of the spline shaft shank circumference diameter (∅d) is easier than in the case of small diameter fits. No profile machines are required to grind the root flank diameter (∅d).

Disadvantages: the bore cannot be hardened, additional calibration heat treatment is required after conventional heat treatment, the differences in concentricity quality can be obtained in the case of calibration broaching, it is difficult to manufacture the spline hole root circle diameter (∅D), and the manufacture of the bore side surface and shaft is time-consuming.

Analysis of Spline Shaft Milling Technology

Once the main shaft has been created, the next step is to mill the splines (Ben et al. 2018; Ying et al. 2022). There are several options for this. Starting with an ordinary horizontal milling machine of the crank type, the base body must be fixed between the centers. The proportional distribution between the splines is ensured by a dividing head.

As shown in Fig. 5, a single spline groove can be made with a disc cutter.

Once it is ready, the tool is removed from the machining area. The distribution between the teeth can be performed by the dividing head, considering the distribution factor depending on the number of slots. Finally, the milling process can be started repeatedly. This process must be repeated depending on the number of perimeter slots. Conversion matrix between tool and spline shaft (2):

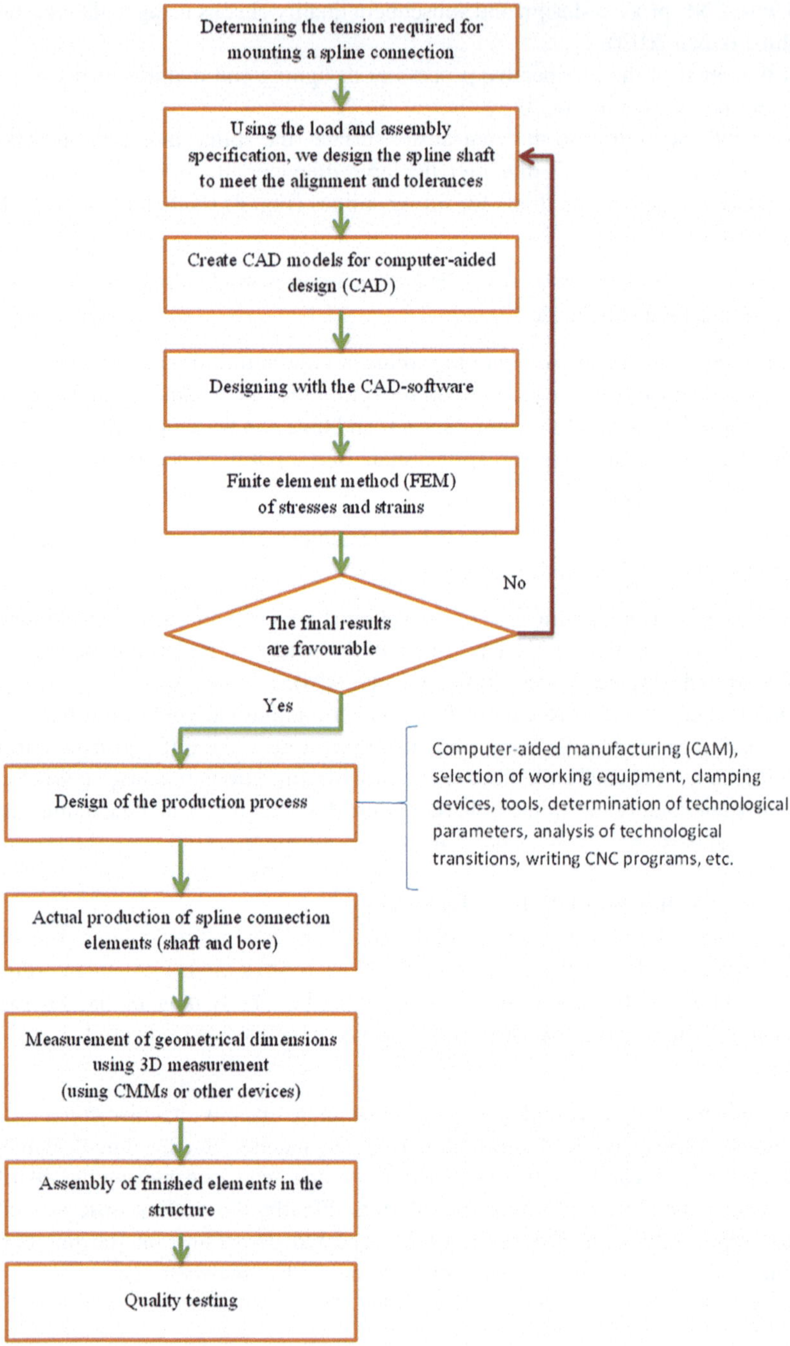

Fig. 3 Diagram of the methodology for ensuring the quality of spline joints

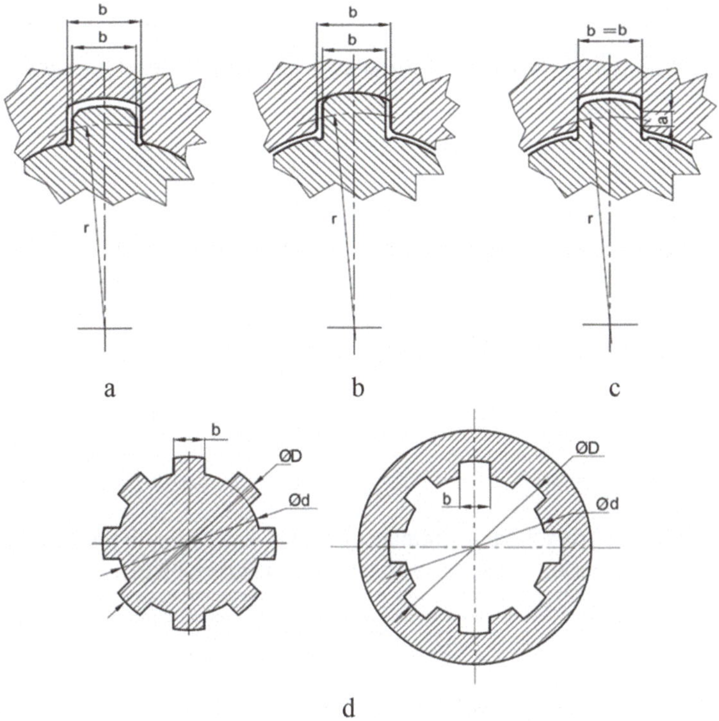

Fig. 4 Different spline fit centres: (**a**) small diameter fit, (**b**) large diameter fit, (**c**) side fit and (**d**) spline shaft and spline bore quantity

Fig. 5 Analysis of spline shaft milling technology (Ying et al. 2022)

$$M_{2R,1R} = \begin{bmatrix} -\cos\varphi_1 \cdot \cos\varphi_2 & \cos\varphi_1 \cdot \sin\varphi_2 & -\sin\varphi_1 & t \cdot \cos\varphi_1 - v_f \cdot \sin\varphi_1 \\ -\sin\varphi_1 \cdot \cos\varphi_2 & \sin\varphi_1 \cdot \sin\varphi_2 & \cos\varphi_1 & t \cdot \sin\varphi_1 - v_f \cdot \cos\varphi_1 \\ \sin\varphi_1 & \cos\varphi_2 & 0 & 0 \\ 0 & 0 & 0 & 1 \end{bmatrix}, \quad (2)$$

where φ_1, φ_2—the angular displacement (°).

The two-parameter vector-scalar function of the cutting edge is given by (3):

$$\underset{r_{1R}}{\rightarrow} = \begin{bmatrix} x_{1R}(\eta, \vartheta) \\ y_{1R}(\eta, \vartheta) \\ z_{1R}(\eta, \vartheta) \\ 1 \end{bmatrix}, \quad (3)$$

where x_{1R}, y_{1R}, z_{1R}—rotational coordinate system associated with the cutting tool.

The normal vector on the surfaces to be joined is given by (4):

$$\underset{n_{1R}}{\rightarrow} = \frac{\partial \underset{r_{1R}}{\rightarrow}}{\partial \eta} \times \frac{\partial \underset{r_{1R}}{\rightarrow}}{\partial \vartheta} = \begin{bmatrix} \underset{i}{\rightarrow} & \underset{j}{\rightarrow} & \underset{k}{\rightarrow} \\ \frac{\partial \underset{x_{1R}}{\rightarrow}}{\partial \eta} & \frac{\partial \underset{y_{1R}}{\rightarrow}}{\partial \eta} & \frac{\partial \underset{z_{1R}}{\rightarrow}}{\partial \eta} \\ \frac{\partial \underset{x_{1R}}{\rightarrow}}{\partial \vartheta} & \frac{\partial \underset{y_{1R}}{\rightarrow}}{\partial \vartheta} & \frac{\partial \underset{z_{1R}}{\rightarrow}}{\partial \vartheta} \end{bmatrix}. \quad (4)$$

The relative velocity between the connected surfaces can be represented as follows (5):

$$v_{1R} \rightarrow = M_{2R,1R} \cdot \frac{d}{dt}\left(M_{2R,1R}\right) \cdot r_{1R} \rightarrow, \quad (5)$$

where

$M_{2R,\,1R}$—the transformation matrix,

$\frac{d}{dt}\left(M_{2R,1R}\right)$—transformation matrix,

$r_{1R}\rightarrow$—is the vector of displacement of a points of a profile curve.

The statement "Connection I" is as follows:

$$\underset{n_{1R}}{\rightarrow} \cdot \underset{v_{1R}}{\rightarrow} = 0, \quad (6)$$

where $v_{1R}\rightarrow$is the relative velocity vector (mm/min$^{(-1)}$).

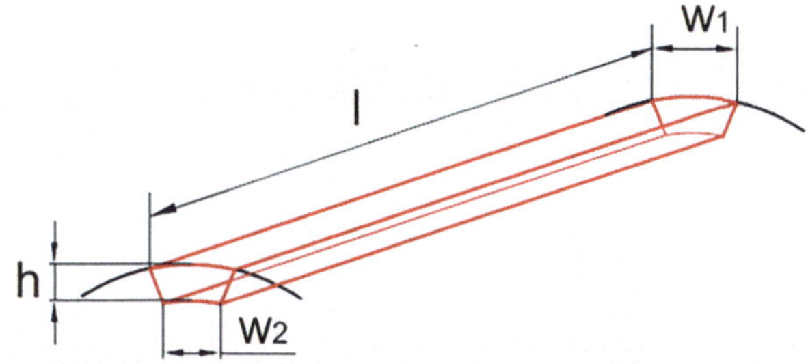

Fig. 6 Geometrical dimensions of a single spline groove on a shaft

The contact points between the tool and the spline shaft can be defined using formulas (2), (3), (4), (5) and (6) (Ben et al. 2018).

The depth t is equal to (7):

$$t = \frac{D}{2} + \frac{d}{2}, \tag{7}$$

where D is the outer diameter of the cutter (mm).

The following initial parameters are required for process design: h, z, l, b_s, d_a, d_f, n and f_z.

The width of the groove w_1 on the diameter of the spline shaft's outer ring is equal to (Fig. 6):

$$\frac{d_a \cdot \pi - b_s \cdot z}{z} = w_1, \tag{8}$$

where

d_a—spline shaft circumference diameter (mm),
b_s—spline shaft width (mm).

$$\frac{d_a \cdot \pi - b_s \cdot z}{z} = w_1, \tag{9}$$

where

w_1—width of the groove on the tip circle diameter (mm),
w_2—groove width on the diameter of the root circle (mm).

The cutting depth a is equal to the slot depth h.

Knowing the outer diameter of the cutter D, we can determine the cutting speed (10):

$$v_? = D \cdot \pi \cdot n_m \tag{10}$$

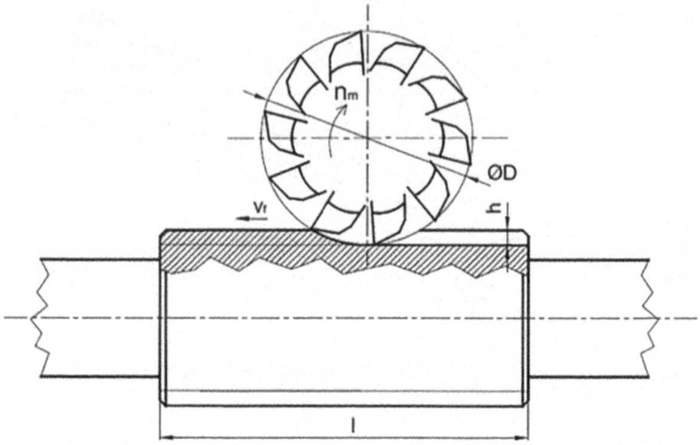

Fig. 7 Diagram of the technological process of milling splines with a disc cutter (Ying et al. 2022)

where n_m is the number of revolutions of the cutting tool ($\min^{(-1)}$).

The feed rate is determined by formula (11):

$$v_f = f_z \cdot z_t \cdot n_{\mathrm{m}}, \tag{11}$$

where

f_z—the feed per edge (mm),
z_t—the number of teeth on the cutter.

The arc of contact between the tool and the detail during the chip formation process is shown schematically in Fig. 7:

$$i = \sqrt{h \cdot D}. \tag{12}$$

The number of contacts can be determined by the formula (13). Number of contacts is the number of working teeth of the cutter along the contact *i* arc

$$\Psi = \frac{i}{t_p} = \frac{\sqrt{h \cdot D}}{\frac{D \cdot \pi}{z}} = \frac{z}{\pi} \cdot \sqrt{\frac{h}{D}}, \tag{13}$$

where

t_p is the step between the teeth (14).

$$\Psi t_p = \frac{D \cdot \pi}{z}. \tag{14}$$

The cutting force per cutting edge is then:

$$F_{c1} = k_c \cdot \underline{h} \cdot w_1, \tag{15}$$

where $\underline{h}$ is the average chip thickness.

The average chip thickness can be determined by formula (16):

$$\underline{h} = f_z \cdot \sqrt{\frac{h}{D}}. \tag{16}$$

Then, based on the formulas (13), (15), and (16), the total cutting force of the milling process can be obtained (17):

$$F_c = \Psi \cdot F_{c1} = \frac{z}{\pi} \cdot f_z \sqrt{\frac{h}{D}} \cdot k_c \cdot w_1, \tag{17}$$

where

F_{c1}—cutting force for one edge (N),
k_c—specific cutting force (N/mm^2).

The cutting power is determined by formula (18):

$$P_c = F_c \cdot v_c. \tag{18}$$

where

F_c—total cutting force (N),
v_c—cutting speed (mm/min).

Then, taking into account the efficiency of the cutting process, the engine power of the working machine should be:

$$P_m = \frac{P_c}{\eta}, \tag{19}$$

where

P_c—cutting power (kW),
η—energy conversion efficiency.

It should be noted that when the cutter is controlled during the cutting process, the center point of the cutting tool is controlled. Thus, having analysed Fig. 7, we conclude that it is necessary to consider an additional distance x (Fig. 8).

$$x = \sqrt{h \cdot (D - h)}, \tag{20}$$

According to the analysis of Fig. 8 and taking into account the excesses (l_1, l_2), we write down the formula for calculating the machine time:

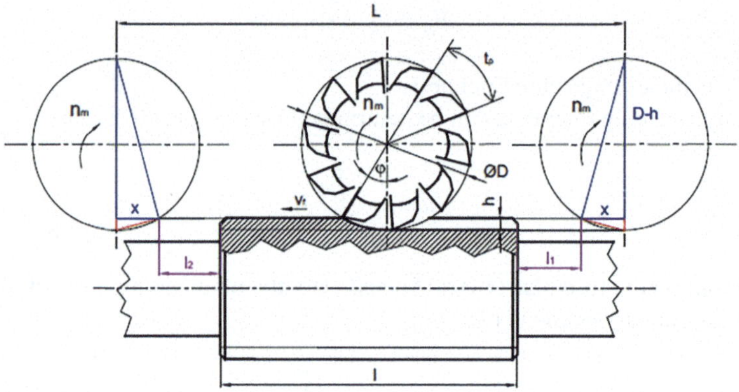

Fig. 8 Revised diagram of the technological process of forming milling cutter slots (Bali 1988)

$$T_m = \frac{L}{v_f} \cdot z = \frac{x + l_1 + l + l_2 + x}{v_f} \cdot z. \tag{21}$$

where

L—the total production length (mm),
l_1, l_2—tool strokes (mm),
l—total length of the connection (mm).

It is advisable to check the adequacy of the proposed models using the finite element method (Figs. 9, 10, 11 and 12).

The spline shaft and tool were modelled for the finite element simulation using the following initial input data. Geometry: shaft diameter d = 40 mm, spline number z = 8, slot length l = 60 mm. Material: structural steel (E = 210 GPa, $\nu = 0.3$, σy = 250 MPa). Boundary conditions: the shaft was fixed at one end, and a torque M = 150 N·m was applied to the opposite end. Contact conditions: friction coefficient μ = 0.15 between tool and spline surface. Mesh: tetrahedral finite elements with average size 1.0 mm, refined to 0.25 mm in the contact zone.

In the same way, we also performed mathematical modelling of groove milling using a worm cutter. This technological method is economical during mass production due to the complexity of the cutting tool and the short calculated machine time.

The analysis of the spline shaft grinding technology has shown many grinding possibilities depending on the type of alignment. The following solutions are recommended if the centre is to be used with a small diameter fit or with a side bearing (Fig. 13).

The abrasive disc performs a rotational and linear motion simultaneously. When one tooth is ready, it can be divided using the dividing head according to the division ratio. The grinding process can then proceed to the second spline.

This process must be repeated for all splines around the perimeter. The side surfaces and the slot width w_2 are subjected to grinding.

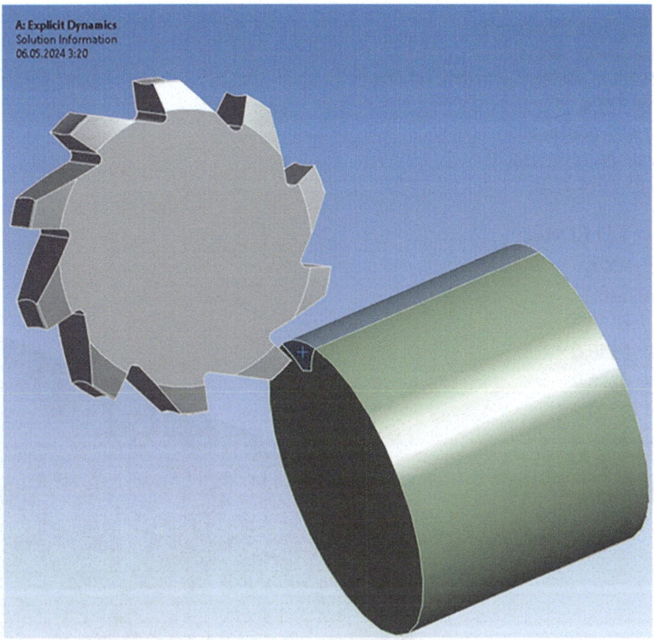

Fig. 9 Construction of the workpiece-tool body system

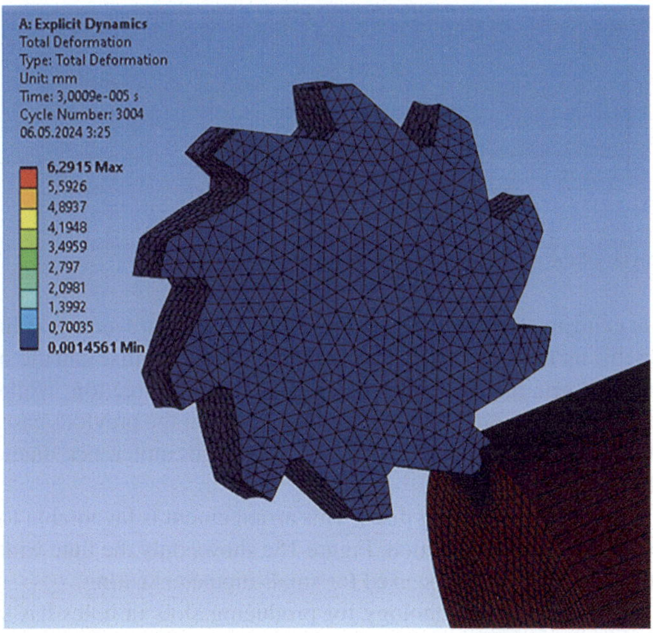

Fig. 10 Milling process simulation step by step

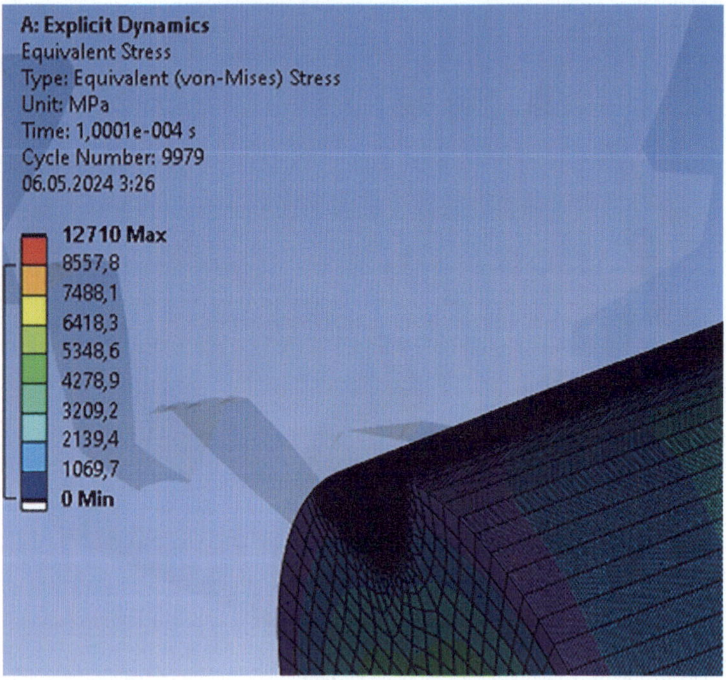

Fig. 11 Refinement simulation of the milling process

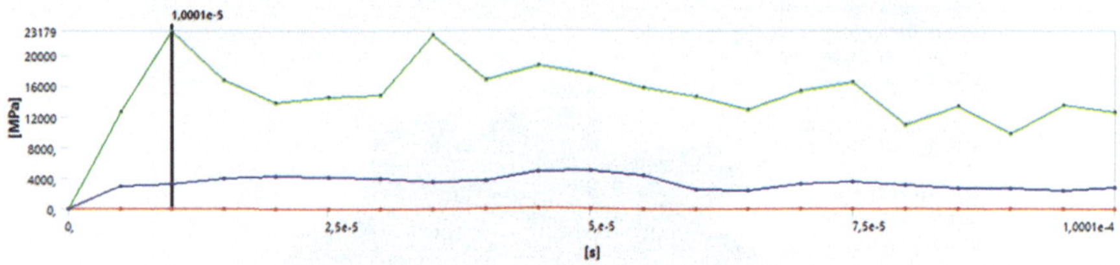

Fig. 12 Analysis of the milling process dynamics

If you are centring a large diameter, we recommend the traverse grinding method (Fig. 14). In this method, the axes of rotation of the abrasive disc and the spline shaft are parallel. The grinding wheel moves in a rotational direction, while the shaft moves linearly. The separation process is the same as in the previous case. It should be noted that this technology involves grinding the circumference diameter of the shaft shank.

Figure 15 shows two grinding discs. This arrangement is favourable for the fit of the side bearing in series production. Figure 15c shows only the flute width $w_{2\ grinding}$ on the root diameter. This can be used for small-diameter seating.

An analysis of slotting technology for producing slots in holes has shown that spline grooves can be produced using slotting technology. The entire depth of cut is

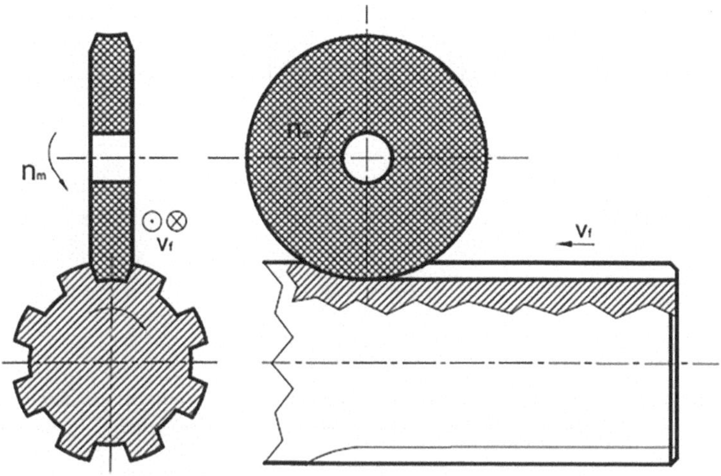

Fig. 13 Schematic of grinding a shaft spline groove for centering with a relatively small diameter or with a side bearing (Bodzás 2020)

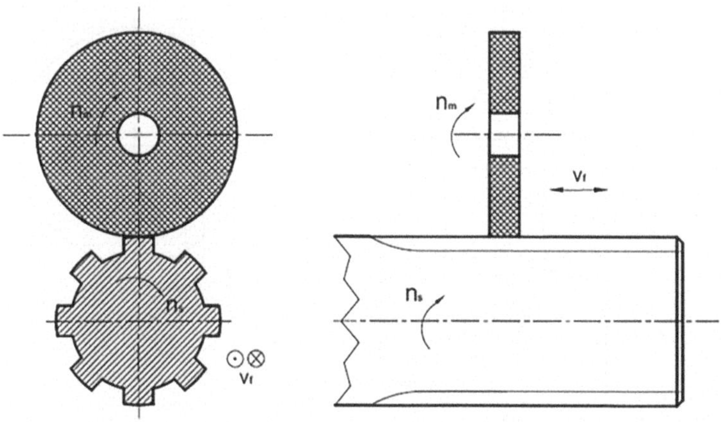

Fig. 14 Diagram of traverse grinding of a spline shaft (Bodzás 2020)

the depth of the groove. The cutting tool makes alternating movements parallel to the workpiece's axis of rotation. Once the cutting depth a_1 has been removed, the process can be repeated until the entire cutting depth (a) has been removed. The workpiece must then be moved to the next slot position, and the process can be repeated (Fig. 16). The total estimated machine time depends on the number of slots:

$$T_m = \frac{L}{v_f} \cdot z = \frac{l_1 + l + l_2}{v_f} \cdot z. \tag{22}$$

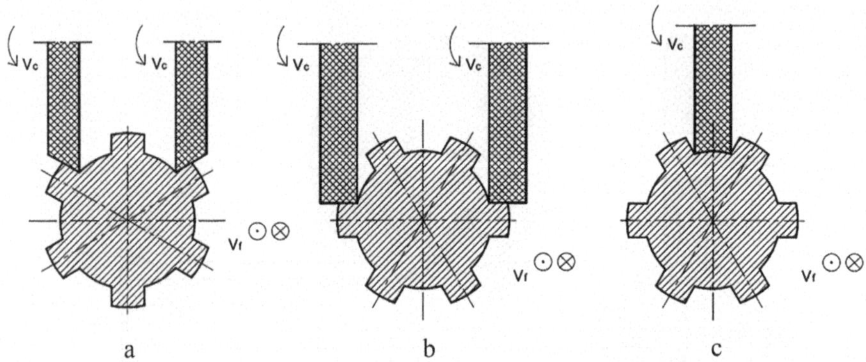

Fig. 15 Schematic of other possible methods for grinding a spline shaft: (**a**) grinding with a double conical grinding wheel, (**b**) grinding with a double disc wheel, (**c**) grinding on the root diameter only

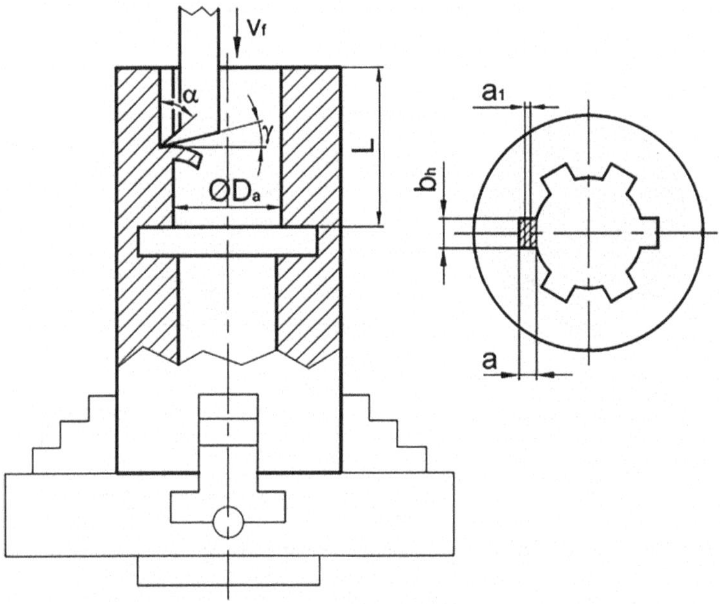

Fig. 16 Making slot grooves using slotting technology

It is promising to analyze the correctness of technological machining modes after checking them by the finite element method (Fig. 17).

After slotting, the slots must be ground based on the alignment options (Fig. 18).

Figure 18a is suitable for grinding spline grooves in large-diameter and side bearing applications. The abrasive disc rotates (vc) around its axis, while the spline is fed linearly (vf).

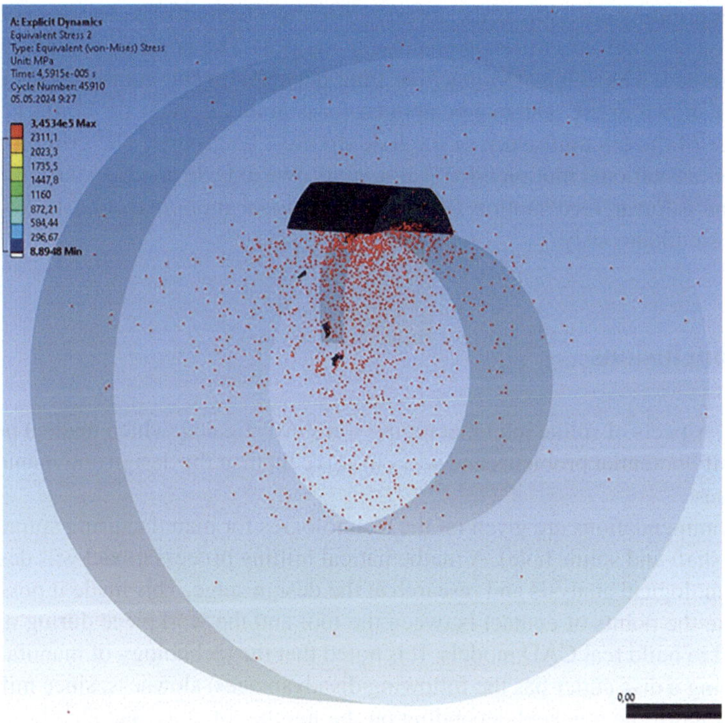

Fig. 17 Analysis of technological parameters by the finite element method (FEM)

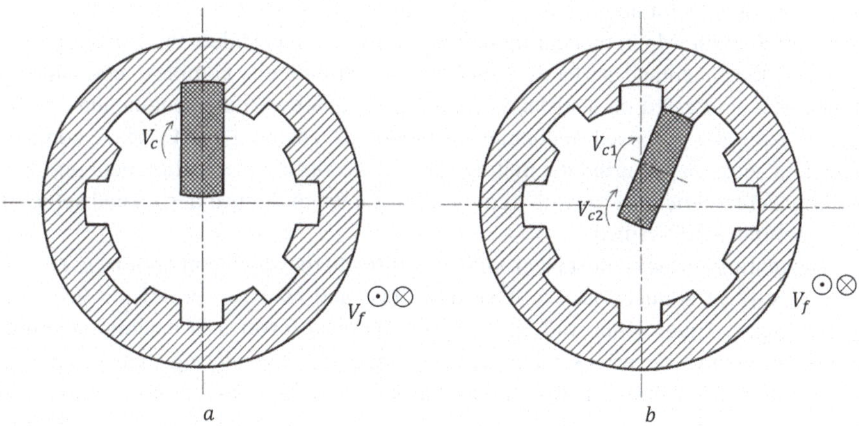

Fig. 18 Diagram of the internal spline grinding technology

After grinding one groove, the abrasive disc must be removed. The slot hole must be moved one step (to a different position in the slot). The grinding process can then be continued. The calculated machine time depends on the number of teeth. The solution shown in Fig. 18b can be applied for a small diameter fit. In this case, the diameter of the circumference of the hole tip must be ground. The grinding wheel performs a rotational motion (v(c1)) around its own axis. In contrast, the spline bore performs a linear feed motion (vf) and a rotational motion around its own axis (v(c2)) simultaneously.

4 Conclusions

The key aspects of spline joint machining were investigated, which made it possible to identify potential problems and ways to solve them at the design and manufacturing stages.

Recommendations are given on the technologies for manufacturing spline joints (spline shaft and spline hole). A mathematical milling process model was described for technological analysis and research at the design stage. This made it possible to calculate the points of contact between the tool and the workpiece during machining, and to build real CAD models. It is noted that the technology of manufacturing slots using a disc cutter has the following disadvantages: slowness, since milling of one slot has to be repeated depending on the number of slots, and significant wear of the cutting tool, since the cutting process occurs on the surface of the cutting edge of the tool.

To increase labour productivity, you can use a worm gear. Due to its complex geometry, this technology requires high tooling costs, so it is economical only in mass production. Mathematical modelling allows us to calculate the contact points between the machined surfaces. Since the worm cutter is complex, new cutting edges will be created on a surface of a different quality. This phenomenon can be traced using the proposed mathematical model. To ensure a good surface connection, it is necessary to grind the mating surfaces according to the alignment options. All the technological parameters required for the technological design and setup of the machine are described.

The mating surfaces must be ground according to the alignment options to ensure a good surface connection. We have described and analysed promising grinding technologies. Thus, while studying the technological features of machining spline joints, we analysed various manufacturing methods, considering their efficiency and economic feasibility. It was found that to ensure high-quality machining, it is necessary to consider not only the technical parameters of the process, but also the choice of material and tool.

Mathematical models of the machining process allow for effective planning and control of the technological process. As a result, this will increase the cost-effectiveness of spline production and reduce reject rates. Optimally match machining strategies to the available machines and tools. It is advisable to check the adequacy of the proposed models using the finite element method.

Acknowledgments The research was partially supported by the International Association for Technological Development and Innovations.

References

Bali J (1988) Forgácsolás. Tankönyvkiadó, Budapest

Basova Y, Dobrotvorskiy S, Balog M et al (2023) Increasing SME supply chain resilience in the face of rapidly changing demand with 3D model visualisation. Int J Mechatron Appl Mech 14:35–47. https://doi.org/10.17683/ijomam/issue14.5

Ben NY, Zhang Q, Lee MG (2018) Effect of floating axial holder on oscillating cold forging of spline shaft. MATEC Web Conf:02001

Bodzás S (2020) Manufacturing of the surfaces of spline fitting connection. Int J Adv Manuf Technol 111(3–4):909–920. https://doi.org/10.1007/s00170-020-06194-w

Chen CH (2013) The common failures of aero-engine mechanical system. Aviation Industry Press, Beijing. ISBN 978-751-650-325-6

Electronic resource (n.d.-a). Accessed at https://shem.com.ua/restavracia-shlicevogo-soedineniya (last accessed 01.04.2024)

Electronic resource (n.d.-b) Accessed at https://at-machining.com/splined-shaft-machining/ (last accessed 03.04.2024)

Gasanov M, Naboka O, Klochko O et al (2019) Simulation modeling in problems of mechanical engineering. NTU "KHPI", Kharkiv

Ivanov V, Evtuhov A, Dehtiarov I, Trojanowska J (2025) Fundamentals of manufacturing engineering using digital visualization. Springer, Cham. https://doi.org/10.1007/978-3-031-74360-3

Jasiulewicz-Kaczmarek M, Antosz K, Zhang C, Ivanov V (2023) Industry 4.0 technologies for sustainable asset life cycle management. Sustainability 15(7):5833. https://doi.org/10.3390/su15075833

Malashchenko VO, Nikolaychuk VV, Tymeychuk OY (2020) Connections of variable stiffness. NUWHP, Rivne. ISBN 978-966-327-452-2

Naboka OV, Klochko OO, Zakovorotny OY (n.d.) A modern approach to the technological support of the manufacture of gear wheels and additive AM-technologies. Bull NTU "KHPI" 1(9):82–91

Pavlyshche VT (2003) Fundamentals of design and calculation of machine parts. Afisha, Lviv. ISBN 966-8013-58-1

Skuratovskyi A, Lesyk D, Stepura O (2021) Connection of machine parts: a visual aid for bachelor's degree applicants in specialities 131 "Applied Mechanics", 133 "Industrial Engineering". Igor Sikorsky Kyiv Polytechnic Institute, Kyiv

Ying SS, Fu CT, Zhang SQ et al (2022) Numerical and experimental investigations on cutting force of broaching internal spline holes. Int J Adv Manuf Technol 120:2803–2814. https://doi.org/10.1007/s00170-022-08739-7

Determining the Machinability of Structural Materials by Cutting

Victor Antonyuk and Sergii Vysloukh

Abstract Based on the analysis of works on determining the machinability of structural materials, it was found that the existing methods are laborious, material-intensive, not reliable enough, and not always suitable for use in production conditions. And most of the accelerated methods for determining technological parameters do not consider the cutting process's real conditions. The developed mathematical models of the parameters of the cutting process and the methods for determining the relative machinability of materials and optimizing cutting modes allow for reducing the volume, time, and costs of tool and structural materials when conducting experimental studies, as well as obtaining mathematical models of the processing process's initial parameters. At the same time, the obtained mathematical models are conveniently stored in the knowledge base for further use in determining the relative machinability of the structural material and the properties of the tool material, both by a separate partial indicator and by a complex one, taking into account expert assessments (weight coefficients) of each of them, and determining the optimal cutting modes taking into account the specific conditions of machining the part according to the corresponding criterion. A method for the mathematical automated determination of the machinability of materials for practical application in real conditions is proposed and is oriented towards further expansion and improvement.

Keywords Machinability · Cutting process · Structural and tool materials · Mathematical model · Optimization

V. Antonyuk (✉) · S. Vysloukh
National Technical University of Ukraine "Igor Sikorsky Kyiv Polytechnic Institute", Kyiv, Ukraine

V. Ivanov (ed.), *Emerging Technologies in Manufacturing and Materials Engineering*, Springer Tracts in Mechanical Engineering,
https://doi.org/10.1007/978-3-032-14914-5_3

1 Introduction

Manufacturing and tool production, like many other industries, use modern structural materials, which, having good operational properties, should provide high machinability by cutting during parts manufacturing with the highest productivity and low cost of finished products (Twardowski and Wieczorowski 2023). However, research into the possibilities of effective machining for difficult-to-process structural materials by cutting is hampered by the complexity of determining technological modes (Klocke 2011).

Today, the main direction is assigning cutting operating modes, regulatory, and reference data. A qualified manufacturing engineer assigns machining modes based on his own professional experience or using automated systems (Ivanov et al. 2023). However, such an approach becomes ineffective with the appearance of new, not yet tested materials in production. At the same time, the terms allocated for their development require the development of automated methods of work of the manufacturing engineer to solve this problem (Astakhov 2014).

As the main method for manufacturing parts, machining materials by cutting is a highly complex physical process that affects the technological system in different ways: machine tool, fixture, cutting tool, and part (Ivanov and Zajac 2017). Therefore, the processed properties of materials cannot be characterized by any single indicator—they contain a set of characteristics (Polishetty et al. 2016).

Such main characteristics include:

- material machinability in relation to the influence of the cutting process on the tool;
- material machinability in relation to the magnitude of cutting forces;
- material machinability in relation to the quality of the product.
- machinability in relation to the nature of chip formation and shape, etc.

In practice, there are no difficulties in choosing the most essential machinability characteristic based on specific requirements for the operation being performed (Liao et al. 2024).

Knowing the absolute machinability of a material is necessary when developing management documents for assigning cutting modes and technical standardization of work (Visloukh 2001). Absolute machinability determines the processing conditions of a structural material (cutting modes, parameters, and geometry of the cutting tool, cooling conditions, etc.) for the corresponding processes (turning, drilling, milling, etc.) based on the mathematical dependences of the initial characteristics of the processing process on the cutting modes and other initial data (Ezugwu et al. 2003).

To obtain data on relative machinability, it is sufficient to set the value of V_{ek} (for example, V60 during turning) or the cutting force P when processing the studied material under certain standard cutting conditions. The ratio of the found value of the studied material parameter to the corresponding characteristic of the material, which is taken as a standard, gives an idea of the relative machinability of the material and can be taken as a basis for assessing its machinability (Astakhov 2014). Similarly, it is possible to determine the relative characteristics of machining properties by a tool material.

Determining relative machinability requires creating the same cutting conditions for the compared materials and conducting new experiments for each pair of materials. This process takes quite a long time, even though, as a result, we only get a partial indicator of the property with a separate comparison parameter.

The use of accelerated research methods is particularly relevant to determining relative machinability. A large number of accelerated research methods based on various principles are known, but their practical applicability is not always effective (Kant and Sangwan 2015).

An accelerated method that fully satisfies the requirements of a convenient and high-quality determination of machinability does not yet exist. As an alternative solution to the specified problem of determining relative and absolute machinability, it is proposed to abandon the "classical" approach and align its mathematical model with the direct cutting process, which fully reproduces it from a technological point of view (Lakic et al. 2011).

To obtain such a model, it is sufficient to conduct short-term experiments and then use their results to obtain mathematical models of the initial indicators of the cutting process as a basis for determining machinability. This will allow the comparison of processed materials according to selected indicators with the specified modes without carrying out additional experimental studies. In contrast, all the necessary information about the processing process can be obtained from direct modeling.

2 Fundamentals and Challenges

The work aims to develop a methodology for determining the cutting machinability of structural materials based on mathematical modeling of technological indicators, to create a methodology for obtaining adequate mathematical models of the initial indicators for the cutting process based on processing the results of experimental studies, which are entered into the knowledge base. To provide an approbation of the methodology for using the obtained mathematical models of the indicators of the processing of structural material with an appropriate tool for determining the relative machinability of the structural material in comparison with the selected standard for individual indicators of the cutting process and for a complex indicator, taking into account their expert assessments and real processing conditions.

The relevance of solving the setting tasks is due to problems with the machinability of construction materials, which was reflected in the works (Antonyuk et al. 2020; Antonyuk et al. 2023; Barandych et al. 2018).

Because absolute and relative machinability are almost not used to optimize the cutting process due to the lack of appropriate application programs, it is therefore advisable to present the determination of the absolute machinability of the structural material by optimizing cutting modes according to the selected criterion using mathematical models of individual indicators of the processing of structural material, which are in the knowledge base (Kim et al. 2024).

For this purpose, it is advisable to obtain adequate mathematical models of individual technological indicators of the cutting process in a unified form for their effective use in determining relative and absolute machinability in automated mode (Antonyuk et al. 2020; Antonyuk et al. 2023).

The task of determining the machinability of construction materials is also problematic when determining the machining properties of new tool materials.

3 Ideas and Solutions

3.1 Modeling the Machinability of Structural Materials by Cutting

The complexity of determining the technological parameters of materials, especially their machinability, lies in the non-stationarity of mechanical processing processes. Therefore, it is practically impossible to reproduce absolutely identical cutting conditions necessary for calculating relative machinability, and the discreteness of registering the values of the necessary characteristics during the experiment and their subsequent graphical interpretation to obtain optimal solutions reduces the probability of determining absolute machinability many times over. In this case, the time factor plays an important role. The significant time costs associated with conducting experimental studies and subsequent processing of the results mainly obtained do not allow technologists to directly apply the theory of material machinability in the production of elements (Lalbondre et al. 2014).

Therefore, a methodology for determining the machinability of structural materials has been proposed, based on the "free" experiment method. The term "free" is the possibility of arbitrary variation of the input parameters set by the experimenter, so the experiment can be considered active, and the object of research (Davim 2014).

This method allows one to obtain maximum information about the object of study, thereby carrying out its higher-quality modeling. The object of study in this case is the machining of structural material (cutting process).

As input parameters when solving the problem set, the system used cutting modes, physical and mechanical characteristics of the processed and tool materials, the values of the design parameters of the tool, processing conditions, etc.

The choice of parameters participating in the experiment depends on the specific requirements for the obtained models and the type of mechanical processing. The optimal option would be to use all indicators in a complex. However, such an approach in practice is impractical because it significantly complicates the processing of results, obtaining mathematical models of the cutting process parameters, and their further analysis and use.

Therefore, when choosing a mathematical model of the parameters and output indicators of the cutting process, it is advisable to represent them as a dependence (Wagri et al. 2022):

$$M_i = f(G_1, G_2, \ldots, G_m), \tag{1}$$

where G_j, $j = 1, 2\ldots, m$,—cutting process parameters that have the greatest impact on the output value M_i, $i = 1, 2, \ldots, n$.

Considering that it is necessary to simultaneously obtain a significant number of mathematical models of various indicators of the cutting process, and taking into account their further storage in the knowledge base, it is advisable to present the models in a unified form, as

$$M_i = C_i \Pi_{j=1}^{m} \left(G_j^{a_{ij}}\right), \tag{2}$$

where C_i, a_{ij}—coefficients and exponents of the powers of mathematical models for each output indicator, the values of which depend on the processed and tool materials, as well as the specific conditions of the cutting process.

This type of mathematical models significantly simplifies the algorithms and programs for their generation, and also presents the task of processing experimental data to obtain mathematical models as an identification task.

The general form of the resulting mathematical models for individual types of processing is presented as follows:

a) turning:

- cutting force components:

$$P_x = C^{x1} V^{x2} S^{x3} t^{x4} HB^{x5} \tau^{x6};$$

$$P_y = C^{y1} V^{y2} S^{y3} t^{y4} HB^{y5} \tau^{y6};$$

$$P_z = C^{z1} V^{z2} S^{z3} t^{z4} HB^{z5} \tau^{z6};$$

- dimensional wear of the cutting tool: $h = C^{h1} V^{h2} S^{h3} t^{h4} \tau^{h5}$;
- roughness of the treated surface: $R_a = C^{r1} V^{r2} S^{r3} HB^{r4} \delta^{r5} r^{r6}$;
- surface treatment error: $\delta = C^{\delta 1} V^{\delta 2} S^{\delta 3} \tau^{\delta 4} P_z^{\delta 5} h^{\delta 6}$;
- cutting tool wear rate at the machining of construction material:

$$J = C^{\alpha 1} V^{\alpha 2} S^{\alpha 3} t^{\alpha 4} \tau^{\alpha 5}$$

b) drilling holes:

- axial cutting force: $P_o = C^{o1} V^{o2} S^{o3} D^{o4} \tau^{o5}$;
- torque: $M_{rot} = C^{M1} V^{M2} S^{M3} D^{M4} \tau^{M5}$;
- dimensional wear of the cutting tool at the machining construction material:

$$h = C^{h1} V^{h2} S^{h3} t^{h4} \tau^{h5};$$

- roughness of the machined hole surface: $R_a = C^{r1}V^{r2}S^{r3}HB^{r4}\delta^{r5}\phi^{r6}$
- hole processing error: $\delta = C^{\delta 1}V^{\delta 2}S^{\delta 3}\tau^{\delta 4}P_z^{\delta 5}h^{\delta 6}$;
- drill cutting edge wear rate: $J = C^{\alpha 1}V^{\alpha 2}S^{\alpha 3}t^{\alpha 4}\tau^{\alpha 5}$

c) surface milling:

- cutting force components:

$$P_x = C^{x1}V^{x2}S^{x3}D^{x4}\tau^{x5}z^{x6}b^{x7};$$

$$P_y = C^{y1}V^{y2}S^{y3}t^{y4}\tau^{y5}z^{y6}b^{y7};$$

$$P_z = C^{z1}V^{z2}S^{z3}t^{z4}\tau^{z5}z^{z6}b^{z7};$$

- dimensional wear of the cutter at the machining construction material:

$$h = C^{h1}V^{h2}S^{h3}t^{h4}\tau^{h5}z^{h6}b^{h7};$$

- roughness of the treated surface: $R_a = C^{r1}V^{r2}S^{r3}HB^{r4}\delta^{r5}r^{r6}z^{r7}$;
- Surface milling error: $\delta = C^{\delta 1}V^{\delta 2}S^{\delta 3}\tau^{\delta 4}P_z^{\delta 5}h^{\delta 6}z^{\delta 7}$;
- cutter wear rate: $J = C^{\alpha 1}V^{\alpha 2}S^{\alpha 3}t^{\alpha 4}\tau^{\alpha 5}z^{\alpha 6}$

d) hole reaming:

- axial cutting force: $P_o = C^{o1}V^{o2}S^{o3}D^{o4}\tau^{o5}$;
- torque: $M_{rot} = C^{M1}V^{M2}S^{M3}D^{M4}\tau^{M5}$;
- dimensional wear of the cutting tool at the machining construction material:

$$h = C^{h1}V^{h2}S^{h3}t^{h4}\tau^{h5};$$

- roughness of the machined surface of the hole: $R_a = C^{r1}V^{r2}S^{r3}HB^{r4}\delta^{r5}\varphi^{r6}$;
- hole machining error: $\delta = C^{\delta 1}V^{\delta 2}S^{\delta 3}\tau^{\delta 4}P_z^{\delta 5}h^{\delta 6}$;
- reamer wear rate at the machining construction material:

$$J = C^{\alpha 1}V^{\alpha 2}S^{\alpha 3}t^{\alpha 4}\tau^{\alpha 5}z^{\alpha 6}.$$

The following symbols are used in the above formulas:

V—cutting speed, m/min;
S—cutting feed, mm/min;
t—cutting depth, mm;
τ—cutting depth, mm;
HB—Brinell hardness of the processed material;
r—radius at the tip of the cutter, mm;
D—diameter of the hole or cutter, mm;
φ—main angle of the cutter in the plan in degrees;
b—milling width, mm;
z—number of cutter teeth.

The above mathematical models must be determined to process the information arrays obtained during experimental studies. During the experiment, the necessary parameters of the cutting process were recorded at certain time intervals (not necessarily the same).

Analysis of methods for identifying mathematical models showed that to obtain them, it is advisable to use the method of group consideration of arguments or the method of least squares. The advantage of the least squares (LS) method is that it does not depend on the number of experiments performed and minimizes the sum of the squares of deviations of the actual results and those calculated using the obtained approximating function. At the same time, the method of modeling group consideration of arguments (MGCA) allows for the acquisition of mathematical models of optimal complexity. A comparative analysis of these methods showed that, in our case, for solving the identification problem, the use of LS is more appropriate because the type of the obtained model is predetermined (Wang 2020).

Thus, it is advisable to model the cutting process parameters based on a unified power dependence (2).

Then the process of determining the technological parameters of the processed materials can be automated, and the studied characteristics can be compared using the developed methodology under any processing conditions.

3.2 *Relative and Absolute Machinability of Structural Materials*

When solving the problem of determining the machinability of materials by cutting, the concepts of relative and absolute machinability were used; therefore, these two types of machinability and their relationship were considered.

It is known that relative machinability is less widely used in production than absolute, due to its laboriousness and low informative value. At the same time, having obtained the value of the relative machinability coefficient, one can only determine the greater or lesser suitability for processing a particular material, but without receiving information about the conditions and cutting modes under which it should be processed (Zheng and Yu 2025).

The proposed method allows changing the attitude to determining relative machinability. With its use, there is no need to conduct control experiments on the material under study. It is enough to have the appropriate mathematical models of the material processing parameters.

All necessary calculations that require a significant amount of time can be performed autonomously, in an automated mode. In this case, it is possible to compare the reference and studied materials for each determined parameter. It is also essential to calculate a complex machinability index.

According to this method, the technologist-researcher only needs to indicate the values of the weight coefficients for the corresponding indicators according to expert assessments, thereby objectively determining their priority (El-Gaddar et al. 2024).

As a result, already at the first stages of research into new materials that have not previously been used in production, it is possible to establish rational or optimal conditions for their processing and recommended areas of their further use (Liao et al. 2024).

The relative machinability of a material is understood as a coefficient equal to the ratio of the indicator values according to which the comparison is made, calculated under the same cutting conditions of the studied and reference materials.

Thus:

$$K_i = \frac{M_{iinv}}{M_{iet}} \tag{3}$$

where

K_i—coefficient of relative processability for the i-th parameter;
$M_{i_{inv}}$ and $M_{i_{et}}$—the value of the initial indicators under given cutting conditions of the tested and reference materials.

Coefficients of processability can be presented both as a percentage and as an absolute value.

In this case, K_i is a positive value that takes into account the properties of the estimated indicators M_i.

The values of the K_i coefficients were determined under the following conditions:

$$\begin{cases} K_i \in (0\ldots1]\, at\, M_{i_{inv}} \le M_{i_{et}}; \\ K_i \in (0\ldots\infty]\, at\, M_{i_{inv}} > M_{i_{et}}. \end{cases}$$

If the coefficient $K_i < 1$ according to the first condition of the indicator $M_{i_{inv}} \le M_{i_{5t}}$, we consider that the researeched material is better in machining according to this indicator than the reference one.

The opposite indicates that the technological processability of the studied material according to the corresponding indicator lower comparing to the reference material.

If the comparison of materials is carried out according to one specific indicator, then the coefficient K_i is also called the partial processability coefficient.

As the indicator M_i, by which the materials are evaluated, you can use:

- the magnitude of the cutting forces arising during the machining process (P_x, P_y, P_z, P_o) or torque (M_{rot});
- cutting tool wear rate on the rear surface at the processing the appropriate construction material ($J_з$);
- roughness of the treated surface (R_a or R_z);
- processing error (δ) etc.

Then the partial processing coefficients for the corresponding indicators will be determined as:

$$K_{P_x} = \frac{P_{x_{inv}}}{P_{x_{et}}}, K_j = \frac{j_{inv}}{j_{et}}, K_{Ra} = \frac{R_{a_{inv}}}{R_{a_{et}}}, K_\delta = \frac{\delta_{inv}}{\delta_{et}} \text{ etc.}$$

In practice, it is often necessary to evaluate a material, especially a new one, not by one, but by several indicators.

For this purpose, the concept of a complex workability coefficient is used, which is determined according to the formula:

$$K_k = \frac{\sum_{i=1}^{n} k_{b1} \cdot K_i}{\sum_{i=1}^{n} k_{b1}} = \frac{k_{b1} \cdot K_1 + k_{b2} K_2 + \ldots + k_{bn} \cdot K_n}{k_{b1} + k_{b2} \ldots + k_{bn}} \quad (4)$$

where

K_k is the complex workability coefficient;
K_i is the workability coefficient of the studied material for the i-th indicator;
k_{bi} is the weighted workability coefficient of the i-th indicator;
n is the number of partial workability coefficients used in determining the complex indicator.

As can be seen from expression (4), K_k consists of a number of partial processability coefficients, to which the weight coefficients k_{bi} correspond.

The role of k_{bi} is reduced to the distribution of priority (importance) between the partial processability indicators K_i for specific production conditions.

Thus, when processing small-sized valuable workpieces, the processing error, the condition of the surface layer and the permissible cutting forces come to the fore. Processing large workpieces requires the priority of the tool stability and the rate of its wear. Therefore, it should be noted that the importance of a particular indicator is not strictly recommendatory, but is determined by the priority series, which is built as a result of a survey of experts, based on their personal experience and knowledge of the production process. Therefore, weighting factors can also be called expert assessments.

Using a priority series allows you to minimize the error in determining relative machinability, eliminate the errors of previous experimental measurements and the subjectivity of their assessment. In exceptional cases, the technologist-researcher has the opportunity to determine the values of the weighting factors himself, but he must take into account in advance the possibility of obtaining results with a certain error.

Thus, the proposed methodology included both stages of determining the relative machinability of materials for the pair "test material – reference material", namely:

- determination of partial workability coefficients;
- calculation of the complex workability coefficient.

The nature of the automation of computational processes and the ability to choose a reference material make this approach more effective for the researcher.

Absolute machinability involves obtaining appropriate mathematical models for different types of processing, i.e. the dependence of the output parameters on the input parameters, with the subsequent solution of the optimization problem according to the selected criteria. These dependencies are obtained for each material separately or for groups of similar materials.

Often, several such dependencies are matched for one material, while breaking down the set of input parameter values into components of its subset. This is due to the increased sensitivity of the processes being modeled to minor changes in external influences.

As previously defined, the type of dependencies of the output indicators on the input parameters according to formula (1) in this method for different types of processing is established by the researcher and does not change throughout the work with it. Therefore, these same dependencies will also be used to determine absolute machinability.

The choice of the optimality criterion when solving the problem of establishing the absolute machinability of materials by cutting is determined based on the real requirements of production (Twardowski and Wieczorowski 2023; Lalbondre et al. 2014).

As such criteria, one can choose:

(a) the criterion of absolute deviation;
(b) processing time (cutting time);
(c) the productivity of the cutting process;
(d) the cost of manufacturing a part for a given type of processing;
(e) the stability of the cutting tool, etc.

These criteria combine the main technical and economic requirements for the production process, namely, obtaining high-quality products (products, parts) in the shortest possible time with minimal expenditure of material and production resources. A universal method for solving such a specific and at the same time quite complex problem has not yet been found. The presented method offers one of the options for its solution.

The technologist is invited to determine the most important criterion of optimality for specific conditions and to establish the absolute machinability of the studied material using it.

Despite the complex and close relationship between the initial indicators Mi, multi-criteria optimization methods allow applying a certain generalized, complex optimization criterion to study the cutting process. In parallel with the problem of choosing a criterion, an equally important problem of choosing an optimization method was solved—a method of determining the largest (smallest) value of the objective function (Sharma et al. 2024; Zheng et al. 2023).

The absolute deviation criterion, like other criteria, involves determining the values of the input parameters G_i at which the objective function reaches its smallest (largest) value.

The objective function F for the absolute deviation criterion has the following form:

$$F = \frac{\sum_{i=1}^{n}\left|\left(\frac{M_{i\,all}}{2} - M_i\right)\cdot k_i\right|}{\sqrt{\prod_{i=1}^{n}\left|\left(\frac{M_{i\,all}}{2} - M_i\right)^2 k_i\right|}} \rightarrow \min, \tag{5}$$

where:

n—the number of initial indicators of the studied process;
$M_{i\,all}$—the permissible value of the i-th indicator;
k_i—the weight coefficients of the i-th indicator.

Expression (5) mathematically describes the requirements for minimizing the absolute deviation of the calculated value from half the permissible value for each of the selected partial optimization indicators.

That is, the determination of the smallest value of the objective function F based on the weight equality of the indicators is implemented.

The values of the weight coefficients k_j were calculated according to the formula:

$$k_i = \prod_{j=1}^{n-1} M_{j\,all}, (j \neq i). \tag{6}$$

At the same time, restrictions were imposed on the values of the calculated indicators, which were determined on the basis of technological requirements for the quality and accuracy of the surface obtained as a result of processing, as well as the degree of tool wear during the cutting process.

These restrictions formed the area of allowable solutions D_1:

$$D_1 : \begin{cases} N = f(Pz, V) \leq N_{all} \\ h = f(V, S, t, \tau) \leq h_{all} \\ R_a = f(V, S, t, \tau, r, HB) \leq Ra_{all} \\ \delta = (V, S, t, \tau, Pz, h) \leq \delta_{all} \end{cases} \tag{7}$$

where:

N_{all}—maximum permissible cutting power, kW;
h_{all}—permissible tool wear on the rear surface, mm;
Ra_{all}—permissible roughness of the machined surface, μm;
δ_{all}—permissible machining error, mm.

In addition, additional boundaries of the region of permissible solutions D_2 are formed by the following restrictions:

$$D_2 : \{ n_{\min} \le n \le n_{\max}; S_{\min} \le S \le S_{\min}; t_{\min} \le t \le z \} \tag{8}$$

where:

n_{mm}, n_{max}, s_{min}, s_{max}, t_{min}—maximum and minimum values of spindle (tool) speed n, feed S and cutting depth t, respectively;
z—machining tolerance.

The resulting search area for the smallest value of the objective function is defined by the intersection of areas D_1 and D_2, i.e. $D = D_1 \cap D_2$.

Thus, the considered criterion allows one to determine the optimal values for the general optimization case, that is, when there is no clearly expressed priority of one of the initial indicators. As an example, let us give the form of the objective function F when optimizing by the criterion of absolute deviation for the turning process:

$$F = \frac{\left|\frac{N_{all}}{2} - N\right| \cdot k_N + \left|\frac{h_{all}}{2} - h\right| \cdot k_h + \left|\frac{R_{a\,all}}{2} - R_a\right| \cdot k_{Ra} + \left|\frac{\delta_{all}}{2} - \delta\right| \cdot k_\delta}{\sqrt{\left|\frac{N_{all}}{2} - N\right|^2 \cdot k_N \cdot \left|\frac{h_{all}}{2} - h\right|^2 \cdot k_h \cdot \left|\frac{R_{a\,all}}{2} - R_a\right|^2 \cdot k_{Ra} \cdot \left|\frac{\delta_{all}}{2} - \delta\right|^2 \cdot k_\delta}} \to \min$$

The processing time criterion is used in the system to determine such cutting process modes, in which the least amount of time is spent on processing a given surface, provided that other initial indicators are maintained within the established limits. This criterion is quite often used in serial and mass production, where a well-established technological process scheme operates, and the resulting products do not belong to the category of high-precision.

Therefore, within permissible limits, other parameters can be neglected, bringing to the fore the issue of reducing processing time. The term "neglect" in this case does not indicate their weakening, but only indicates the possibility of changing within the specified limits, which allows you to expand the optimization area and, therefore, find the appropriate optimal solution.

Taking this into account, the general form of the objective function will be:

$$F = f(V, S, t) \to \min \tag{9}$$

To determine the type of analytical dependence (objective function) $\tau = f(V,S,t)$, as an example, consider the process of turning a cylindrical workpiece.

Then the expression for determining the processing time can be written as:

$$\tau = \frac{L}{n \cdot S} \tag{10}$$

where

L—length of the machined surface, mm;

n—speed of the machine spindle, rpm;
S—longitudinal feed of the cutter, mm/min.

The tolerance z, which is taken for processing, can be removed in one or several passes i, depending on the proposed requirements for the accuracy and surface quality of the resulting part. If the tolerance is removed in several passes, then the processing time τ increases accordingly.

Taking this into account and the dependence of cutting speed on the speed of rotation of the machine spindle, the expression for determining the time spent on turning will ultimately look like this:

$$\tau = \frac{L \cdot z \cdot \pi \cdot d}{1000 \cdot V \cdot n \cdot t \cdot S} \tag{11}$$

where d—processed surface diameter.

The obtained dependence $\tau = f(V,S,t)$ can be applied not only to turning, but also to other types of processing. The restrictions imposed on the search area, as when using the previous criterion, were determined by the size of the area D, which is given by constraints (7) and (8).

The productivity of the machining process of parts on metal-cutting machines is a quantity that is the reciprocal of the surface machining time. Therefore, in this case, formulas are used that are similar to the previous optimization criterion. When solving the optimization problem, cutting modes are determined that deliver the maximum value of labor productivity in the area represented by constraints (7), (8) and the objective function (9).

Practically any machine-building and instrument-making production faces the problem of extending the service life of the tools used. One of the options for its solution is the design of new tools with wear-resistant coatings and improved design and technological characteristics (Patel et al. 2021). However, this option requires significant expenditure of material and time resources for its implementation.

Therefore, it is proposed to extend the service life of the existing tool due to the rational conditions of their use. In the proposed methodology, such an approach is reflected in the determination of machinability in the form of a criterion for the stability of the cutting tool.

The need to abandon the further use of the cutting tool and its subsequent regrinding is determined by the value of the maximum permissible h_{all} wear, which in all cases is taken as the largest width (mm) of the worn contact area on the back surface of the tool. In this case, it is worth distinguishing between the concepts of maximum permissible and permissible wear. The permissible wear determines the boundaries of the optimization area D and only in some cases can it be equal to the maximum permissible

$$h_{all} \leq h_{max\ all}.$$

The expression that defines the objective function F for the cutting tool stability criterion has the following form:

$$F = \frac{h_{maxall}}{J} \to \max \tag{12}$$

where

J—cutting tool wear rate, mm/min;
$h_{max\ all}$—maximum permissible wear value, mm.

As a result of solving the optimization problem according to criterion (12), in contrast to the previously given criteria, we will obtain not only the values of the cutting parameters that ensure the high-quality operation of the tool until the maximum permissible wear is achieved, but also its main characteristic—durability, which is presented in minutes. This will allow us to plan the production cycle in advance, evenly load the tool services and, most importantly, extend the service life of the cutting tool without design and technological improvements. Here, the area of search for the optimal solution remains area D. The next criterion when solving the problem of determining absolute machinability is the criterion of the cost of the manufactured product. It most fully covers the entire production process and allows us to approach the assessment of machinability from a technical and economic point of view, because the basis of the organization of any production is the principle of obtaining profit.

Technological cost C (UAH/piece), which in this case acts as the objective function F, which is determined by the following expression

$$F = \text{B}_n + \tau \cdot \left(a_T + \frac{M\left(K_V\right)}{T} \right) \to \min; \tag{13}$$

$$b_n = t_{ar} \cdot \left(a_{cwd1} + a_w\right) + \frac{Tp.f}{Q} \cdot b_{cw}; \quad b_{cw} = a_{cwd1} + a_w + a_{aw};$$

$$\tau = \frac{\pi \cdot d \cdot L}{10^3 \cdot V \cdot S} \cdot i \quad a_T = a_{mh} + a_w,$$

where:

t_{ar}—time of auxiliary movements when processing one part, min;
a_{cwdl}—costs associated with machine downtime, UAH/min;
a_w—worker's hourly wage (with accruals), UAH/min;
$T_{p.f}$—preparatory and final time, min;
Q—number of parts that must be manufactured in a given period of time (batch size);
a_{aw}—hourly wage of the adjuster (with accruals), UAH/min;
a_{mh}—cost of machine-minute of the main machine time including overhead costs, UAH/min;
d—diameter of the machined surface (tool diameter), mm;

L—length of the machined surface, mm;
V—cutting speed, m/min;
S—feed, mm/min;
i—number of passes when machining the surface of the part, which is determined by the expression:

$$i = \begin{cases} k_p & if\left(z - k_p t\right) < t_{all} \\ k_p + 1 & other \end{cases}; \quad k_p = \left[\frac{z}{t}\right] \tag{14}$$

z—the amount of tolerance for processing, mm;
t—cutting depth, mm;
t_{all}—the maximum allowable cutting depth, mm;
$M(K_t)$—mathematical expectation of the specific cost of the tool, UAH/min;

The mathematical expectation of the specific cost of the tool is determined by the formula:

$$M\left(K_t\right) = \left\{t_{cn} \cdot b_{cw} \cdot \left[m_1 + M\left(m_2\right)\right] + a_g \cdot t_r \cdot m_1 + B \cdot \left[1 + M\left(m_2\right)\right]\right\} \cdot m_1^{-1} \tag{15}$$

$$t_{cn} = t_{cn} + \frac{h_{all}}{\delta - \Delta} \cdot t_{da} \quad h_{np} = h \cdot tg\ \alpha + \Delta T_i;$$

$$M\left(m_2\right) \frac{m_1 \cdot T}{t_{MTBF}} \quad t_{MTBF} = \int_0^{\infty} p\left(t_{pb}\right) dt$$

where

m_1—the number of tool changes due to its wear or the permissible number of regrindings;
$M(m_2)$—the mathematical expectation of the number of tool changes due to its breakage or chipping;
a_g—the cost of the machine-minute of the main time of the grinding machine taking into account overhead costs, UAH/min;
B—the cost of the tool, UAH.
t_r—the time spent on regrinding the tool, min; time spent on regrinding the tool
t_{cf}—the time spent on changing the tool, min; tool change time
h_{all}—the limit value of the dimensional wear of the tool, mm;
δ—the tolerance for the obtained precision parameter, mm;
Δ—the total processing error, mm;
t_{da}—the time for dimensional adjustment and adjustment of the technological system, min;
h—the value of the wear of the cutting tool on the rear face, mm;
α—the rear angle of the cutting tool in degrees;
ΔT_i—error caused by temperature deformation of the tool, mm;
T—stability period of the cutting tool, min;
t_{MTBF}—mean time between failures (mean time between failures) of the tool, min;

$p(t_{pb})$—probability of destruction of the cutting tool.

To solve the optimization problem, it is necessary to specify the values of economic indicators, which relate to conditionally constant information of a regulatory and reference nature and are determined on the basis of the current economic situation in the production sector. The method for determining the absolute machinability of a structural material by the criterion of the technological cost of processing allows for interactive editing of economic indicators for the corresponding type of mechanical processing.

As a result of optimization by the cost criterion, we obtain the values of the cutting elements (V, S, t, τ) that correspond to the minimum of the objective function F, as well as the actual cost of the corresponding processing process, presented in hryvnias.

Thus, the proposed method allows not only to determine the absolute machinability in its classical sense, but also to obtain additional information about the cutting process (Antonyuk 1998).

To determine the machinability of materials by cutting based on modeling of technological parameters and the developed method, an algorithm for its implementation has been created.

According to the developed algorithm, the database of structural materials and the knowledge base containing mathematical models of indicators for the corresponding pair "structural - processed material" according to the selected cutting method are opened. In the next step, the name of the studied material is entered and mathematical models of its parameters are loaded from the knowledge base.

In the case of solving the problem of determining the relative machinability of a structural material, the name of the material—the standard is entered at the system request, with which the initial indicators of the studied material will be compared. According to the name of the material—the standard, the corresponding mathematical models of the initial parameters of the cutting process are selected from the knowledge base. To determine the relative machinability indicators, the processing conditions are entered in the dialog mode, namely: the dimensions of the processed surface, the parameters of the cutting tool, cutting modes, etc.

Based on the selected mathematical models of the studied and standard materials and the entered processing conditions, partial indicators are determined according to formulas (1) and (2).

In order to determine the complex indicator of the machinability of the material by the set of partial indicators, according to formula (4), weight coefficients (expert estimates) of each partial indicator are entered in the dialog mode, which are established as a result of a survey of experts.

The values of the complex indicator of the relative machinability of the studied structural material are printed. This completes the work of the algorithm for determining the relative machinability of a material by cutting.

When solving the problem of determining the absolute machinability of a material by cutting, which consists in optimizing the processing conditions of a structural material (most often determining the optimal cutting modes), the optimality criterion, processing conditions and parameters of the processed surface are entered in the dialog mode. The area of allowable solutions is formed by specifying the necessary parameters of the quality of the processed surface, the limit values of the machine spindle speed and feed, the stability period of the cutting tool, etc. Based on mathematical models of individual indicators of the cutting process for the processed and tool materials used, a general mathematical model of the cutting process is formed. This model includes an expression of the objective function and constraint functions that create the area of allowable solutions.

The solution of the optimization problem is carried out by one of the methods of multidimensional nonlinear programming (Zheng et al. 2023; Kumar et al. 2022).

Thus, the proposed methodology for determining the machinability of materials and optimizing cutting modes allows:

- to reduce the volume and consumption of tool and structural materials, as well as the time for conducting experimental research;
- to improve the quality of the obtained mathematical models of the initial indicators of the cutting process;
- to determine the relative machinability of the structural material both by a separate partial indicator and by a complex one, taking into account expert estimates (weight coefficients) of each partial indicator;
- to determine the optimal cutting modes taking into account the specific conditions of processing the part according to the selected optimality criterion.

3.3 Practical Determination of Machinability of Structural Materials

The relative machinability of all studied materials was calculated using the developed methodology for determining the machinability of materials by cutting. In this case, previously obtained mathematical models for each type of mechanical processing, which are presented in the knowledge base, were used.

For example, mathematical models were obtained for turning X40Cr13 steel with a through cutter and a VK8 hard alloy insert. The system of mathematical models of the turning process consisted of seven models that determine each of the initial parameters of the cutting process: cutter wear on the rear surface, wear rate, cutting forces, roughness of the machined surface and machining accuracy.

Each model corresponds to its own set of coefficients and power indices, constituent elements of the knowledge base for the studied machining method.

Experimental research to obtain mathematical models was performed on a machine Compact 400 V tool using a cutting speed V in the range of 50…250 m/min, feed S = 10…200 mm/min and cutting depth t = 0.1…1.5 mm.

The experimental research results were conducted by computer technology with programs created by the authors using multifactor statistics analysis methods.

These models have the following form:

– tool wear rate on the rear surface:

$$J = 0,392 \cdot 10^{-3} \cdot V^{0.9633} \cdot S^{-0.19803} \cdot t^{-0.04244} \cdot \tau^{1.2345}, (\text{MM} / \text{XB}); \quad (16)$$

– tool wear on the back surface at the processing the corresponding construction material:

$$h = 3,7 \cdot 10^{-9} \cdot V^{3.78} \cdot S^{-0.9269} \cdot t^{-1.45779} \cdot \tau^{0.08718}, (\text{MM} / \text{XB}); \quad (17)$$

– axial component of cutting force:

$$P_x = 0,9055967 \cdot 10^{-5} \cdot V^{-1.07956} S^{-0.351} t^{1.1266324} \tau^{0.14709}, (\text{H}); \quad (18)$$

– radial component of cutting force:

$$P_y = 0,10364822 \cdot 10^{6} \cdot V^{-1.9517} \cdot S^{0.93243} \cdot t^{1.39113} \cdot \tau^{0.14582}, (\text{H}); \quad (19)$$

– tangential component of the cutting force:

$$P_z = 0,2960037 \cdot 10^{4} \cdot V^{-0.3773} \cdot S^{-0.0675} \cdot t^{-0.8892} \cdot \tau^{0.07957}, (\text{H}); \quad (20)$$

– roughness of the treated surface:

$$R_a = 0,9289302 \cdot 10^{5} \cdot V^{-2.113358} \cdot S^{2.5475} \cdot t^{0.29935} \cdot \tau^{-0.0619} \cdot r^{40.06077} \times HB^{1.65077} \cdot \sigma_B^{1.05706} \varepsilon^{7.9631} \quad (21)$$

– processing error:

$$\delta = 1,46 \cdot 10^{-1.2205} \cdot V^{1.2205} \cdot S^{1.66102} \cdot t^{-6.7039} \cdot \tau^{0.2904} \cdot J^{0.4359} \quad (22)$$

According to the Fisher criterion, the provided mathematical models are adequate within the limits of the used cutting modes with a confidence probability of 0.95.

As an example, we will define the machinability of structural steel X40Cr13. This calculation of machinability requires obtaining the values of partial coefficients and the complex coefficient of relative machinability, which are determined under the same processing conditions for both the studied and the reference materials. In this case, as a reference, it is worth taking a material close to the studied one in terms of physicochemical characteristics (Barandych et al. 2018), the cutting processes of which under different conditions are well studied and have wide application in real production. In most cases, the reference material is steel 45: firstly, based on the above requirements, and, secondly, because this material was used as a reference in other works when developing the basics of determining machinability (Astakhov 2014; Liao et al. 2024).

The initial parameters of the turning process of steel X40Cr13 and steel 45 with a cutting tool made of VK8 hard alloy was determined based on the relevant mathematical models available in the knowledge base for the specified materials.

In this case, the following cutting process modes were selected: speed *V*—90 m/min; feed *S*—60 mm/min; cutting depth *t*—0.2 mm; and cutting time τ—2 min.

The values of the initial indicators of the turning processes of the studied and reference materials, the values of the partial machinability indicators, and the complex indicator under the condition of equal weighting of the partial indicators are given in Table 1.

Analysis of the obtained results shows that the relative machinability of steel X40Cr13 has different values for individual indicators ranging from 0.61 to 2.87. Steel X40Cr13 has the best machinability in terms of machining accuracy, and the worst in terms of cutting forces.

Definition of Absolute Machinability The most important stage in studying the technological parameters of structural materials is the determination of absolute machinability. Its determination is presented as an example of assessing the absolute machinability of a structural material.

The main task solved using absolute machinability is establishing optimal processing modes for each material under study. At the same time, the values obtained for cutting modes did not always correspond to the optimal ones. They did not coincide with the specific features of a particular production process.

The search for optimal cutting modes is based on the obtained mathematical models, based on the selected optimization criterion. The choice of one or another criterion depends exclusively on production requirements, thereby eliminating one of the main disadvantages of previous methods (Kumar et al. 2022).

It is possible to optimize technological parameters for each of the materials entered into the database using the proposed methodology.

Table 2 presents the results of the optimization analysis of the machining of X40Cr13 steel with VK8 hard alloy, on the example of which the relative machinability was previously determined according to various criteria, namely the criterion

Table 1 Results of determining the relative machinability of steel X40Cr13 relative to steel 45

	Output parameter		
Output parameter	Steel X40Cr13	Steel 45	Partial processability coefficients K_i
h	0.0336	0.0237	1.42
J	0.0230	0.0199	1.56
Px	110.928	38.2493	2.87
Py	85.2876	42.9238	1.98
Pz	103.7889	77.7709	1.33
Ra	24.3446	18.5448	1.31
δ	0.0483	0.740	0.61
Complex machinability coefficient $K_к$			1.58

Table 2 Results of optimization of technological parameters of the pair of steel X40Cr13 - VK8 during turning

Pair: Workpiece material—tool	Optimization criteria															
	Absolute deviation				Processing productivity				Tool stability				Cost			
	V	S	t	τ	V	S	t		V	S	t	τ	V	S	t	τ
Steel X40Cr13-VK8	80	31	0.4	1.5	247	30	0.2		40	11	0.2	1.2	242	122	1.0	1.4

of absolute deviation, machining productivity, the value of the stability of the cutting tool, and the cost of surface treatment.

The solution to the optimization task was performed using the original program, and the algorithms created by the authors used the sequential multidimensional nonlinear programming method.

As can be seen from Table 2, the main elements (parameters) of the machining process *(V, s, t, τ)* have different values for each of the optimization criteria. This is due to the fact that the values of these parameters, which are optimal for one criterion, do not satisfy the requirements of the second and vice versa. Therefore, if it is necessary to simultaneously take into account all the criteria, the optimal values of the parameters are determined as the arithmetic mean for each of the initial indicators of the investigated process.

For turning steel X40Cr13 with a through cutter with a VK8 carbide insert, the optimal values of the cutting speed, feed, depth and cutting time parameters will be as follows:

$$V = 227{,}25\,\text{m} / \text{min}, s = 194\,\text{mm} / \text{min}, t = 0{,}45\,\text{mm}, \tau = 1{,}33\,\text{min}.$$

Thus, the examples given for solving the problem of determining the relative and absolute machinability of structural materials have shown the effectiveness of using the proposed method for any type of material processing by cutting, especially when researching new structural materials and establishing optimal conditions for their processing according to various optimization criteria.

4 Conclusions

Based on the analysis of works devoted to determining technological parameters of machining materials, the methodology developed for obtaining mathematical models and determining machinability, and software implementation, the following conclusions can be drawn.

The available methods for determining technological parameters of machining materials are laborious, material-intensive, not sufficiently reliable, and not always suitable for use in production conditions; most of the accelerated methods for determining technological parameters do not consider the real conditions of the cutting process.

The proposed accelerated assessment of the parameters of machining materials in an automated mode is one of the advantages of the proposed methodology, which is advisable to carry out based on a unified dependence.

The mathematical models obtained using the provided methodology are conveniently stored in a knowledge base for further use in determining the machinability of structural materials and the machined properties of tool materials, as well as for optimizing the cutting process parameters. In contrast, the methodology is focused on data processing directly based on research results and the use of other sources.

The mathematical models obtained using the provided methodology are conveniently stored in a knowledge base for further use in determining the machinability of structural materials and the machinable properties of tool materials, as well as for optimizing the cutting process parameters.

The provided mathematical modeling methodology is universal, that is, it can be used for any methods of cutting materials, determine the relative machinability of a structural material both by a separate partial indicator and by a complex one taking into account expert assessments (weight coefficients) of each partial indicator, and also determine the optimal cutting modes taking into account the specific processing conditions of the part according to the selected optimality criterion.

Acknowledgments The research was partially supported by the International Association for Technological Development and Innovations.

References

Antonyuk VS (1998) Formation of wear-resistant antifriction composite coating reinforced with ultradispersed diamonds. Sverkhtverd Mater 4:72–76

Antonyuk V, Barandych K, Vysloukh S (2020) Multicriteria optimization of the part's finishing turning process working in the conditions of alternating loadings. Adv Manuf Proc II:491–501. https://doi.org/10.1007/978-3-030-68014-5_48

Antonyuk V, Barandych K, Gladskyi M, Vysloukh S, Voloshko O (2023) Cyclic durability modeling of parts during finish-turning. CEUR Workshop Proc 3628:365–369. https://ceur-ws.org/Vol-3628/

Astakhov V (2014) Machinability: existing and advanced concepts. In: Machinability of advanced materials. https://doi.org/10.1002/9781118576854.ch1

Barandych KS, Vysloykh SP, Antonyuk VS (2018) Ensuring fatigue life of parts during finish turning with cubic boron nitride tools. J Superhard Mater 40:206. https://doi.org/10.3103/S1063457618030085

Davim JP (2014) Modern mechanical engineering: research, development and education. Springer. https://doi.org/10.1007/978-3-642-45176-8

El-Gaddar A, Azab A, Baki MF (2024) Batch-sizing and machinability data systems for milling operations: an optimal sustainable cost of quality approach. Manuf Lett 41:19–30. https://doi.org/10.1016/j.mfglet.2024.09.006

Ezugwu EO, Bonney J, Yamane Y (2003) An overview of the machinability of aeroengine alloys. J Mater Process Technol 134(2):233–253. https://doi.org/10.1016/S0924-0136(02)01042-7

Ivanov V, Zajac J (2017) Flexible fixtures for CNC machining centers in multiproduct manufacturing. EAI Endor Trans Indus Netw Intell Syst 4(12):e4. https://doi.org/10.4108/eai.10-1-2018.153552

Ivanov V, Vashchenko S, Pavlenko I, Hatala M, Kolos V, Andrusyshyn V (2023) Automated subsystem for cutting modes calculations. In: Balog M, Iakovets A, Hrehova S (eds) EAI international conference on automation and control in theory and practice. EAI ARTEP 2023, EAI/springer innovations in communication and computing. Springer, Cham, pp 299–313. https://doi.org/10.1007/978-3-031-31967-9_23

Kant G, Sangwan KS (2015) Predictive modelling and optimization of machining parameters to minimize surface roughness using artificial neural network coupled with genetic. Proc CIRP 31:453–458. https://doi.org/10.1016/j.procir.2015.03.043

Kim T, Kim K-K, Song E (2024) Selection of the most probable. Best Oper Res. https://doi.org/10.1287/opre.2022.0343

Klocke F (2011) Manufacturing processes, cutting. Springer, New York. https://document.kdu.edu.ua/info_zab/131_1946.pdf

Kumar MSK, Gurudatt B, Reddappa HN, Suresh R (2022) Parametric optimization of cutting parameters for micro-machining of titanium Grade-12 alloy using statistical techniques. Int J Lightweight Mater Manuf 5(1):74–83. https://doi.org/10.1016/j.ijlmm.2021.10.003

Lakic GG, Sredanovic B, Nedic B (2011) Development of mathematical model of universal material machinability. J Balkan Tribol Assoc 17(4):501–510. https://www.researchgate.net/publication/269703388

Lalbondre R, Krishna P, Mohankumar GC (2014) An experimental investigation on machinability studies of steels by face turning. Procedia Mater Sci 6:1386–1395. https://doi.org/10.1007/s00170-012-4703-1

Liao Z, Schoop JM, Saelzer J, Bergmann B, Priarone PC, Splettstößer A, Bedekar VM, Zanger F, Kaynak Y (2024) Review of current best-practices in machinability evaluation and understanding for improving machining performance. CIRP J Manuf Sci Technol 50:151–184. https://doi.org/10.1016/j.cirpj.2024.02.008

Patel NS, Parihar PL, Makwana S (2021) Parametric optimization to improve the machining process by using Taguchi method. Mater Today Proc 47(12). https://doi.org/10.1016/j.matpr.2021.03.005

Polishetty A, Littlefair G, Goldberg M, Nomani J (2016) Current trends in machinability research. Int J Mater Form Mach Process 3(1). https://doi.org/10.4018/IJMFMP.2016010101

Sharma AK, Aravindan N, Ignatia KMJ et al (2024) Hybrid multi criteria decision making methodology for improving sustainability of the manufacturing sector. Int J Interact Des Manuf (IJIDeM). https://doi.org/10.1007/s12008-024-01846-7

Twardowski P, Wieczorowski M (2023) Monitoring of cutting process and tool condition of metal and metal composite. Materials 16:3660. https://doi.org/10.3390/ma16103660

Visloukh SP (2001) Determination of process parameters of new tool. J Superhard Mater 23(5):65–69

Wagri NK, Petare A, Agrawal A, Rai R, Malviya R, Dohare S, Kishore K (2022) An overview of the machinability of alloy steel. Mater Today Proc 62(6):3771–3781. https://doi.org/10.1016/j.matpr.2022.04.457

Wang, D.: Probability theory and mathematical statistics. Beijing Institute of Technology Press, Beijing (2020)

Zheng M, Yu J (2025) Parametric optimization and determination in machining processes by means of probabilistic multi-objective optimization. Mech Eng Adv 3(1):1950. https://doi.org/10.59400/mea1950

Zheng, M., Teng, H., Yu, J., et al.: Probability-based multi-objective optimization for material selection. Springer Nature Singapore (2023). https://doi.org/10.1007/978-981-99-3939-8

Prediction of Turning Accuracy of Large-Pitch Lead Screws with ACME Profile

Oleh Onysko and Volodymyr Kopei

Abstract Running kinematic pairs: screw-nut occupy a large niche of essential parts in mechanisms and machines. In the process of performing screw surfaces, the turning tool is often positioned at an angle of inclination relative to the perpendicular to the thread axis. This is necessary to ensure the cutter's operability for a long time. However, this position of the cutter provokes a significant deviation from the trapezoidal profile of the thread it produces. This is especially true for threads with a large pitch. The purpose of the work is to develop an analytical model of turning lead screws and create and use an interactive program for predicting the accuracy of the cut thread. As a result of using the developed program, it was established:—if a maximum pitch of 1 inch is used for a 3.5 inch diameter lead screw according to the ASME standard, the deviation of the half-profile angle of thread will be 8 arcmin, which is 67% of the tolerance; – the maximum permissible deviations due to ISO standard for lead screws with a diameter of 25.4 mm (1 inch) are from 0.37° to 0.45°, when a deviation of 0.2° was calculated, which is approximately half the tolerance; – in the case of a negative angle of inclination the deviation of the half-profile angle is insignificant.

Keywords Thread profile · Convolute screw surface · Lathe tool · Inclination angle

1 Introduction

The kinematic pair of a nut and a screw is probably the most common mechanical transmission. Various operational requirements are imposed on such mechanisms. For example, transmission reliability and thread profile accuracy are required in

O. Onysko (✉) · V. Kopei
Ivano-Frankivsk National Technical University of Oil and Gas, Ivano-Frankivsk, Ukraine
e-mail: oleh.onysko@nung.edu.ua

V. Ivanov (ed.), *Emerging Technologies in Manufacturing and Materials Engineering*, Springer Tracts in Mechanical Engineering,
https://doi.org/10.1007/978-3-032-14914-5_4

areas where high tightness is provided (hydraulic shut-off valves of water management, oil and gas pipe valves), contributing to high environmental safety (Hryniuk and Arkhypova 2018). Laboratory and measuring devices containing high-precision screw gears are widely used in production and scientific activities (Mazur et al. 2022), as well as in the educational and innovative industries (Shkitsa et al. 2024; Shkitsa et al. 2017). For the quality of the screw, it is essential to use the most optimal technology for grinding its thread (Ropyak et al. 2021). The accuracy of the screw surface depends on the predicted and real radial displacements (Tutko et al. 2021). Chemical coatings and metallization significantly improve threaded surfaces' wear resistance, especially in aggressive operating conditions (Dubei et al. 2022). Power lead screws for jacks must be both strong and precise enough to withstand rapid wear (Hurey et al. 2020). The accuracy of shafts and critical parts, including screws, largely depends on the technological aspects of their production (Karpus et al. 2019; Karpus and Ivanov 2012).

Special requirements for accuracy apply to lead screws used in machine tool drives, robots, and 3D printers. Such kinematic screws are usually made with a large pitch and from wear-resistant materials. As the demand for such drives is constantly growing, technology, efficiency, and stable performance of their production must play a significant role (Kusyi et al. 2024; Kusyi and Stupnytskyy 2020).

2 Fundamentals and Challenges

Lead screws with a large pitch are usually manufactured by turning (Li et al. 2020). However, whirling is used when necessary to increase productivity (He et al. 2019). In (Li et al. 2020), the effect of vibration on thread cutter wear is investigated, which can affect the accuracy of the resulting thread. Accuracy is affected by temperature deformations. Research (He et al. 2019) focuses on analytical modeling of temperature fields during whirling. Several works are related to thread turning on workpieces made of various materials, including difficult-to-machine materials (Costa and Polli 2021; Krawczyk et al. 2022). In (Costa and Polli 2021), the effect of thread turning methods on cutting forces in stainless steel workpieces was investigated. The condition of the thread surface after turning it from a special, difficult-to-machine material, Inconel 718, was investigated in (Krawczyk et al. 2022). It is worth emphasizing that there are special scientific studies on technological damageability in manufacturing (Kusyi 2021). Studies of the technology of manufacturing screw surfaces (Boral and Gołębski 2022; Kawasaki and Tsuji 2023) concern cylindrical worms, not lead screws. Both of these works investigate the accuracy of the thread obtained by the less common and less productive method of milling with an end mill. Ensuring the accuracy of the thread profile in the turning process is investigated in (Fu et al. 2023). This work ensures the stability of the thread cutter profile under vibration conditions. The mathematical model of the helicoid surface obtained as a result of whirling is presented in (Merticaru et al. 2022). The emphasis is on micro-geometric accuracy. The work (Dudás 2016) theoretically predicts the accuracy of the screw obtained in the turning process, but without taking into account the parameters of the turning cutter. The closest to the topic of our research is the study (Onysko et al. 2024), which considers the influence of the geometric parameters of the turning cutters on the accuracy of the thread. However, a conical thread with a triangular profile was investigated, rather than a trapezoidal one, which is used for kinematic lead screws. An analytical study and modeling of the geometry of the profile of kinematic screw surfaces was considered in (Kacalak et al. 2021). However, this work considers exclusively the accuracy of the profile of worm shafts obtained by grinding, which is not typical for the manufacture of most lead screws. This study is interesting because the dependences of the profile deviations from straightness (k and k_2) on the helix angle were analytically derived (Fig. 1).

However, in the presented dependencies, the parameters are the average diameter, thread pitch, and module, which are typical for gear and worm gears. In addition, the equations proven in (Kacalak et al. 2021) do not indicate the functional dependence of the profile of the screw surface on the parameters of the manufacturing process. Common in the papers (Onysko et al. 2024; Kacalak et al. 2021) is the consideration of the screw as a convolute surface. The study (Onysko et al. 2024) indicates that all threads (triangular and trapezoidal) are convolute, and not Archimedean screws, as declared by the standard (Fig. 1). In particular, in (Kacalak et al. 2021), convolute involute worms were described.

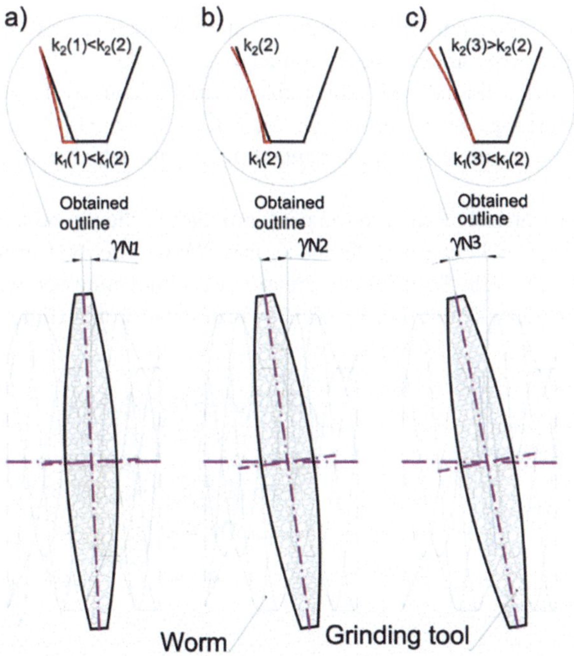

Fig. 1 Influence of the grinding wheel axis inclination angle on the shape of the obtained axial outline of conical helical surface and deviations from straightness k and k_2 (Kacalak et al. 2021)

Research objective to develop an analytical model for turning lead screws with a trapezoidal profile and a large pitch, which takes into account the inclination angle of the cutter, and create an interactive program to predict the accuracy of the resulting thread profile.

3 Ideas and Solutions

3.1 Theoretical Fundamentals

A gap must be provided between the cutting insert's side surfaces and the thread's sides to ensure uniform tool wear and obtain high-quality threads. Therefore, the cutter, or its cutting insert, should be inclined at an angle λ (inclination angle) to obtain the maximum symmetrical gap relative to the sides of the thread. This angle is actually equal to the thread lead angle ψ (Thread turning, n.d.) (Fig. 2):

$$\lambda = \psi = \arctan\frac{P}{\pi D}, (\text{rad}) \quad (1)$$

However, in (Onysko et al. 2024) it is proved that the thread profile will not correspond to the standard one when using a thread cutter with a cutting edge that is not aligned with the axial plane of the thread. It is known that the profile of a trapezoidal thread consists of two mutually symmetrical straight sections, located at an angle $\varphi/2$ relative to the perpendicular to the thread axis (Fig. 3).

To produce such a thread with conventional cutters from well-known manufacturers (Thread turning, n.d.), only when the cutting edge (1-2-3-4) entirely coincides with the axial plane of the fundamental thread cone, which is marked in yellow in Fig. 4.

If the cutter is rotated relative to the perpendicular to the thread axis by an angle λ, an ideal fundamental cone with the generatrix *BN* will not be formed. The cutter (red) rotated by an angle λ causes the creation of a convolute screw, rather than a standard Archimedean screw. The result is a curvilinear profile (lime, Fig. 5).

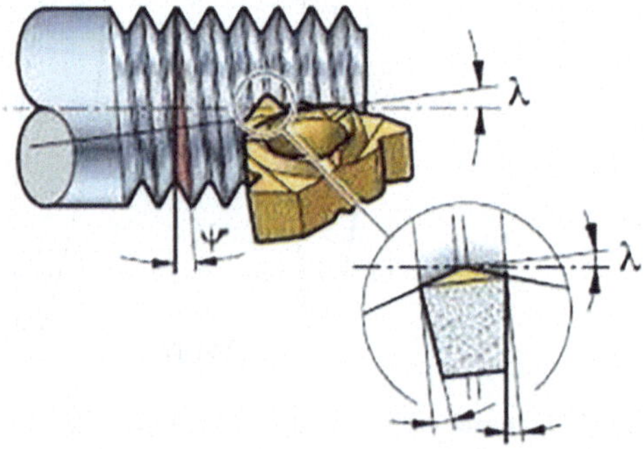

Fig. 2 Scheme of installing the cutting insert of the thread cutter at an angle λ to ensure its uniform wear (Thread turning, n.d.)

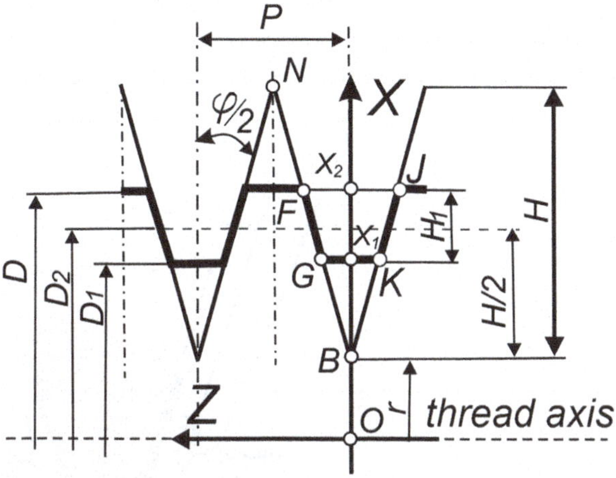

Fig. 3 Schema of metric trapezoidal thread according to a GOST and ISO 2904 standard in ZX Cartesian coordinate system: D—major diameter, D_1—minor diameter, D_2—pitch diameter, P—pitch, H—height of fundamental triangle ($H = 1.866P$), H_1—depth of thread ($H_1 = 0.5P$)

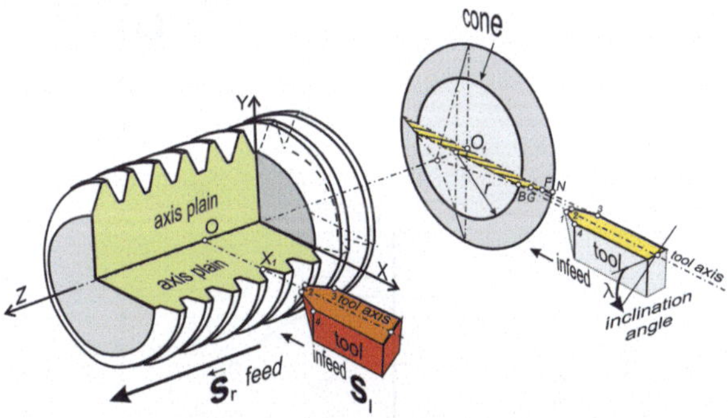

Fig. 4 Scheme of layout of the threading turning tool layout at an angle of inclination λe

This curvilinear profile is created by the edge 3-2 of the cutter (section *AD* of the thread) and the edge 1-4 (section *EL* of the thread) (Fig. 5). In the case of a straight-line interpolation of the curve *AD* by its two extreme points *A* and *D*, as well as the curve *EL* by its two extreme points *E* and *L*, a new profile angle $\varphi + \delta$ can be determined, where the value δ is a positive or negative deviation from the standard thread profile angle φ.

According to the study (Onysko et al. 2024), the resulting profile of the convolute screw is defined in *ZX* coordinates (Fig. 6):

$$Z(x) = \tan(\sigma)x\frac{\sin\tau}{\sin\eta} - \frac{P}{2\pi}\tau, \tag{2}$$

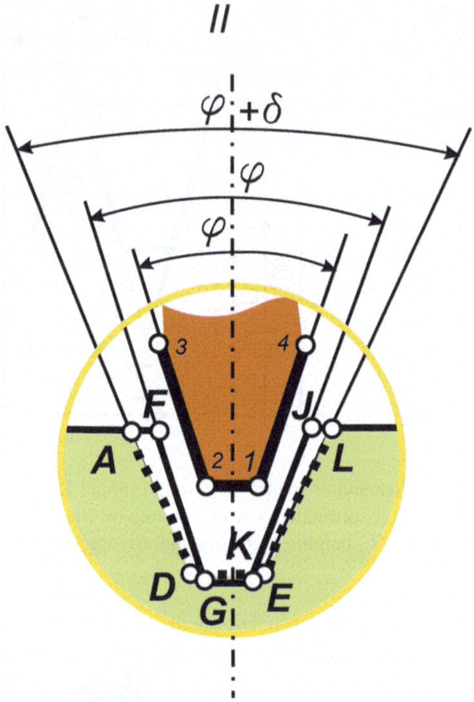

Fig. 5 Diagram of a standard and tool-formed thread profile

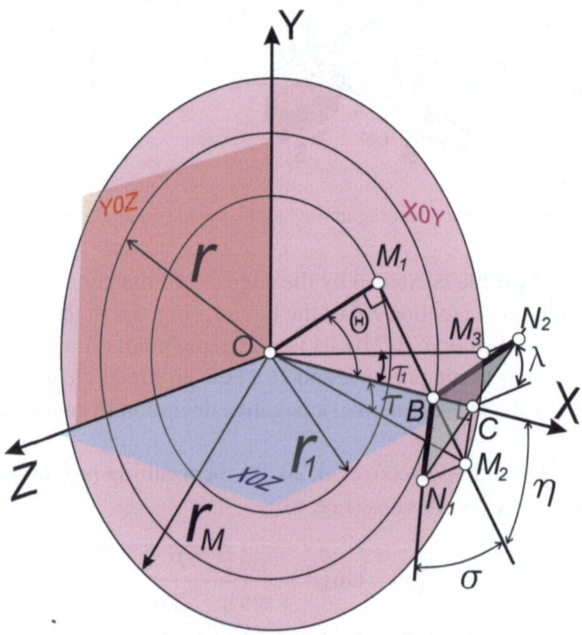

Fig. 6 Scheme for calculating the thread profile obtained by a cutter with a non-zero cutting edge inclination angle

where

$$\tau = \eta - \arcsin\left(\frac{r sin\eta}{x}\right)(\text{rad}); \tag{3}$$

$$\eta = \arctan\left(\frac{P sin\lambda}{2H}\right)(\text{rad}); \tag{4}$$

$$\tan(\sigma) = \frac{P cos\lambda}{2H}\cos\left(\arctan\left(\frac{P sin\lambda}{2H}\right)\right). \tag{5}$$

The quantities σ, τ, η are determined using cylindrical coordinates (with axis *OZ*, polar axis *OX*, polar radius OM_2 and polar angle τ) that are combined with Cartesian coordinates *OZYX* (Onysko et al. 2024) (Fig. 6).

The radius *r* according to the method (Onysko et al. 2024) corresponds to the distance from the thread axis to the vertex of the fundamental triangle (point *B*) (Fig. 6). The left side of the fundamental triangle is denoted by the segment *BN* (Fig. 3), i.e. it is the generator of the fundamental starting cone of the standard screw (Fig. 4).

To determine the value of *r*, use the diagram in Fig. 3:

$$r = \frac{D}{2} - \frac{H_1}{2} - \frac{H}{2} = D/_2 - 0.25P - 0.933P = D/_2 - 1.183P. \tag{6}$$

The profile curves *AD* and *EL* of a trapezoidal thread obtained by a cutter placed at an angle λ should obviously be determined in the range from X_1 to X_2, which are the coordinates of points *G*, *F*, and also *K*, *J*.

The diagram in Fig. 6 allows us to interpret equation (2–5). On it, the straight line BN_1 contains the cutting edge of the cutter (1–4), the straight line BN_2 contains the cutting edge of the cutter (2–3). It can be seen from the diagram that the polar angle τ corresponds to a clockwise rotation around the *Z* axis, and τ_1—counterclockwise rotation. Therefore, in calculating the profile of the thread section *EL* obtained by turning, formula (4) with the "–" sign should be used.

Based on Eqs. (1)–(6), an interactive program was created to construct the cutter profile.

3.2 Program Description

Developed in Python program (Kopei and Onysko 2025) is designed to build the profile of the lead screw thread, which is formed by turning with $\lambda \neq 0$. To construct the profile, the program uses the dependence (2). The program also calculates the following profile parameters:

- axial deviation from the ideal profile y_0 with $\lambda \neq 0$ (in the program it is denoted as d$y = y_0 - y$), where $y = Z(x)$ (2);
- angle of the tangent to the profile (α);
- profile curvature (c), calculated by numerical differentiation;
- angle of the line constructed through the extreme points of the profile y (α_{12}).

To get the results, you need to execute all the cells ("Run all") and enter the input data (Fig. 7): diameter d (mm), pitch P (mm), thread profile angle φ (°), inclination angle λ (°), right axis of the graph "*Raxis*", variable "*variable*", its step "*step*" and number of curves n. After changing these data, the graph is automatically rebuilt. Two values of the profile angle φ are provided (30° and 29°). If λ = "*optimal*", then the program calculates the value of λ according to formula (1). If $n > 1$, then the program builds several curves for the selected in the "*variable*" list value (d, P, ϕ, λ). The first curve is built for the value specified in the fields "*d*", "*P*", "*fi*", "*lmbda*". To build the second curve, the *step* value is added to this value. To build the third curve, 2**step* is added to it. And so on. The results are plots (Figs. 8, 9 and 10).

3.3 *Calculation Results for ISO-GOST Threads*

The results (Figs. 8, 9 and 10, Tables 1 and 2) show that increasing the thread diameter leads to a decrease in dy deviations, profile angle deviations and an increase in profile curvature. Increasing the thread pitch leads to an increase in dy deviations, profile angle deviations and profile curvature. Increasing the nominal profile angle leads to a decrease in dy deviations, an increase in profile angle and curvature deviations. Increasing the angle λ leads to an increase in dy deviations, profile angle deviations and an increase in profile curvature. Negative values of λ lead to an increase in deviations not as sharply as positive ones. The profile curvature almost does not change for values of +λ and -λ.

Fig. 7 Program input data

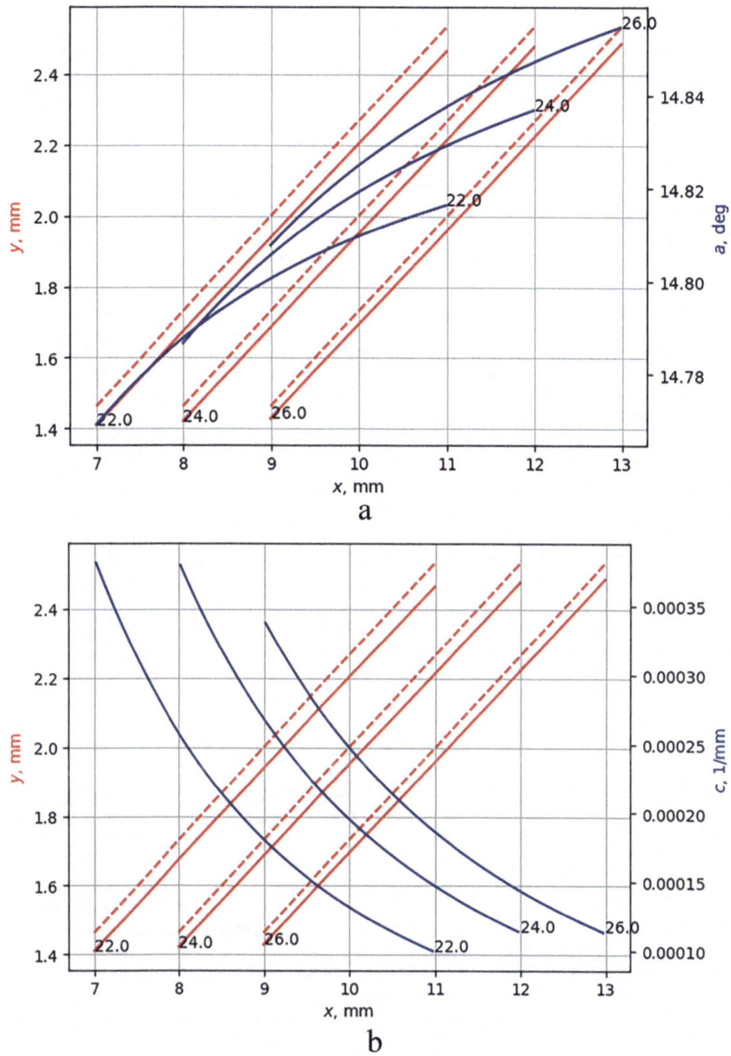

Fig. 8 Thread profiles y of lead screws $d \in \{22, 24, 26\}$ (mm), $P = 8$ mm: optimal $\lambda \neq 0$ (solid red); $\lambda = 0$ (dashed red); tangent angle α (blue, **a**), curvature c (blue, **b**)

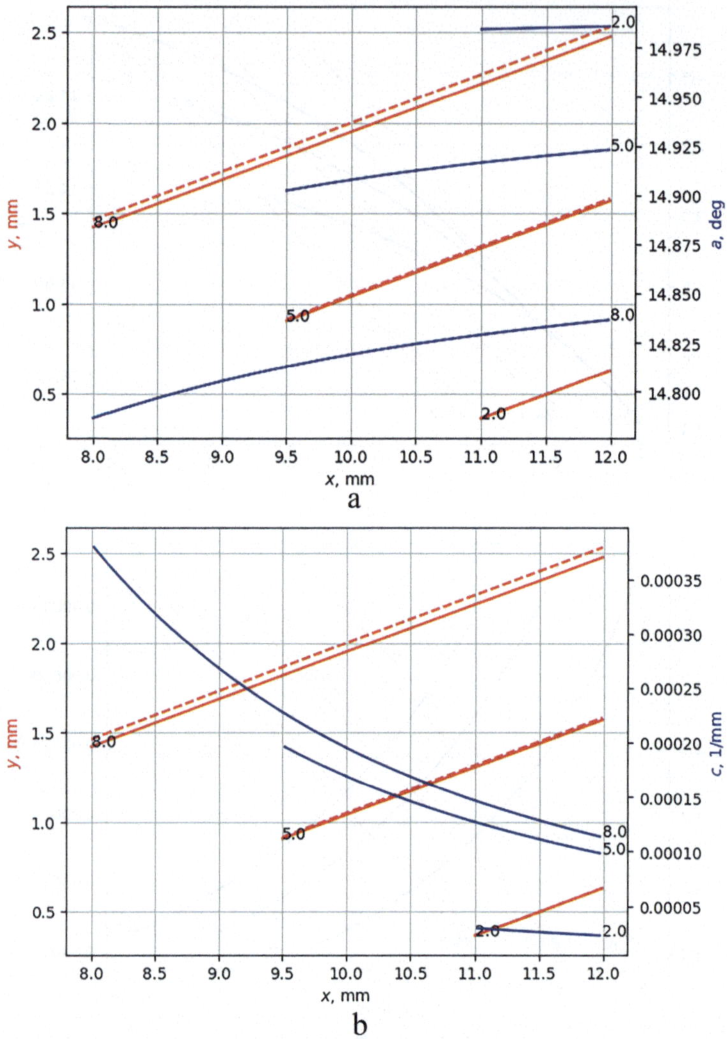

Fig. 9 Thread profiles y of lead screws $P \in \{2, 5, 8\}$ (mm), $d = 24$ mm: optimal $\lambda \neq 0$ (solid red); $\lambda = 0$ (dashed red); tangent angle α (blue, **a**), curvature c (blue, **b**)

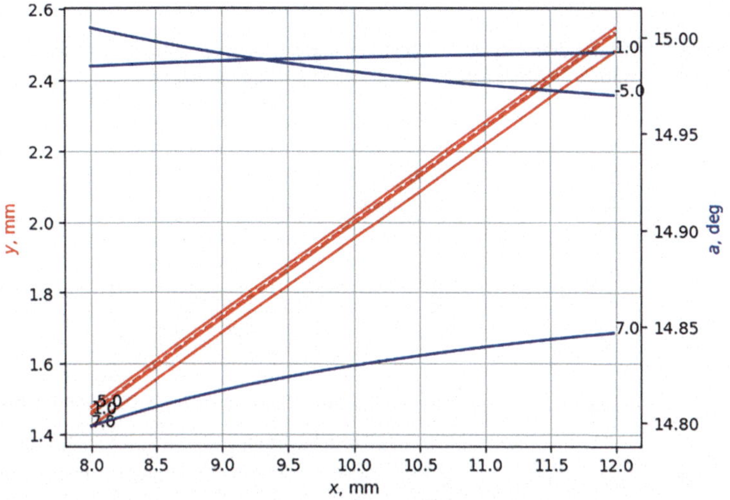

a

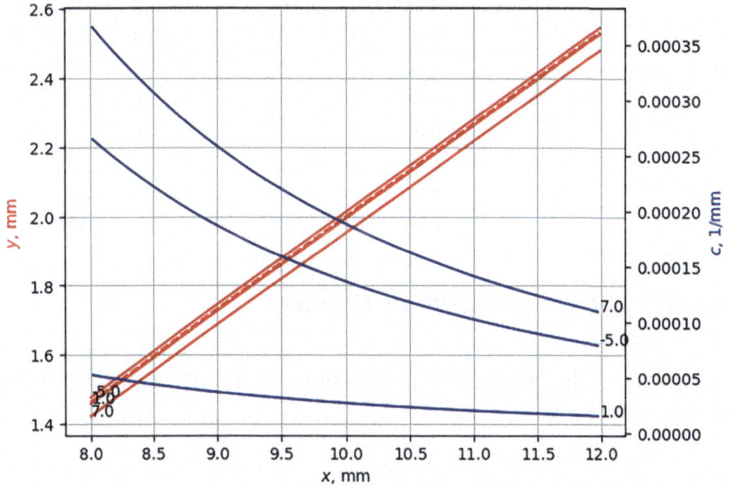

b

Fig. 10 Thread profiles y of lead screws $\lambda \in \{-5°, 1°, 7°\}$, $d = 24$ mm, $P = 8$ mm: $\lambda \neq 0$ (red solid); $\lambda = 0$ (red dashed); tangent angle α (blue, **a**); curvature c (blue, **b**)

Table 1 Results of calculations of half-profile angle values (ISO, GOST)

Thread pitch P, mm	Major diameter d, mm	Inclination angle λ, °	Semi-profile angle of the resulting thread $\varphi/2°$
8	22	8.052259	14.797943
8	24	7.256083	14.816678
8	26	6.602550	14.835449
2	24	1.585492	14.985184
5	24	4.233630	14.914047
8	24	−5	14.983876
8	24	1	14.989257
8	24	7	14.827294

Table 2 Tolerances and calculated deviations of a 14.5° flank angle for ASME B1.5-1997

				14.5° tolerance			14.5° deviation	
d, in.	Threads/in.	P, in.	λ, deg	deg	arcmin	a_{12}, deg	deg	arcmin
0.25	16	0.0625	5.196	0	47	14.388	0	7
0.375	12	0.08333	4.550	0	41	14.408	0	6
0.5	10	0.1	4.046	0	39	14.424	0	5
0.625	8	0.125	4.046	0	35	14.424	0	5
0.75	6	0.16667	4.550	0	30	14.408	0	6
1	5	0.2	4.046	0	27	14.424	0	5
1.375	4	0.25	3.643	0	25	14.436	0	4
2.25	3	0.33333	2.915	0	22	14.456	0	3
3.5	1	1	6.057	0	12	14.361	0	8

3.4 ASME Thread Calculation Results

The results of the calculations of the deviations of the ASME thread half-profile angle are shown in Table 2.

3.5 Discussion

For the study, screws with a diameter of 22, 24, 26 mm were selected precisely because the ISO and GOST standards provide for them with pitch options of 2, 5, 8 mm. For these screws, the lead angle ψ varies from 1.6° to 8°. For a diameter of 22 mm, the largest pitch is 8 mm, which causes a lead angle of 8° and leads to a significant deviation of the semi-profile angle (0.2°). A similar situation occurs when considering screws with a diameter of 30 mm, for which the maximum pitch is 10 mm and the lead angle is 7.25°. However, the ISO and GOST standards do not regulate the permissible deviations for the half-profile angle.

Considered ASME lead screws that have a deviation in the half-profile angle within their tolerance. The largest deviation (8 arcmin) is observed for a screw with a diameter of 3.5 inches, a pitch of 1 inch, and a lead angle of 6°. This is 67% of the tolerance of 12 arcmin.

4 Conclusions

For lead screws with large pitches, a mandatory technological norm is to use an inclination angle equal to the helix lead angle. This approach increases the cutter stability and increases the productivity of the thread turning process, but can lead to deviations in the profile of the resulting trapezoidal thread. The profile becomes curvilinear with insignificant curvature and acquires a deviation in the profile angle.

A significant profile deviation is a decrease in the half-profile angle by up to 0.2°, which is predicted to occur if there is turning of lead screws with a diameter of 22–26 mm and a maximum pitch of 8 mm. Such a change can be considered acceptable, since according to the ASME standard, the maximum permissible deviations for screws with a diameter of 25.4 mm (1 inch) are from 0.37° to 0.45°. That is, 0.2° is approximately half the tolerance.

If a maximum pitch of 1 inch is used for a 3.5-inch diameter lead screw according to the ASME standard, the deviation of the semi-profile angle will be 8 arcmin, which is 67% of the tolerance. Additional machining errors in real production conditions can lead to defects.

If a negative cutting edge inclination angle is in use −5°, the deviation of the half-profile angle is insignificant and is 0.02°, equivalent to 4% of tolerance only.

In the near future, studies are planned to check the results of this article experimentally.

Acknowledgments The authors are grateful to the Ministry of Science and Education of Ukraine for the grant to implement the project (RK 0124U000654). The research was partially supported by the International Association for Technological Development and Innovations.

References

Boral P, Gołębski R (2022) Technology of Manufacturing of ZC cylindrical worm. Materials 15:6412. https://doi.org/10.3390/ma15186412

Costa CE, Polli ML (2021) Effects of the infeed method on thread turning of AISI 304L stainless steel. J Braz Soc Mech Sci Eng 43:253. https://doi.org/10.1007/s40430-021-02978-7

Dubei OY, Tutko TF, Ropyak LY, Shovkoplias MV (2022) Development of analytical model of threaded connection of tubular parts of chrome-plated metal structures. Metallofiz Noveishie Tekhnol 44(2):251–272. https://doi.org/10.15407/mfint.44.02.0251

Dudás I (2016) The extension of the general mathematical model developed for helicoidal surfaces to the whole system of manufacturing technology and production geometry (ProMAT). Int J Adv Manuf Technol 86:1557–1572. https://doi.org/10.1007/s00170-015-8233-5

Fu X, Li K, Li Z et al (2023) A SVM-based design method for cutting edge profile stability of large-pitch thread turning tool considering vibration. Int J Adv Manuf Technol 125:4529–4547. https://doi.org/10.1007/s00170-023-10985-2

He Y, Liu C, Wang Y, Li Y, Wang S, Wang L, Wang Y (2019) Analytical modeling of temperature distribution in lead-screw whirling milling considering the transient un-deformed chip geometry. Int J Mech Sci 157–158:619–632. https://doi.org/10.1016/j.ijmecsci.2019.05.008

Hryniuk VI, Arkhypova LM (2018) Regularity of effects of climatic changes on quality indicators of surface water of the Dniester basin. Naukovyi Visnyk Natsionalnoho Hirnychoho Universytetu 3:125–133. https://doi.org/10.29202/nvngu/2018-3/17

Hurey I, Hurey T, Gurey V (2020) Wear resistance of hardened nanocrystalline structures in the course of friction of steel-grey cast iron pair in oil-abrasive medium. In: Advances in design, simulation and manufacturing. DSMIE 2019, Lecture notes in mechanical engineering. Springer, Cham, pp 572–580. https://doi.org/10.1007/978-3-030-22365-6_57

Kacalak W, Majewski M, Budniak Z, Ponomarenkow J (2021) Worm gear drives with improved kinematic accuracy. Materials 14:7825. https://doi.org/10.3390/ma14247825

Karpus VE, Ivanov VA (2012) Locating accuracy of shafts in V-blocks. Russ Engin Res 32:144–150. https://doi.org/10.3103/S1068798X1202013X

Karpus, V., Ivanov, V., Dehtiarov, I., Zajac, J., Kurochkina, V.: Technological assurance of complex parts manufacturing. In: Ivanov, V., et al. Advances in design, simulation and manufacturing. DSMIE 2019. Lecture notes in mechanical engineering, pp. 51–61. Springer, Cham (2019). https://doi.org/10.1007/978-3-319-93587-4_6

Kawasaki K, Tsuji I (2023) Manufacturing method for large cylindrical worm gear set of ISO type I on universal CNC machine tools. J Manuf Mater Process 7:53. https://doi.org/10.3390/jmmp7020053

Kopei V, Onysko O (2025) Interactive program threading_dev3.ipynb. https://colab.research.google.com/drive/1SStV-wu6LcEfNX2oPD29kfmTGcemNNNP. last accessed 2025/02/02

Krawczyk B, Szablewski P, Mendak M, Gapiński B, Smak K, Legutko S, Wieczorowski M, Miko E (2022) Surface topography description of threads made with turning on Inconel 718 shafts. Materials 16(1):80. https://doi.org/10.3390/ma16010080

Kusyi Y (2021) Technological inheritability of product material using the criterion of technological damageability. In: Ivanov V, Trojanowska J, Pavlenko I, Zajac J, Peraković D (eds) Advances in design, simulation and manufacturing IV. DSMIE 2021, LNME, vol 1. Springer, Cham, pp 52–61. https://doi.org/10.1007/978-3-030-77719-7_6

Kusyi Y, Stupnytskyy V (2020) Optimization of the technological process based on analysis of technological damageability of castings. In: Ivanov V, Trojanowska J, Pavlenko I, Zajac J, Peraković D (eds) Advances in design, simulation and manufacturing III. DSMIE 2020, Lecture notes in mechanical engineering. Springer, Cham, pp 276–284. https://doi.org/10.1007/978-3-030-50794-7_27

Kusyi Y, Stupnytskyy V, Kostiuk O, Onysko O, Dragašius E, Baskutis S, Chatys R (2024) Control of the parameters of the surface layer of steel parts during their processing applying the material homogeneity criterion. Eksploatacja i Niezawodnosc-Maint Reliab 26(3):187794. https://doi.org/10.17531/ein/187794

Li Z, Fu X, Li J et al (2020) Establishment of vibration wear model for turning large-pitch thread tools and its wear suppression method. Int J Adv Manuf Technol 109:857–876. https://doi.org/10.1007/s00170-020-05403-w

Mazur TM, Slyotov MM, Prokopiv VV, Slyotov OM, Mazur MP (2022) Light emitters based on II-VI chalcogenides with nanostructured surface. Mol Cryst Liq Cryst 752(1):95–102. https://doi.org/10.1080/15421406.2022.2091276

Merticaru V, Nagîț G, Dodun O, Merticaru E, Rîpanu MI, Mihalache AM, Slătineanu L (2022) Influence of machining conditions on micro-geometric accuracy elements of complex helical surfaces generated by thread whirling. Micromachines 13:1520. https://doi.org/10.3390/mi13091520

Onysko O, Kopei V, Barz C, Kusyi Y, Baskutis S, Bembenek M, Dašić P, Panchuk V (2024) Analytical model of tapered thread made by turning from different machinability workpieces. Machines 12:313. https://doi.org/10.3390/machines12050313

Ropyak LY, Vytvytskyi VS, Velychkovych AS, Pryhorovska TO, Shovkoplias MV (2021) Study on grinding mode effect on external conical thread quality. IOP Conf Series Mater Sci Eng 1018(1):art. no. 012014. https://doi.org/10.1088/1757-899X/1018/1/012014

Shkitsa LY, Panchuk VG, Kornuta VA (2017) Innovative methods of popularizing technical education. IOP Conf Series Mater Sci Eng 200(1):art. no. 012023. https://doi.org/10.1088/1757-899X/200/1/012023

Shkitsa L, Yatsyshyn T, Kornuta V, Deinega I, Boiko N (2024) Innovative educational and production laboratories at the Technical University. In: Shchokin R, Iatsyshyn A, Kovach V, Zaporozhets A (eds) Digital technologies in education. Studies in systems, decision and control, vol 529. Springer, Cham, pp 67–81. https://doi.org/10.1007/978-3-031-57422-1_6

Thread turning (n.d.). https://www.sandvik.coromant.com/en-gb/knowledge/threading/thread-turning, last accessed 2025/01/30

Tutko T, Dubei O, Ropyak L, Vytvytskyi V (2021) Determination of radial displacement coefficient for designing of thread joint of thin-walled shells. In: Advances in design, simulation and manufacturing IV. DSMIE 2021, LNME, Lecture notes in mechanical engineering. Springer, Cham, pp 153–162. https://doi.org/10.1007/978-3-030-77719-7_16

Innovative Approaches to the Restoration of Engineering Products

Ihor Yakovenko and Olexander Permyakov

Abstract A key trend in modern engineering is the reduction of product life cycles and a corresponding decrease in the emphasis on repair and restoration activities. However, this trend does not extend to unique and high-value equipment and machinery, where repair and modernization remain critical. This chapter explores approaches to enhancing the efficiency of repair and restoration operations by leveraging advancements in information technologies, CALS methodologies, and the "Industry 4.0" paradigm. The discussion focuses on two principal areas: the reverse engineering of machine components and using portable machinery to repair and modernize engineering systems, mainly when disassembly and transportation are challenging or impractical. The chapter provides a detailed analysis of the organizational framework for reverse engineering processes, examining factors influencing the efficiency and precision of creating 3D models of restored objects. The selection of manufacturing methods, whether additive or subtractive, for producing replacement parts and the associated software development are also thoroughly evaluated. Key characteristics of portable machine tools are reviewed and compared to traditional machine tool layouts, emphasizing their unique operational features. These distinctions inform optimal strategies for deploying portable equipment. The study concludes that designing and manufacturing portable machine tools based on the aggregate-modular principle is generally the most effective approach. This methodology not only reduces design and production timelines but also enhances the operational efficiency of individual modules.

Keywords Reverse engineering · 3D model · Repair pieces · Portable machine tools · Modular construction

I. Yakovenko (✉) · O. Permyakov
National Technical University "Kharkiv Polytechnic Institute", Kharkiv, Ukraine
e-mail: ihor.e.yakovenko@khpi.edu.ua

V. Ivanov (ed.), *Emerging Technologies in Manufacturing and Materials Engineering*, Springer Tracts in Mechanical Engineering,
https://doi.org/10.1007/978-3-032-14914-5_5

1 Introduction

Due to the rapid development of scientific and technological progress, modern production can meet a wide range of people's needs for essential goods in various fields, including mechanical engineering, which causes a steady increase in the production volume of goods. On the other hand, scientific and technological progress allows for the rapid creation of more functional, economical, and technologically advanced products, which leads to faster obsolescence of consumer goods and the desire of consumers to replace old products with new, more advanced ones (a striking example of this is the production of computer and household appliances, smartphones, etc.) (Basova et al. 2023).

Thus, according to (Manoochehri et al. 2022), at present, the average period obsolescence of equipment in mechanical engineering has decreased from 10–15 years in the 1990s to 4–6 years in developed countries and countries with less developed economies, such as some countries in Asia, Africa or Latin America, the period of obsolescence may remain above than 10–15 years and, accordingly, significant repairs of such equipment remain more popular. All this impacts the trends that are taking place in the field of technical maintenance and repair of equipment. On the one hand, the steady growth of production volumes and product compliance principles (10 R) of the increasingly popular circular economy requires an increase in the volume of various aspects of maintenance (Bondar et al. 2024a). On the other hand, obsolescence and, in many cases, the economic inexpediency of repairs lead to a decrease in the repair work associated with restoring the operability of products.

According to (Marcus et al. 2020), the volume of the repair market in developed countries has decreased by 15–20% over the past two decades. Still, the demand for modernization services has grown by approximately 30% (mainly due to heavy engineering facilities) (Basova et al. 2023). This ensures stabilization or even a slight increase in the volume of work in the maintenance and repair industry (up to 2–4% in various countries). However, this does not apply to expensive, unique machines whose service life exceeds moral aging. In such cases, the scope of repair work increases due to the need to modernize and adapt equipment to modern requirements (Bondar et al. 2024b). In such industries as energy, mining, aviation, shipbuilding, and others, where heavy equipment is used, the share of repair work is 40–60% since the acquisition and installation of new equipment are associated with enormous investment costs. Examples include main pipelines, hydroelectric power plants, mining equipment, etc.

2 Fundamentals and Challenges

Repair work in the mechanical engineering industry can be conditionally divided into four groups.

1. Repairs and possibly upgrades of the product are performed through disassembly and assembly work at the manufacturing plant or branches (certified representatives) after the product is delivered to the plant (branch) for repairs. In this case, certified elements manufactured similarly to the original replace damaged parts and units. Additional work may be performed during upgrades or metalwork during repairs using the plant's equipment (Bondar et al. 2024a).
2. Repairs or upgrades are performed at specialized enterprises not part of the manufacturer's network. In this case, elements are replaced using parts ordered from other manufacturers, or, if there are appropriate production capacities, a non-original part is manufactured directly at the enterprise using the available drawings, technical requirements, and manufacturing technology.
3. Repair and modernization are carried out by enterprises providing various services specializing in restoring parts and elements for various purposes. Moreover, in many cases, there is no design documentation, technology, or technical requirements for manufacturing the failed part (Bondar et al. 2024a).
4. Repair and modernization are performed directly at the place of installation (location) of the product. For products that do not require additional mechanical operations, repair is carried out using assembly and disassembly work using metalworking tools (mainly manual). In cases where additional mechanical processing is necessary, special equipment is used. In modern mechanical engineering, portable machines are increasingly used for these purposes. This is especially typical for large-sized products; the dismantling and transportation are impractical or even impossible (Bondar et al. 2024b).

The issues of organizing repair work at enterprises of the first and second groups in modern development conditions and by the provisions of the circular economy concept are widely covered in technical literature (Fischer et al. 2023; Bradley and Persson 2022; Ntsondé et al. 2023; Breteler, n.d.). We want to describe some aspects of organizing work at enterprises of the second and third groups.

3 Ideas and Solutions

Developing new materials, object modeling systems, and modern methods for creating and restoring parts, such as additive technology, plasma spraying, electric spark methods, etc., have contributed to a whole new direction in the mechanical engineering industry—reverse engineering (Bondar et al. 2024b). The improvement of these areas leads to an intensive growth in the number of enterprises (especially small and medium-sized) that work in this area (Global Reverse Engineering Service Market 2024), which leads to software development for this industry. According to estimates (Reverse engineering market report overview 2024), the global reverse engineering software market was valued at approximately US$1.5 billion in 2023 and is projected to reach US$3.8 billion by 2032, indicating a compound annual growth rate (CAGR) of 10.9%.

The development of reverse engineering in mechanical engineering is currently proceeding in four directions (What Is Reverse Engineering? 2024): rapid prototyping; additive and subtractive manufacturing when replacing inoperable parts; modernization of heavy engineering facilities; virtualization of objects and creation of digital twins.

3.1 Restoration of Parts Outside the Product Location Area

The process of restoring parts is a complex design and technological task. In addition to geometric dimensions, the solution partially changes some characteristics of the part laid down by the designer: the material of individual sections, physical and mechanical properties, surface roughness, etc. In this regard, the issue of organizing the production process of restoring parts, during which the problems of the quality of restoring parts are resolved, is of particular importance.

As noted in their work, Pham and Hieu "Reverse engineering (RE) is generally defined as the process of analyzing an object or an existing system (hardware and software) to identify its components and their relationships and to investigate how it works, to reverse engineer or create a copy without access to the design from which it was made" (Pham and Hieu 2007). Initially, this process was based on scanning and converting the scanned data into a three-dimensional surface or solid model (Raja 2007). Later, as technology and computer-aided design systems developed, various methods of digitizing the physical model were considered in detail, such as the use of photogrammetric measurement systems (Bonitz and Krzystek 1997), laser rangefinder scanners (Page et al. 2007), and the use of fast tactile or optical sensors to define free-form surfaces and display geometric and topological information (Roth-Koch 1996). Subsequently, these approaches to data acquisition, surface-adjustment during reconstruction, and 3D object modeling were advanced under the Industry 4.0 paradigm across sectors involved in manufacturing highly complex geometries (Skytt 1996; Kyaw et al. 2023; García-Molina et al. 2023; Valerga et al. 2023), including artificial intelligence (Galvez et al. 2023).

Any process of restoring an object without design and technological documentation begins with the design of its geometric model, which determines the production methods and technology.

Under these circumstances, reverse engineering should be understood as a process that includes traditional engineering steps alongside reverse-engineering procedures. This set contains heuristic methods, design calculation methods, and metrological procedures to create an adequate model of a real object with possible implementation using traditional subtractive or additive automated manufacturing technologies.

In general, a formalized methodology for technological support of reverse engineering tasks includes a sequence of actions (Fig. 1, the dark arrows show the process of coupling restoration using additive technology, the light arrows show possible options for using subtractive technology): analysis of the part; creation of

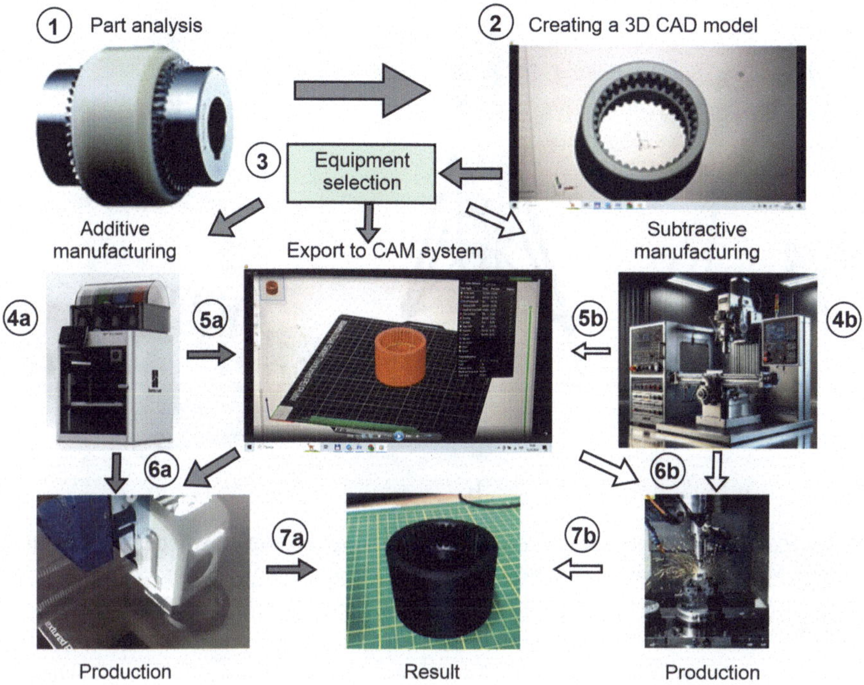

Fig. 1 Structural diagram of technological support for reverse engineering tasks, on the example of coupling manufacturing

3D CAD models; export this model to the CAM system; equipment selection; obtaining a control program or slider; equipment preparation/adjustment (CNC machine or 3D printer); production of the part.

The method of restoring a product (additive or subtractive) is determined primarily by the functional purpose of the part, the conditions of its operation, and, accordingly, the geometric parameters of the part and the physical and mechanical properties of the material used. In this case, the economic indicators of one method of obtaining an analog part are also considered.

At the initial stage of object restoration, its geometric shape is analyzed. In modern conditions, this involves 3-D scanning of a real object or damaged part. Currently, contact and non-contact scanning methods are used.

Depending on the direction of its activities (dimensions of the restored object, its location, geometric accuracy, and complexity of the surface configuration), an enterprise has the opportunity to choose equipment (using Hexagon models (Hexagon 2024) as an example) from hand-held laser scanners of the hand type (Fig. 2a) and portable trackers for measuring large-sized parts (Fig. 2b) to coordinate measuring machines of various types (Fig. 2c). Table 1 shows the main characteristics of the measuring equipment.

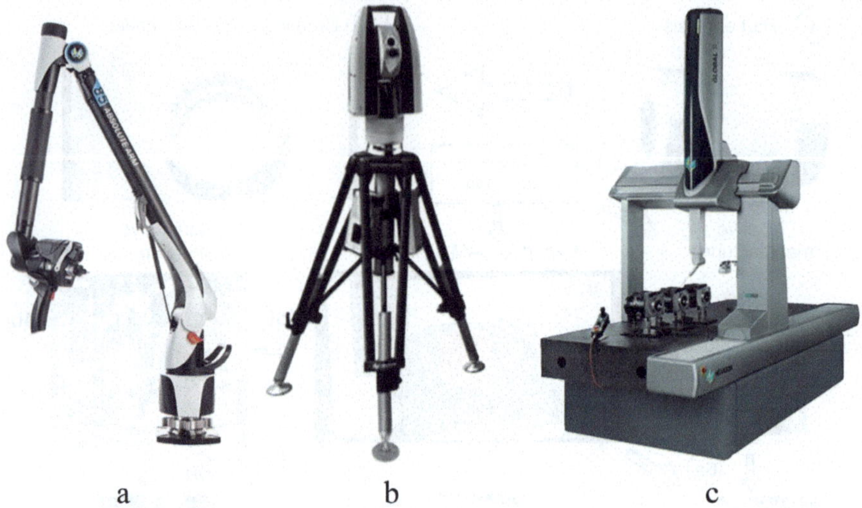

Fig. 2 Examples of object scanning equipment: (**a**) Hexagon Absolute Arm CMM; (**b**) LEICA AT500 CMM; (**c**) DEA GLOBAL CMM

Table 1 Main characteristics of the measuring equipment

Equipment model	Scanning method	Measuring range	Measurement accuracy,	Scanning speed	Recognition	File format
Hexagon absolute arm CMM 7 axis	Blue or red lasers	2,0…4,5 m	From 39 μm	Up 1, 2 million points/s	From 0,001 mm	STL, OBJ, DXF, X_T
Leica AT500 CMM	Touch	To 300 m	From 15 μm +6 μm/m	From 100 points/s	From 0,001 mm	X_T, DXF
DEA GLOBAL CMM	Touch, or blue laser	To 2×4×1,5 m	From 1,2 μm	1000 points/s (touch) 600,000 points/s (laser)	From 0,0001 mm	STL, OBJ, DXF, X_T

Processing of measurement results and construction of a 3D model of the object is carried out by software tools that are built into the CMM by manufacturers, third-party programs for analyzing measurement results (Geometric Design X, PolyWorks/InnovMetric, 3DReshaper RealityCapture, etc.), and extensions with a reverse engineering function for the most popular CAD systems (ScanTo3D for SolidWorks, Autodesk Fusion 360, Siemens NX, CATIA, etc.).

When creating a 3D model, much attention is paid to correcting measurement results (eliminating "noise") and developing a parametric model. This is especially important when creating a model based on analyzing the parameters of a part that has lost its original shape (due to wear, breakage, deformation, etc.). The software allows you to take into account damaged segments, the symmetry of the part elements depending on their functional purpose, and the formation of the surface contour due to geometric primitives in cases where this does not contradict the functional purpose of complex-profile parts (for example, the profile of turbine blades is built by smoothing without the use of traditional geometric primitives). Thus, Fig. 3(a) shows the result of scanning a real part, and Fig. 3(b) shows a computer model after eliminating image noise and correcting surfaces.

Figure 4 shows the result of processing the measurements of the worn out “Sprocket” part in the Geometric Design X system when creating a 3D model.

After creating a 3D model, the question arises about how to recreate the object. Modern methods of restoring parts offer two main options: additive and subtractive.

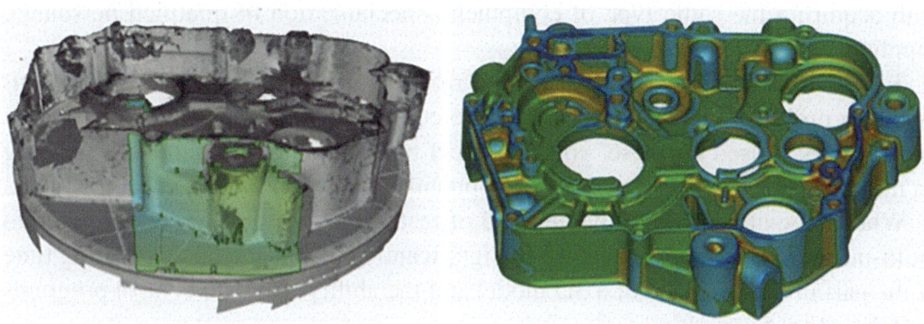

a b

Fig. 3 Comparison of a scanned part before (**a**) and after noise correction (**b**)

Fig. 4 Processing of scanning data of a fundamental part “Sprocket” during 3D model creation (Permyakov et al. [24] 2023)

The additive method is widely used in reconstructing parts from various polymers and, somewhat less frequently, in reconstructing parts from metal. This is because equipment for additive work with metals is quite expensive, a limited set of materials for work requires special preparation of the source material, high requirements for working conditions, the need for subsequent heat treatment to relieve stress, fire hazard, etc.). Therefore, when restoring metal products, traditional subtractive technology with the removal of a layer of material is most often used.

The work method is specified based on the selected direction of the part's restoration for additive technologies.

When working with polymers, various technologies that work with different materials provide different surface quality, manufacturing speed, the ability to combine colors, etc. As the analysis showed, repair companies usually use one, less often, several options of equipment for additive technologies, which is associated with acquiring the same type of equipment, specialization of qualified personnel, suppliers of materials, etc.

Table 2 shows the operating principles of equipment for various additive manufacturing processes, and Table 3 shows the comparative characteristics of additive technologies. Based on these, you can select product creation methods depending on the company's area of activity, development concept, and financial capabilities.

When choosing a subtractive method of restoring a part, preference is given to multi-purpose CNC equipment, as this significantly reduces the manufacturing time of the part in the presence of a 3D model and the ability to create control programs for CNC (Fig. 5) quickly.

3.2 Restoration and Modernization of Heavy Engineering Products without Dismantling

Complex heavy engineering products have a much longer service life than ordinary ones, so they require repair or modernization. However, moving them to the production base is difficult or impossible (quarry excavators, mining equipment, power plant turbines, pipelines, etc.). As noted earlier, portable machines are increasingly used to solve such problems in modern mechanical engineering. Portable machines have some features that have a significant impact on ensuring the accuracy of processing.

Firstly, instead of moving the part to a fixed machine, the machine is brought to the part, often transported in sections for easy handling.

Secondly, the part is not clamped on the machine; instead, the machine module—with its tool and locating system—is anchored to an existing structure feature. Even for the same surface, the referencing approach can differ and is selected based on numerous factors.

Thirdly, portable machining units are generally tailored for a single, narrowly defined operation.

Table 2 Technology of additive manufacturing methods

Technology	Principle works	Peculiarities
FDM/FFF (Fused Deposition Modeling/Fused Filament Fabrication)	The extruder heats the polymer thread (filament) and lays it in layers on the platform.	Low cost of the product, rough texture
SLA (Stereolithography)	Curing liquid photopolymer resin layer by layer.	Smooth surface, long manufacturing time
DLP (Digital Light Processing)	Curing photopolymer resin in one layer using a digital projector.	Smooth surface, long manufacturing time
SLS (Selective Laser Sintering)	The laser fuses the polymer powder in layers.	Durable details, rough texture
MJF (Multi Jet Fusion)	Thermal energy fuses the polymer powder, and jet heads apply bonding agents.	Typical for mass production
PolyJet	Jet application of photopolymer resin followed by UV curing.	Multicolored products, smooth surface
DLS (Digital Light Synthesis)	UV light and oxygen to control the polymer curing process.	Durable serial products
SLM/DMLS (Selective Laser Melting/Direct Metal Laser Sintering)	The laser beam fuses or sinters the metal powder layer by layer.	High density and strength of products (comparable to traditional methods).
EBM (Electron Beam Melting)	The electron beam melts metal powder in a vacuum layer by layer.	Working with refractory metals, relatively high productivity
DED (Directed Energy Deposition)	The metal powder/wire is fed into a melting zone created by a laser, electron beam, or plasma	Creation of large imprecise parts, the need for subsequent processing
Binder Jetting	A binder is applied to the metal powder to form layers, then sintered.	High productivity requires additional heat treatment (sintering)
LMD (Laser Metal Deposition)	The laser melts metal powder or wire fed onto the surface layer by layer.	High durability of products, possibility to combine different metals
CS (Cold Spray)	Metal powder is "packed" onto the surface under high pressure and speed without melting.	High production speed sometimes requires additional processing.
Metal Extrusion	A polymer thread filled with metal powder is laid in layers and then sintered.	Low cost, requires additional heat treatment (sintering)

Fourthly, because the unit is mounted directly on the part, its architecture must support compact transport and fast, straightforward assembly and teardown.

Fifthly, to ensure the required processing accuracy, the units and elements of the portable machine must form a sufficiently high structural rigidity after installation at the processing site while minimizing their dimensions and weight.

Analyzing the design features of portable machines from leading manufacturers (Mirage 2024; Serco 2024; Tomco Industrial Ltd. 2024; Climax Portable Products

Table 3 Comparative characteristics of additive manufacturing methods

Technology	Accuracy	Speed	Materials
FDM/FFF	50–300 μm	**10–50 mm³/s**	ABS, PLA, PETG, nylon
SLA	25–50 μm	**up to 10 mm³/s**	Photopolymers (liquids and resins)
DLP	25–50 μm	**up to 10 mm³/s**	Photopolymers (liquids and resins)
SLS	50–120 μm	**10–50 mm³/s**	Nylon, PA, composites
MJF	50–80 μm	**50–200 mm³/s**	Polyamides, TPU
PolyJet	16–50 μm	**10–50 mm³/s**	Composites, elastomers
DLS	50 μm	**200–1000 mm³/s**	Elastomers, solid polymers, polyurethanes
SLM/DMLS	20–100 μm.	**10–30 cm³/hour**	Aluminum, titanium, stainless steel, cobalt-chromium, heat-resistant alloys
EBM	50–200 μm	**30–70 cm³/hour**	Titanium and its alloys, cobalt-chromium
DED	100–300 μm	**50–150 cm³/hour**	Wide range of metals (powder or wire)
Binder jetting	50–200 μm	**up to 500 cm³/ hour**	Stainless steel, aluminum, copper, bronze
LMD	100–300 μm	**–**	Steel, aluminum, titanium, cobalt-chrome.
CS	–	**up to 1000 cm³/ hour**	Aluminum, copper, titanium, stainless steel.
Metal extrusion	100–200 μm.	**10–40 cm³/hour**	Stainless steels, bronze, copper.

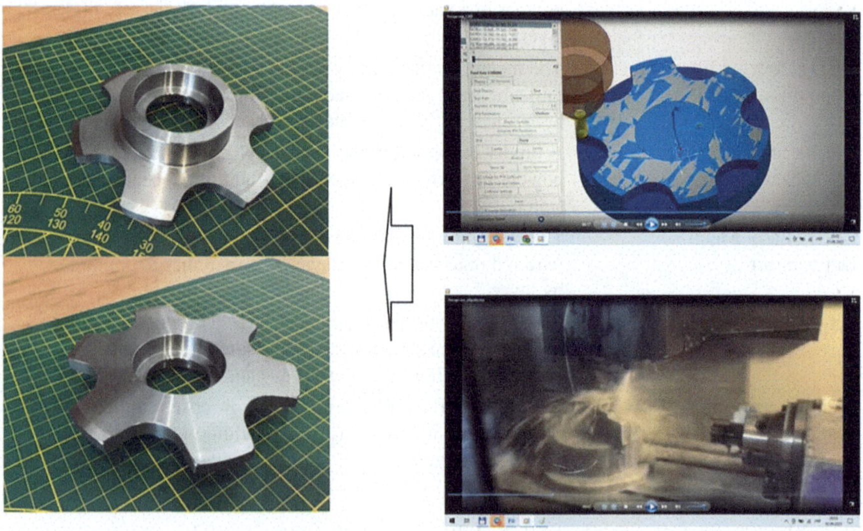

Fig. 5 Results of restoring the genuine part "Sprocket" on the multi-purpose machine using the previously created 3D model (Permyakov et al. [24] 2023)

2024; Protem 2024; Metrom 2024) allows us to conclude that the main factor influencing the layout and design of the machine is the kinematics of the formation of the workpiece surface.

The most typical technological operations for portable machines are the processing of flanges, diametrical planes, holes of various diameters, and grooves of multiple profiles; machines for processing external cylindrical surfaces of sufficient length are scarce, and abrasive processing operations are practically absent (mainly manual execution). The conducted analysis of the largest manufacturers allows us to classify the variants of kinematic schemes for machines of this group (Yakovenko et al. 2023a). This will enable us to significantly reduce the time spent searching for the necessary variants of layout solutions for a portable machine in automated design.

The most common kinematic schemes of this classification are given in Table 4.

We want to note the portable machines of group 7 with hexapod kinematics for working with an end tool, which allows for processing many surfaces on one side of the part. This design enables the tool to be positioned in a specific range of surface sizes and angles of inclination of the processing axis. The most popular on the market are portable groups 1 and 4 machines.

Therefore, leading manufacturers produce such machines in series in a specific size, which allows for the processing of not a particular surface with a given size but a group of surfaces with a specific range of geometric dimensions and a particular option for basing the machine on the processing object (Fig. 6). Moreover, for boring holes, the machine basing scheme is very often used directly on the processed hole using center finder device in mechanical, contact or laser design, which allows for reasonably precise installation of the machine during installation.

The use of such machines is possible both by repair services of large corporations for their facilities and by smaller companies for carrying out repair and modernization orders at the facilities of various enterprises.

However, creating a portable machine to perform a specific operation by basing and fixing a particular object, i.e., an original design, is often necessary. In this case, further use of this machine in the original configuration becomes unlikely after the work is completed. The authors found that in this case, it is advisable to use the aggregate-modular principle of equipment layout (Yakovenko et al. 2023b), i.e., after dismantling the equipment (and such dismantling is mandatory to ensure the operation of the main object), standardized units and assemblies can be used to create a new machine. Such units are power plants that provide the main and auxiliary movements of the tool during processing, basing systems, spindle groups, etc. In this case, the originality of each machine is ensured by the design of the base system (this can be called the body for the traditional machine layout), which connects all other elements, and the machine mounting system on the processing object.

In this case, the design process for such equipment should be updated according to the following scheme (Fig. 7).

During the concept phase, as in the classical scheme, based on the geometric parameters of the product and the formed surface, the processing technology is developed, and the basing system and the kinematic scheme for ensuring the cutting process are determined. Then, the design layout of the portable machine is formed.

Table 4 Kinematic schemes of surface shaping for portable machines

№	Scheme	Buttstock	Surfaces
1	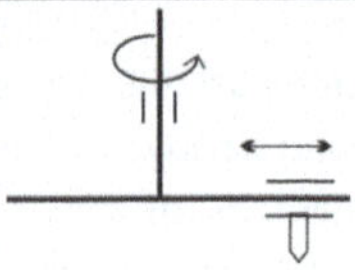		<u>Climax</u> End faces, flanges, grooves
2	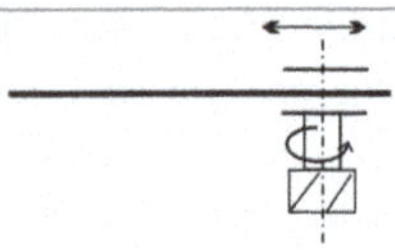	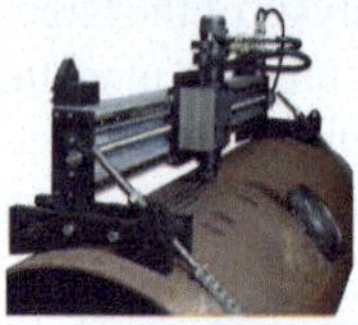	<u>Mirage</u> Planes, grooves of a rectilinear profile.
3	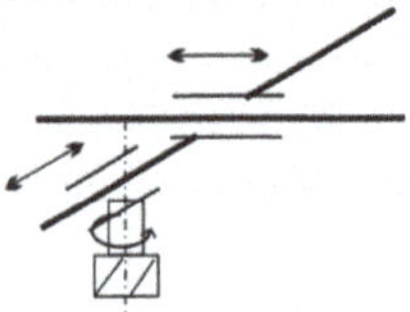		<u>Mirage</u> Planes, grooves of free profile.
4	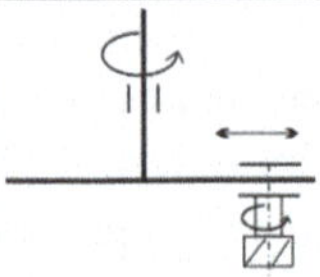		<u>Climax</u> Ends, diametrical planes, ledges.
5	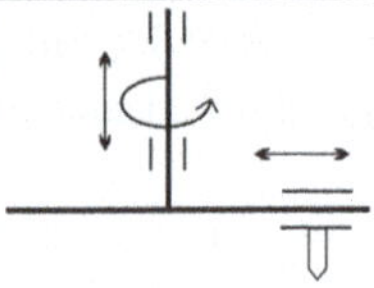		<u>Serco</u> Large diameter holes.
6		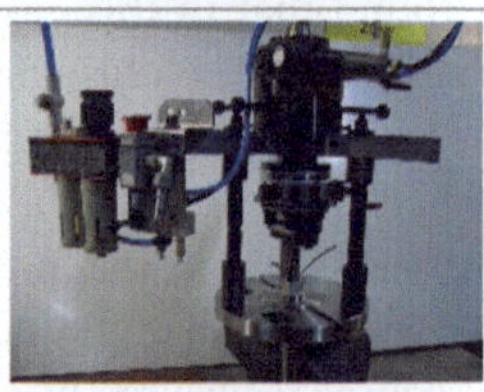	<u>Protem</u> Processing holes with an axial cutting tool.
7	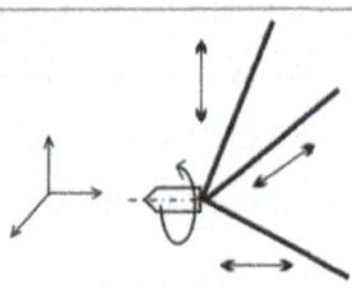		<u>Metrom</u> Processing of a large number of identical holes on one sawn at different angles

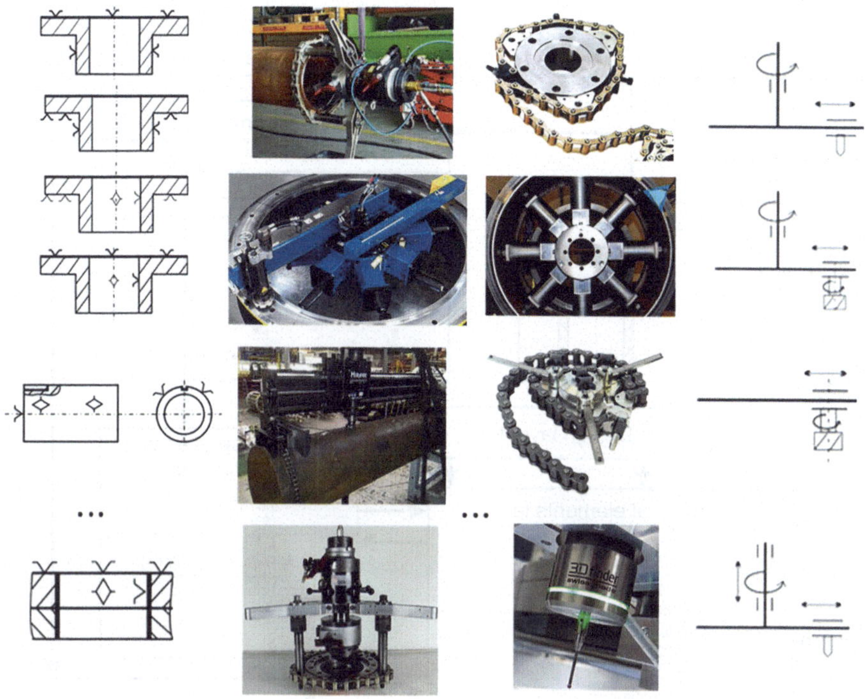

Fig. 6 Most common portable machine tools, basing and clamping schemes, and design implementation options

The scheme proposed by the authors differs from the classical design by expanding the next design stage—the design layout, due to the selection of modules and units from the database of such elements. The database of unified units of portable machines consists of two sections. The first block contains information on units obtained after dismantling portable machines that have completed their production tasks and are available at the enterprise warehouse. Moreover, information is taken into account not only on the models and technological capabilities of the module or unified element, but also additional characteristics (rigidity, vibration resistance, etc.), which can be obtained on special test benches. If the required units cannot be found in the first section, the second block is accessed, where information is stored on units that can be purchased from related companies. Based on the structural layout and the selected set of units, original units and elements are designed to be manufactured at the enterprise.

This information is necessary at the next stage of design—the construction and analysis of a 3D model of the processing process in modern computer-aided design systems (we used the Solid Works system). This analysis allows us to control the future accuracy of the shaping process and make the necessary adjustments to the design layout of the portable machine (selection of other technological modules,

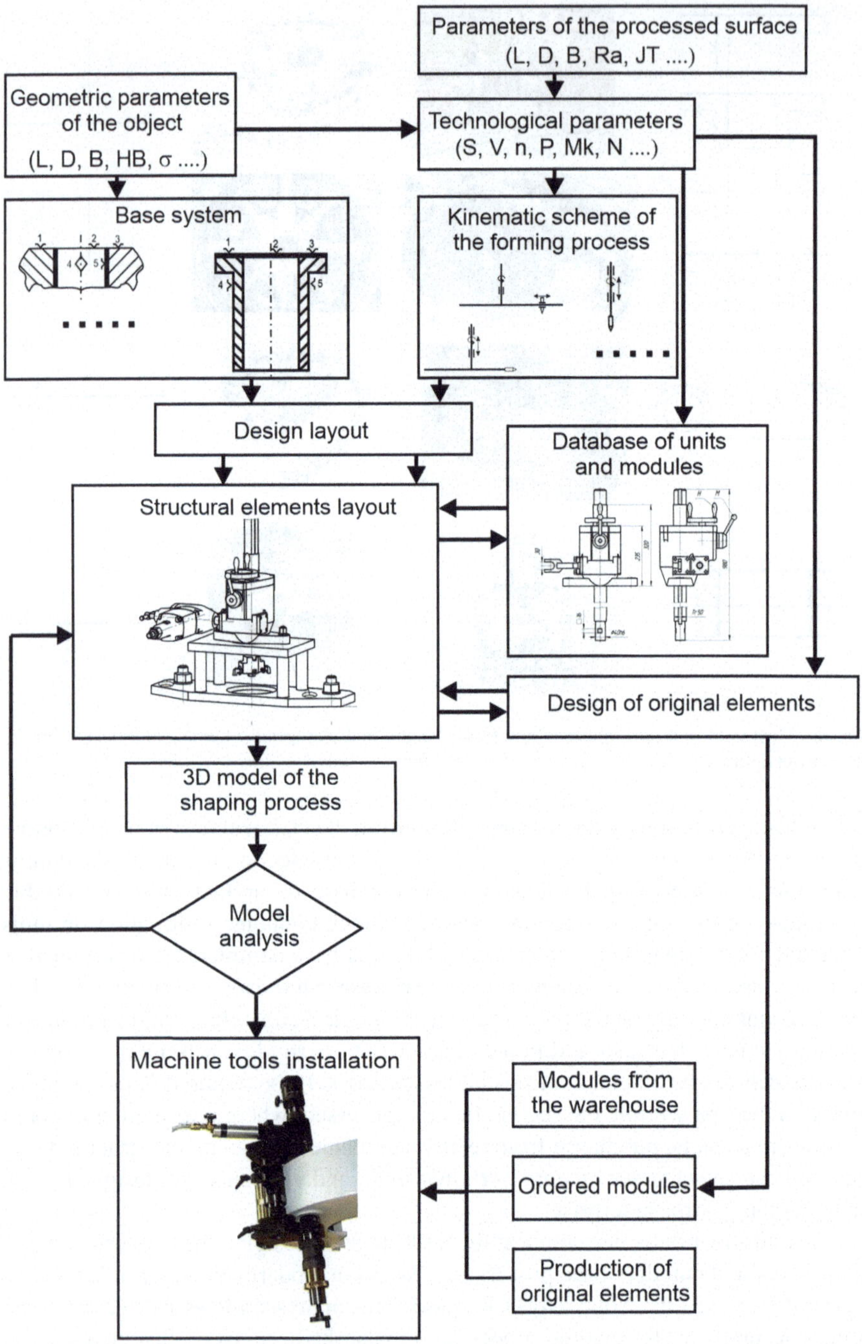

Fig. 7 Design concept for modular portable machine tools

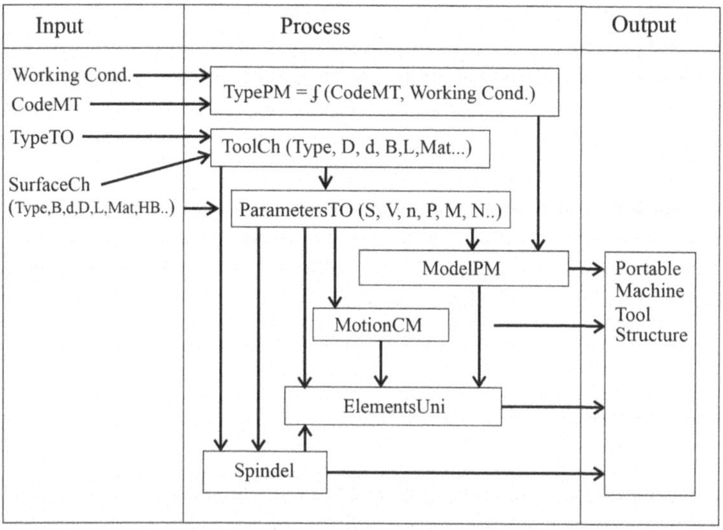

Fig. 8 HIPO diagram for conceptual layout synthesis of portable machine tools

changes in the designs and parameters of the housing and original elements, etc.) before starting the production and installation process.

This design model pays special attention to creating a design layout for a portable machine (Fig. 8).

Before detailing the selection of components, the working context (Working Cond) and the machine layout code (CodeMT) are used to determine the portable machine class (TypePM). The remaining layout parameters are specified later, during the project's development.

At the same time, the technological approach to processing the workpiece (i.e., the technological operation, TypeTO) is determined taking into account the technical and dimensional constraints of the problem (SurfaceCh): the material of the part and its physical and mechanical properties, the geometry and size of the target surface, the rigidity of the unit, and other factors. For the adopted processing method, a cutting tool is selected (or, if necessary, designed) and its characteristics (ToolCH) are determined: type; material grade (HSS, carbide, ceramic, etc.); geometric parameters (dimensions of the cutting part, angles, etc.); other characteristics.

The next step is determining the selected operation's cutting process parameters (ParametersTO). Typical parameters include feed (S), cutting speed (V), spindle speed (n), cutting force (P), torque (M), and power consumption (N). These values can be obtained automatically from standard data sets stored in the system or set by the process engineer and entered into the database.

Using the collected cutting process data and the previously defined layout option, the design team starts creating the Portable Machine Tool Structure (PMTSS): selecting the models of power units (ModelPM), standardized motion conversion mechanisms (MotionCM), spindle units (Spindle), and other standardized portable machine tool modules (ElementsUni). The decision-making procedure allows optimizing several criteria—typically cost, dimensions, and weight—to achieve an efficient solution. In practice, cost is often considered the dominant criterion, while the other characteristics impose constraints on feasibility. Suitable units can be obtained from an internal catalogue (including our production stocks and products from our closest partners) and an extended database of proven serial suppliers, which is especially relevant for pneumatic motors and associated drives. Without a suitable standardized module, the required units and assemblies are designed, and special interfaces or devices are developed to ensure the required positioning accuracy and fixation of the machine during installation.

4 Conclusions

Reverse engineering significantly reduces the labor intensity of repair work regardless of the methods of restoring parts—additive or subtractive, ensuring the necessary accuracy and functionality of the restored product. At the same time, it is possible to create sets of models and documents to carry out the same restoration work in the future. To restore the image of a damaged part and make its 3D model, the authors suggest using neural networks and artificial intelligence in their subject area.

The use of portable machines in the modernization and repair of unique and other large-sized non-dismantling equipment significantly reduces the time of work and commissioning, considerably increasing the quality and accuracy of the work performed compared to manual operations. Therefore, this area is one of the most promising in machine tool building. When creating such equipment for performing less common work, it is necessary to use the aggregate-modular principle of constructing a portable machine, which saves time both in its design and in production and installation at the customer's site. This approach allows you to significantly reduce financial investments due to the multiple uses of the main elements of the portable machine.

The recommendations articulated by the authors for organizing repair and restoration work on engineering products are primarily aimed at small and medium-sized enterprises that are engaged in or plan to engage in such work in the future. It is also possible to use the proposed methodology by engineering enterprises specializing in producing large-sized and unique products when organizing divisions focused on servicing, repairing, and modernizing manufactured products.

Acknowledgments The authors express their deep gratitude to the management of the "Ukrainian Energy Systems Enterprise", which helped to test the proposed methodology in creating portable machines for the modernization of energy facilities in Ukraine and made valuable comments during the system's operation. The authors would like to convey their deep appreciation to the management of the "HERMES TECHNIC Enterprise" (https://www.hermes-technic.pl) in Poland for actively discussing the methods for constructing 3D models of worn parts based on object scanning data.

The research was partially supported by the International Association for Technological Development and Innovations.

References

Basova Y, Dobrotvorskiy S, Balog M et al (2023) Increasing SME supply chain resilience in the face of rapidly changing demand with 3D model visualisation. Int J Mechatron Appl Mech 14:35–47. https://doi.org/10.17683/ijomam/issue14.5

Bondar D, Basova YV, Vodka O (2024a) Automation of quality control for used parts based on 2D images and nonvisual information: An approach to extending the product life cycle. Bull NTU KhPI Series Dynam Stren Mach 2:60–66. https://doi.org/10.20998/2078-9130.2024.2.316341. NTU "KhPI", Kharkiv (in Ukrainian)

Bondar D, Basova Y, Vodka O (2024b) Analysis of the efficiency of quality control algorithms for modified surfaces for high-error processes based on 2D miniatures and non-visual information. In: Cioboată DD (ed) International conference on reliable systems engineering (ICoRSE) - 2024, Lecture notes in networks and systems, vol 1129. Springer, Cham, pp 35–50. https://doi.org/10.1007/978-3-031-70670-7_3

Bonitz P, Krzystek P (1997) Reverse engineering in combination with digital photogrammetry (ICEM SURF/ICEM PHOTO). In: Pratt MJ, Sriram RD, Wozny MJ (eds) Product modeling for computer integrated design and manufacture, IFIP advances in information and communication technology. Springer, Boston, MA. https://doi.org/10.1007/978-0-387-35187-2_20

Bradley K, Persson O (2022) Community repair in the circular economy—fixing more than stuff. Local Environ 27(10–11):1321–1337. https://doi.org/10.1080/13549839.2022.2041580

Breteler FHR (n.d.) Enhancement of the process of reusing building products. https://pure.tue.nl/ws/portalfiles/portal/205362293/Breteler_1490729_CME_Vries_d.pdf, last accessed 2024/12/12

Climax Portable Products: Climax Portable Products homepage, https://www.climaxportable.com/products, last accessed 2024/11/21

Fischer M et al (2023) The future of sustainable business: the circular economy. In: Sustainable business, Springer briefs in business. Springer, Cham, pp 129–140. https://doi.org/10.1007/978-3-031-25397-3_9

Galvez A, Iglesias A, Fister I, Fister I (2023) Industrial artificial intelligence approach for shape reconstruction in quality assessment of digital data from manufactured workpieces. In: 2023 4th international conference on industrial engineering and artificial intelligence (IEAI), Chiang Mai, Thailand, pp 86–93. https://doi.org/10.1109/IEAI59107.2023.00020

García-Molina DF, Valderrama-Zafra JM, Montalvo-Gil JM, Rubio-Paramio MÁ (2023) Optimization of modeling and 3D printing of architectural elements in historical buildings through the systematization of Boolean solid operations. In: Cavas-Martínez F, Marín Granados MD, Mirálbes Buil R, de-Cózar-Macías OD (eds) Advances in design engineering III. INGEGRAF 2022, Lecture notes in mechanical engineering. Springer, Cham, pp 765–777. https://doi.org/10.1007/978-3-031-20325-1_59

Global Reverse Engineering Service Market (2024) Market size, status & forecast to 2031. Verified Market Research, 103 p

Hexagon: Portable scanning. Hexagon homepage, https://hexagon.com/products, last accessed 2024/12/12

Kyaw AC, Nagengast N, Usma-Mansfield C, Fuss FK (2023) A combined reverse engineering and multi-criteria decision-making approach for remanufacturing a classic car part. Proc CIRP 119:222–228. https://doi.org/10.1016/j.procir.2023.02.133

Manoochehri S, et al. (2022) An overview of Europe's repair sector. In: European topic centre on circular economy and resource use, ETC CE report, 73 p

Marcus JS et al. (2020) Promoting product longevity. Study for the Committee on the Internal Market and Consumer Protection. Policy Department for Economic, Scientific and Quality of Life Policies, European Parliament, Luxembourg, 46 p

Metrom: Metrom homepage, https://www.metrom.com, last accessed 2024/11/21

Mirage: Mirage homepage, https://www.miragemachines.com, last accessed 2024/11/21

Ntsondé J, Bergmann M, Steux C, Micheaux H, Beulque R (2023) How can mandatory repair scores foster repair activity-based circular business models? In: New business models conference proceedings 2023. Maastricht University Press. https://doi.org/10.26481/mup.2302

Page D, Koschan A, Abidi M (2007) Methodologies and techniques for reverse engineering–the potential for automation with 3-D laser scanners. In: Reverse engineering. Springer, London, pp 11–32. https://doi.org/10.1007/978-1-84628-856-2_2

Permyakov A, Yakovenko I, Skyba O, Kalinichenko V, Jozkowicz P (2023) Reverse engineering and implementation of modern methods and means of control operations. Bull Nat Tech Univ Kharkiv Polytech Inst Series Tech Mach Indus 2(8):91–99. https://doi.org/10.20998/2079-004X.2023.2(8).10. NTU "KhPI", Kharkiv (in Ukrainian)

Pham DT, Hieu LC (2007) Reverse engineering–hardware and software. In: Reverse engineering. Springer, London, pp 33–70. https://doi.org/10.1007/978-1-84628-856-2_3

Protem: Protem homepage, https://www.protemusa.com, last accessed 2024/11/21

Raja V (2007) Introduction to reverse engineering. In: Reverse engineering. Springer, London, pp 1–9. https://doi.org/10.1007/978-1-84628-856-2_1

Reverse engineering market report overview. Source: https://www.businessresearchinsights.com/market-reports/reverse-engineering-software-market-112729, last accessed 2024/12/12

Roth-Koch S (1996) 3D-shaping in CAD systems by the steps creation—Inspection—optimization. In: Reverse engineering. Vieweg+Teubner, Verlag, pp 125–132. https://doi.org/10.1007/978-3-322-84819-2_10

Serco: Serco products homepage, https://www.serco-tools.com/products, last accessed 2024/11/21

Skytt V (1996) Two methods for surface reconstruction from given point distributions. In: Reverse engineering. Vieweg+Teubner, Verlag, pp 159–168. https://doi.org/10.1007/978-3-322-84819-2_14

Tomco Industrial Ltd.: Tomco Industrial Ltd. Homepage, https://www.tomco.ca/tomco-industrial-ltd/machine-tool, last accessed 2024/11/21

Valerga AP, Cabrera-Revuelta E, Alonso-Garcia M, Fernandez-Vidal SR (2023) Photogrammetry applied to evaluation of the geometry changes on cutting tools Wear. In: Cavas-Martínez F, Marín Granados MD, Mirálbes Buil R, de Cózar-Macías OD (eds) Advances in design engineering III. INGEGRAF 2022, Lecture notes in mechanical engineering. Springer, Cham, pp 583–594. https://doi.org/10.1007/978-3-031-20325-1_46

What Is Reverse Engineering? Source: https://astromachineworks.com/what-is-reverse-engineering, last accessed 2024/12/12

Yakovenko I, Shepeliev D, Sharlay V, Permyakov A, Slipchenko S, Havryliuk Y (2023a) Analysis and synthesis of mobile portable machine tools layouts. In: Cioboată DD (ed) International conference on reliable systems engineering (ICoRSE) - 2022. ICoRSE 2022, Lecture notes in networks and systems, vol 534. Springer, Cham, pp 160–171. https://doi.org/10.1007/978-3-031-15944-2_16

Yakovenko I, Permyakov A, Basova Y, Kotliar A, Rudenko O (2023b) Ensuring accuracy during processing of heavy machinery objects with portable machines. Bull Nat Tech Univ KhPI Series Eng CAD 1:160–166. https://doi.org/10.20998/2079-0775.2023.1.15. NTU "KhPI", Kharkiv (in Ukrainian)

Quality Assurance and Measurement Systems

Quality Management in Agile Manufacturing: Best Practices for Effective Reporting and Continuous Improvement

Angelina Iakovets and Bohdan Haidabrus

Abstract Within the Manufacturing 4.0 and 5.0 paradigm, integrating quality management systems (QMS) with agile methodologies has become a critical strategy for improving operational efficiency and product quality. Agility is a key aspect of enterprise competitiveness. This article examines best practices for effective reporting and continuous improvement in agile manufacturing environments. Synthesizing current literature and empirical research, we identify key principles that underlie successful agile quality management in modern manufacturing. We examine how these principles facilitate real-time quality assessment and foster a culture of continuous improvement, enabling organizations to respond quickly to market demands while maintaining high-quality standards. In addition, we discuss the challenges associated with implementing agile methods in traditional manufacturing settings and propose a concept for an agile quality management system. The results highlight the importance of adaptive quality frameworks that support flexibility and improve product reliability and customer satisfaction. This review aims to provide a comprehensive understanding of how agile methodologies can effectively optimize quality management processes, ultimately facilitating innovation and competitiveness in the manufacturing sector.

Keywords Manufacturing · QMS · Agile · Quality · SME · Sustainability · CMMI

A. Iakovets (✉)
Technical University of Košice, Prešov, Slovakia
e-mail: angelina.iakovets@tuke.sk

B. Haidabrus
Accenture Baltics, Riga, Latvia

Institute of Information Technology, Riga Technical University, Riga, Latvia

V. Ivanov (ed.), *Emerging Technologies in Manufacturing and Materials Engineering*, Springer Tracts in Mechanical Engineering,
https://doi.org/10.1007/978-3-032-14914-5_6

1 Introduction

The rapid development of production and the spread of Industry 4.0 influenced the approach to those technologies that were actively promoted. The huge arrays of data that enterprises began to collect and the energy crisis in the EU (Hebda and Mišík 2024) and other countries influenced the formation of the Industry 5.0 concept. The new concept began to welcome human-oriented technologies (Avsec et al. 2024) and eco-friendly technologies to reduce energy consumption and promote sustainability. Industry 5.0 actively supports human-oriented technologies, including, for example, the use of exoskeletons, collaborative robots, AI-driven systems, and other technologies that make work easier for people. All these technologies should ensure that human expertise remains at the core of production. To sum up, the main goal of Industry 5.0 is to unite people, digital technologies, and robots to achieve sustainable production (Ivanov 2022). The concept of sustainable energy efficiency, waste reduction, and introducing environmentally friendly technologies expands the scope of quality control. In this case, it is essential to refer not only to traditional metrics of product performance and defect reduction but also to introducing additional quality indicators that must be considered at each stage of production and testing of the product. Quality 5.0 is an extension of the Industry 5.0 concept and covers not only quality aspects but also environmental sustainability and ethical considerations. In this case, manufacturing must ensure high-quality outcomes while minimizing resource consumption, reducing waste, and prioritizing emphasis on the long-term impact on the environment and society. This approach means an urgent need to integrate specialists from many areas to solve comprehensive production problems.

While Industry 5.0 outlines the vision for an interconnected, sustainable, and human-driven production ecosystem, Quality 5.0 provides the framework for maintaining high standards within this ecosystem. However, implementing these concepts in real-world manufacturing requires adaptive management models that balance flexibility, efficiency, and sustainability. Excluding a person, training systems to think like specialists in various fields is a laborious and expensive process, which, at the stage of debugging the system, can entail associated costs. As a key criterion, the cooperation of specialists will allow a comprehensive response to the quality of the work performed and the product's sustainability in relation to the environment.

This study examines how to integrate agility into quality management models that align with Industry 5.0 and Quality 5.0 principles. The research explores best practices for continuous improvement, effective reporting, and cross-disciplinary collaboration, providing insights into how manufacturing enterprises can optimize quality while addressing emerging sustainability and technological integration challenges.

2 Fundamentals and Challenges

Traditionally, quality control was carried out from the inspection of incoming material in the warehouse to the final inspection before sending the goods to the consumer.

Quality management, like the industry, went through five stages, from Quality 1.0 to 5.0 (Fig. 1).

Each era of Quality can be briefly characterized:

- Quality 1.0 was aimed at detecting defects. The results were summed up at the end of the production cycle. Quality control was not integrated, and defect elimination and debugging were carried out after the products were released.
- Quality 2.0 is the era of standardization and statistical process control (SPC). Quality was part of the company's strategy, so making decisions before the end of production was essential.
- Quality 3.0 is the era of cost reduction and efficiency improvement. At this time, Lean, Six Sigma, and Kaizen appeared. The production tried to adapt quality

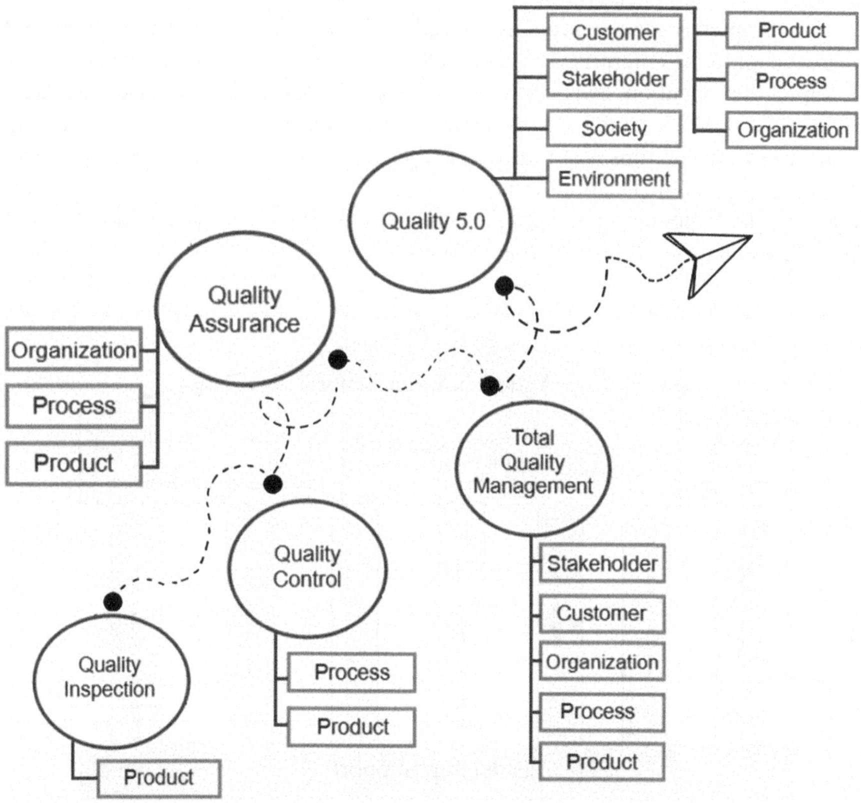

Fig. 1 Evolution of Quality (Fiałkowska-Filipek and Dobrowolska 2024)

with the help of the European Foundation for Quality Management (EFQM) and Baldrige to reach an agreement with the customer (Jan and Grudowski 2023).

- Quality 4.0 arose under the influence of Industry 4.0. This era is characterized by the analytics of large data sets collected using IIoT (Goknil et al. 2023) to predict quality management.
- Quality 5.0 is about integrating human experience with technology. This mostly manifests in collecting feedback and training AI systems to process data while humans engage in creative tasks (Fiałkowska-Filipek and Dobrowolska 2024; Jan and Grudowski 2023; Goknil et al. 2023).

Quality 5.0 is a new branch, and its methodology and approaches to quality management are not yet clear. The main unresolved problem, according to some scientists, is sustainability. The interpretation of "sustainability" and "environmental friendliness" is sometimes considered identical and sometimes completely different concepts. Most are inclined to believe that this is not the bottom and the same thing. Sustainability is a more rational approach that thinks about environmental friendliness in the long term (Dobrotvorskiy et al. 2023). It is essential to understand that when introducing "green" manufacturing (Sun et al. 2023), the problem of adapting the enterprise to a radically new work organization may arise.

The complexity of the problem of full-fledged Industry 5.0 and Quality 5.0 should adapt to the enterprise's readiness level.

Rapid digitalization and smartization of production have affected industrial enterprises' work structure and organization. The industry began to cooperate with IT more closely and thus was forced to adopt the experience of new paradigms and project management. The traditional structure of a manufacturing enterprise underwent a gradual integration of Agile, which led to the emergence of a hybrid form of Agile Manufacturing (Gunasekaran 1998) (Fig. 2).

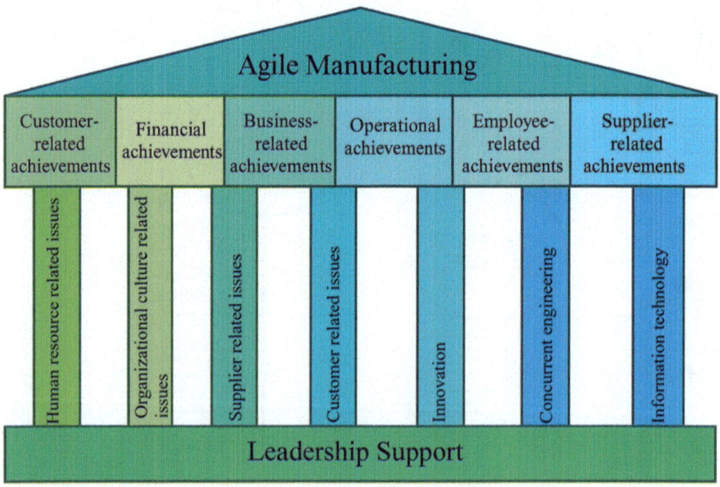

Fig. 2 Agile Manufacturing house of pillars (Kumar et al. 2019)

The first integrations of Agile and Lean in 2019 promised to successfully unify approaches to managing large manufacturing enterprises (Qamar et al. 2019). The formation of small working groups operating based on the Scrum methodology allows for the effective implementation of lean manufacturing, the rapid involvement of personnel in the change process, and the identification of change leaders. According to the experience of 140 automotive manufacturing firms from the UK (Qamar et al. 2019), it became clear that full integration of Agile into all types of enterprise activities is impossible at this stage. With the transition to Industry 5.0, enterprises still could not overcome some barriers, such as a multidisciplinary approach to problem-solving, which was actively promoted in the Agile concept. The solution for many enterprises was to distribute such problems among external consulting firms. The race for technology has led to the fact that the main message of Agile, which was embedded in Industry 5.0 and, likewise, in Quality 5.0 (Dehghan et al. 2025), has been lost. We are talking about "be one step ahead of the competition, Conduct regular training, Have a "plan B" in case of changes" (Banáš and Hrablik Chovanová 2023). The widespread integration of AI, IoT (Jasiulewicz-Kaczmarek et al. 2023), and other technologies has affected the client's perception of the uniqueness of companies. In this case, the concept of Sustainability and human-centeredness is lost (Fig. 3).

The concept of Agile Manufacturing is very suitable for the concept of Industry 5.0 due to its employee-related achievements and flexibility to customer needs.

Based on the barriers to fully implementing all fundamental modern concepts, scientists are increasingly inclined to combine traditional approaches with new ones (Jadoon et al. 2020; Ali and Wasim 2022). Studies focus on selected aspects of production reconfiguration and process organization, forgetting about the life stages of the enterprise and its internal and economic readiness (Jadoon et al. 2020; Ali and

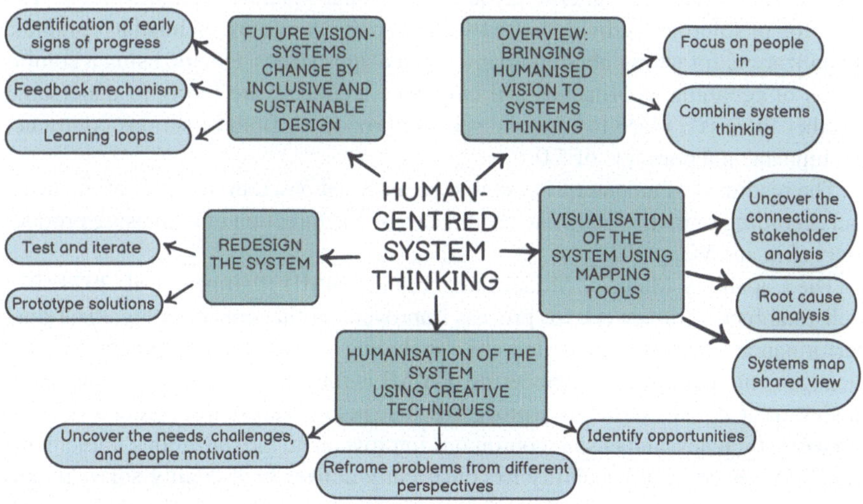

Fig. 3 Human-centered system thinking (Avsec et al. 2024)

Wasim 2022). The role of the joint approach (economic and non-economic methods) was emphasized by Holzner et al. (2023). The study involved the development of a CANVAS-based assessment model for determining the readiness of enterprises for the digital business models of Industry 4.0. The authors used 45 assessment criteria organized into nine categories according to the CANVAS business model. The proposed approach has proven effective in the enterprise studied. The proposed approach takes place only in the preparatory part of the project, but not as a permanent tool for a continuous cycle of management and improving the quality of the Plan Do Check Act cycle (PDCA) (Li et al. 2025). PDCA dates back to the 60 s of the twentieth century but remains firmly at the core of Agile, Lean, and Industry 4.0. In parallel, the FMEA risk assessment method emerged in the 1960s. FMEA, in various forms, analyzes risks within the framework of modern concepts, just like PDCA, but as a form of reporting and summarizing the collected data. The effectiveness and relevance of these approaches are emphasized by Longyu Li (Pekarčíková et al. 2023), who integrates these principles into the risk assessment of the Digital Twin Project Operation.

An optimal hybrid approach should include models that improve organizational performance based on metrics of their economic effect per modern trends in Industry and Quality. The proposed approach can and should use specific traditional approaches, as they find their application even in the era of high digitalization (Li et al. 2025; Pekarčíková et al. 2023).

3 Ideas and Solutions

Analysis of modern trends and practical studies of selected authors showed that the main problem areas of enterprises are operational planning and flexibility. The direction of solutions is divided into two main parts. Some are inclined to complete digitalization, forgetting about human-centered approaches. Others, using a combination of economic approaches and employee feedback, are trying to understand whether there is a place to improve specific processes. Both directions contradict the fundamental concepts of 5.0.

The readiness of an enterprise should be assessed based on its level of maturity, and according to it, the enterprise should choose the direction of increasing productivity (Lin and Wang 2021).

The Capability Maturity Model Integration (CMMI) (Laitinen et al. 2024) is a well-established framework for process improvement that enhances organizational performance across multiple domains, including software development, service management, and product engineering. CMMI is inherently a quality management model, as it defines structured processes that help organizations reduce defects, improve efficiency, and ensure continuous improvement. Organizations implementing CMMI improve their ability to consistently deliver high-quality software and services by following well-defined process areas and maturity levels, aligning closely with established quality management principles such as Total Quality

Management (TQM), Six Sigma, and ISO 9001. The key aspects of Quality Management in CMMI are:

1. Defect prevention at the early stage to identify and mitigate defects through structured development processes.
2. Process standardization as consistent and repeatable way to reduce variability and enhance predictability.
3. Continuous improvement through regular evaluation and optimization of processes to enhance performance and efficiency.
4. Risk Management as a formalized process to identify, perform qualitative and quantitative risk analysis, and create risk register (CIO.com Staff 2023).

The primary relationship between CMMI and quality management is its ability to institutionalize best practices across software development and service delivery. Organizations that achieve higher CMMI maturity levels experience fewer defects, improved customer satisfaction, and enhanced operational efficiency. CMMI defines five maturity levels (ISACA 2025), each reflecting an organization's ability to manage and continuously improve its processes:

- Level 1: Initial—Ad-hoc, unstructured processes with unpredictable outcomes.
- Level 2: Managed—Basic project management processes are established.
- Level 3: Defined—Organization-wide process standardization is achieved.
- Level 4: Quantitatively Managed—Quality metrics and data-driven decision-making (Muzylyov et al. 2021) are implemented.
- Level 5: Optimizing—Continuous process improvement is embedded in the culture [22–24].

In Manufacturing 4.0/5.0 CMMI aligns with a process-driven approach, emphasizing structure, governance, and compliance to ensure quality. However, Agile advocates for a flow-driven approach, prioritizing continuous delivery, automation, and fast feedback loops. These two approaches differ significantly in managing quality (Table 1). Considering the aspects of both methods, the optimal combination was colored green. The decision was made based on barriers to the full-fledged introduction of Agile principles in manufacturing.

Table 1 Comparison of CMM and Agile driven-processes

Aspect	Process-Driven (CMMI)	Flow-Driven (Agile, Scrum, Kanban)
Quality control	Formalized quality gates, audits, and compliance reviews	Automated testing, CI/CD, real-time monitoring
Defect prevention	Risk management, structured verification	Shift-left testing, fast feedback loops
Governance	Strict documentation and approval workflows	Automated compliance checks, policy-as-code
Delivery speed	Slower due to process adherence	Faster due to automation and incremental releases
Flexibility	Low, with strict process standardization	High, with adaptability to change

The QMS under Quality 5.0 and Industry 5.0 should be based on three main principles:

- sustainability,
- human-centric systems,
- flexibility,
- intelligence.

These main pillars were highlighted by Mariana Palhau, et al. (Palhau et al. 2024) during the performance of a new way of economic growth of manufacturing enterprise by establishing new quality management principles. The QMS principles were presented via three Key Performance Indicators of sustainable modern QMS: economics, environment, and social.

CMMI will ensure the "economics" and sustainability aspects by evaluating the economic level and readiness of the enterprise. CMMI will ensure stricter measurement of risks during the continuous improvement of processes under Agile Manufacturing per the principles of Industry 4.0/5.0. The CMMI will provide quality management with a more structured framework for the decision-making process while leveraging automation and continuous improvement practices to ensure Agility.

The Agility will be fully integrated into the decision-making process under quality management. The collaboration, communication, and creativity of Subject Matter Experts (SME) will provide a new approach to quality assessment and planning (Haiderzai et al. 2025). In this way, Agile provides a human-centered approach that can serve to train and develop better AI-powered systems.

Collaboration and decision-making based on Agile principles will enable the system's flexibility to meet customer requirements. Flexibility would be ensured by continuous collection and processing of production data based on the principles of the PDCA cycle.

Use the PDCA cycle and continuously improve quality based on data collection and its analysis immediately using traditional statistical approaches, as well as anomaly analysis and its analysis by different specialists in cooperation. This approach will provide data processing before the manufacturing process ends, and also predict the accumulation of big data that does not bring value to the enterprise.

SME using the Pareto principle identify anomalies in the external environment that may arise and implement them in adaptive AI to warn of an impending anomaly.

The manufacturing process is similar to the Input, Process, Output (IPO) model that describes data flow during the process. This principle works today because this model was founded by software engineers. The IPO model has undergone several transformation stages as the world has gradually changed and everything has moved towards Industry 4.0. (Fig. 4).

The main approach of each modern manufacturing process is to collect data about all the steps of manufacturing and non-manufacturing processes. A simplified principle is shown in Fig. 5.

The "Visual inspection" is a special process of visual quality control of the material before production. Visual inspection refers to pre-production control. This type

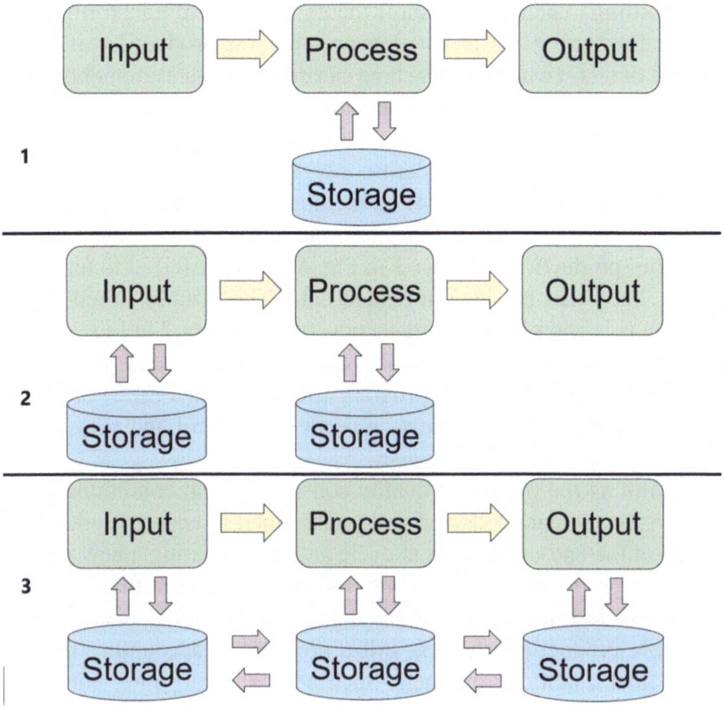

Fig. 4 Evolution of IPO model

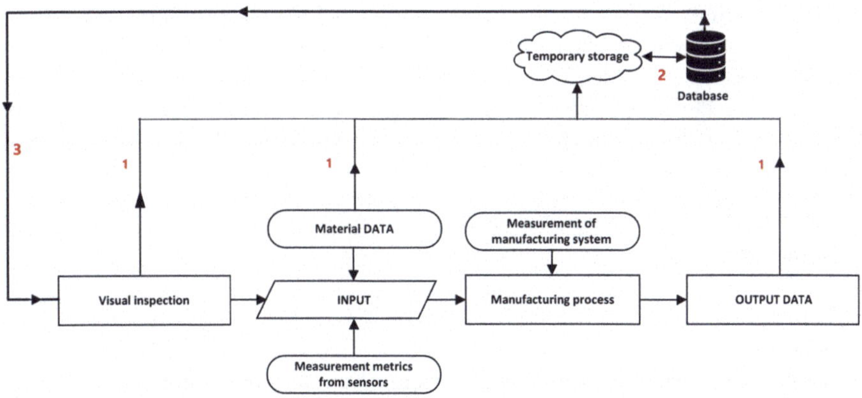

Fig. 5 Manufacturing and quality inspection

of control is carried out to eliminate obviously defective or damaged material before starting the production line.

The "Input" includes the material, supplementary documentation, and data for setting up and adjusting equipment for the quality and parameters of the processed material. According to general economic theory, this type of transformation process includes data for setting the production program for production machines, data about possible defects (which the machine can detect) ("Measurement metrics from sensors" Fig. 5), accompanying documentation and material, working instructions, maintenance documentation, and so on. At this stage, a person can still make adjustments before processing. Still, modern concepts of Industry 4.0 and 5.0 want to exclude human intervention (Fig. 4) to facilitate work and speed up the quality control process. "Material data" and other data about the manufacturing process collected during production (number 1 in Fig. 5) are transferred to temporary storage facilities, like other production-related data. This data is stored there until a batch of products is completed and then converted (number 2 in Fig. 5) into reports and data for debugging production and quality management. Normalized data returns to the new manufacturing program (number 3 in Fig. 5) are returned as instructions and technical documentation needed to control and adjust the production process.

When examining the origins of quality control and real conditions in manufacturing enterprises, the data collected and transformed to ensure production quality are used only at the end of the current cycle and for preparing a new one. The data collected during production are used only for the remote control of machines. The question arises as to how adequately the data is taken into account. In real conditions of flexible production, there is still a problem of accounting for anomalies affecting the final quality of products. Such anomalies are rare, but in this case, these anomalies cause more harm than long-known defects. In this case, it is important to apply the Pareto principle by transforming it: 20% of quality anomalies bring 80% of financial losses associated with their elimination, but 80% of everyday defects bring only 20% of financial losses associated with their elimination.

A typical example: A batch of products was launched on 3D printers working in parallel (Fig. 6).

Before the launch of printing, the main parameters affecting the print quality were identified. Based on the features of this technology, the following critical parameters were identified:

- layer height, (Lishchenko et al. 2022)
- retraction,
- inflow,
- blowing,
- adhesion,
- temperature (of the print head and surface), (Kalman et al. 2024)
- print speed.

Based on previous experience, it was decided to use sensors to monitor the height and distance between the filament layers (Fig. 7).

This approach ensured better print quality of parts, for example, by changing the printing conditions according to the identified bottlenecks. The Rz parameter (average height between the peaks and troughs of a profile) was improved by 3 times.

Fig. 6 3D printers in the laboratory

The problem of removing the filament supporting the main part remained unresolved. However, this is a category of "eliminable" defects, and therefore, the methods and manual for elimination have already been integrated into the production plan.

Series after series, the company traditionally improves its 3D models to optimize the known problem with the filament support layer.

Let's imagine a situation where a sharp temperature drop unexpectedly occurred due to a draft during the production of a thin-walled product. Such an anomaly affected the rate of filament hardening, which led to several changes invisible to sensors and the quality control operator (Budnik et al. 2015).

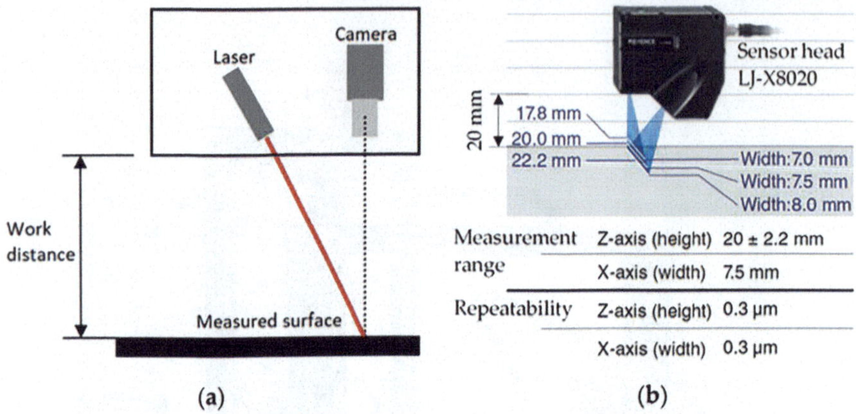

Fig. 7 Measurement schemes using laser profilometry (Haiderzai et al. 2025)

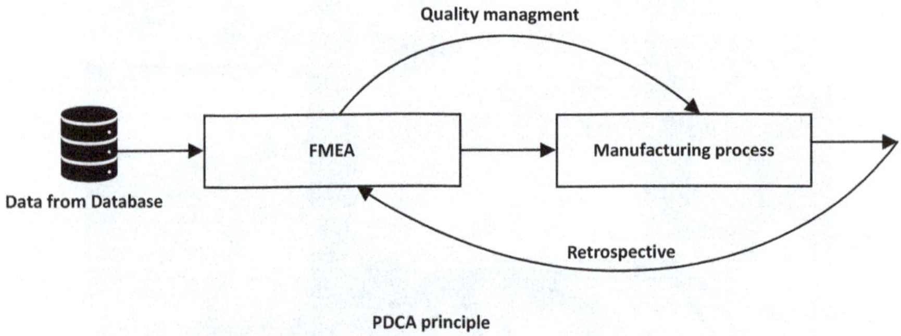

Fig. 8 Integration of the principles of PDCA and retrospective into the processing of quality data

The traditional approach of quality management does not anticipate these anomalies. In this way, during the final strength check of the design, critical flaws will be discovered and will not be eliminated.

If some of the products of this batch reach the consumer, the costs will be critical. This situation entails costs such as payment of a penalty to the client, costs associated with production since the client must get his money back, costs associated with laboratory research and investigations, and costs of improving the production process.

Instead of the traditional approach, using the Agile way of creating the team will be more effective, and CMMI will show the readiness for changes.

Approaches based on technology or production line bottlenecks are often used in practice. Analysis of all anomalies, data spikes, and changes is post-factum. In the proposed study, we propose to combine the experience of the past for a sustainable transition to modern trends.

Figure 8 shows a concept that combines the principles of PDCA and Agile, where the key role will be played by a person engaged in making complex decisions.

The main tool for future training of the quality management and control system will be the results of the FMEA report, which can be: Concept—FMEA of a conceptual proposal, Design—FMEA of a design, System—FMEA of a system, Process—FMEA of a production process, Product—FMEA of a product, Service—FMEA of service, Software—FMEA of software.

If previously Agile principles were applied only to the organization of logistics and new production projects, then the prospect of gradual implementation in quality management opens up.

The main idea is to look at quality control from different angles and solve anomalies not only from the point of view of the design of production units and technology, but also from the point of view of specialists in other fields who will solve the properties of materials under the influence of abnormal factors. Such an approach can help train innovative systems (Wieczorowski et al. 2023) to promptly warn about possible anomalies.

4 Conclusions

A study of trends in Quality Management, Agile in Manufacturing, and approaches to enterprise readiness assessment methods for new trends revealed the need to integrate structured process improvement methodologies. The proposed study proposes a combination of best practices to improve a manufacturing enterprise's QMS to meet the key principles of Industry 5.0 and Quality 5.0. Traditional process-based quality management models provide stability, risk mitigation, and regulatory compliance, but often lack the speed and flexibility required in today's rapidly changing manufacturing environments. Conversely, although responsive and iterative, pure Agile approaches can overlook the need for standardization and robust risk management. A hybrid model incorporating both structured and flow methodologies represents an optimal solution. Integrating Agile principles into quality management enables continuous monitoring, real-time anomaly detection, and adaptive decision-making while maintaining the disciplined structure provided by CMMI. The proposed approach ensures that enterprises will not only meet current quality standards but will also be able to actively identify and eliminate potential defects before they develop into costly failures. The proposed study offers a new perspective on the organization of QMS, combining traditional and new methods to achieve sustainability and development of enterprises per the main pillars of modern trends (Denysenko et al. 2020). A balanced approach and Agile principles facilitate decision-making and human-oriented flexibility, while CMMI ensures sustainability, economic feasibility, and structured risk assessment. Continuous improvement based on data. Integrating the PDCA cycle with real-time monitoring and data analysis from different perspectives and updating data using FMEA will increase the ability to predict and prevent anomalies before they affect product quality. FMEA data and the experience of the SME team will help train intelligent systems based on AI (Budnik et al. 2015), leading to more adaptive and sustainable production processes.

Acknowledgments This work was supported by the Slovak Research and Development Agency under the contract no. APVV-23-0591 and by the projects VEGA 1/0704/22, KEGA 022TUKE-4/2023 granted by the Ministry of Education, Research, Development and Youth of the Slovak Republic. This research was partially supported by the International Association for Technological Development and Innovations.

References

Ali A, Wasim A (2022) Innovative framework for assessing the impact of agile manufacturing in small and medium enterprises (SMEs). Sustainability 14:11503. https://doi.org/10.3390/su141811503

Avsec S, Jagiełło-Kowalczyk M, Żabicka A, Gil-Mastalerczyk J, Gawlak A (2024) Human-centered systems thinking in technology-enhanced sustainable and inclusive architectural design. Sustainability 16:9802. https://doi.org/10.3390/su16229802

Banáš D, Hrablik Chovanová H (2023) Agile manufacturing vs. lean manufacturing. Research Papers Faculty of materials science and technology, Slovak University of Technology 31(52):58–67. https://doi.org/10.2478/rput-2023-0007

Budnik AF, Rudenko PV, Berladir KV, Budnik OA (2015) Structured nanoobjects of polytetrafluoroethylene composites. J Nano-Electron Phys 7(2):02022

CIO.com Staff (2023) CMMI explained. CIO. https://www.cio.com/article/274530/cmmi-explained.html, last accessed 2025/01/31

Dehghan S, Sattarpanah Karganroudi S, Echchakoui S, Barka N (2025) The integration of additive manufacturing into industry 4.0 and industry 5.0: a bibliometric analysis (trends, opportunities, and challenges). Machines 13:62. https://doi.org/10.3390/machines13010062

Denysenko Y, Ivanov V, Luscinski S, Zaloga V (2020) An integrated approach for improving tool provisioning efficiency. Manag Prod Eng Rev 11(4):4–12. https://doi.org/10.24425/mper.2020.136115

Dobrotvorskiy S, Balog M, Ruzmetov A, Basova Y, Hrdzelidze S, Dobrovolska L (2023) Increasing the sustainability of manufacturing processes in the conditions of SMEs enterprises by predicting, their information intensity. In: Balog M, Iakovets A, Hrehova S (eds) EAI international conference on automation and control in theory and practice. EAI ARTEP 2023, EAI/springer innovations in communication and computing. Springer, Cham. https://doi.org/10.1007/978-3-031-31967-9_16

Fiałkowska-Filipek M, Dobrowolska A (2024) Quality 5.0: towards sustainable quality improvement in organizations. Ekonomia 29:49–61. https://doi.org/10.19195/2658-1310.29.3.4

Goknil A, Nguyen P, Sen S, Politaki D, Niavis H, Pedersen KJ, Suyuthi A, Anand A, Ziegenbein A (2023) A systematic review of data quality in CPS and IoT for industry 4.0. ACM Comput Surv 55:1–38

Gunasekaran A (1998) Agile manufacturing: enablers and an implementation framework. Int J Prod Res 36(5):1223–1247. https://doi.org/10.1080/002075498193291

Haiderzai MD, Dakić P, Stupavský I, Aleksić M, Todorović V (2025) Pattern shared vision refinement for enhancing collaboration and decision-making in government software projects. Electronics 14:334. https://doi.org/10.3390/electronics14020334

Hebda W, Mišík M (2024) In search of energy security: nuclear energy development in the Visegrad group countries. Energies 17:5390. https://doi.org/10.3390/en17215390

Holzner M, Rauch E, Orzes G, Matt DT (2023) A CANVAS based assessment model to evaluate SMEs readiness for digital business models. In: Ivanov V, Trojanowska J, Pavlenko I, Rauch E,

Piteľ J (eds) Advances in design, simulation and manufacturing VI. DSMIE 2023. Lecture notes in mechanical engineering. Springer, Cham. https://doi.org/10.1007/978-3-031-32767-4_4

ISACA, CMMI Performance Solutions, https://cmmiinstitute.com/, last accessed 2025/01/31

Ivanov D (2022) The industry 5.0 framework: viability-based integration of the resilience, sustainability, and human-centricity perspectives. Int J Prod Res 61(5):1683–1695. https://doi.org/10.1080/00207543.2022.2118892

Jadoon G, Ud Din I, Almogren A, Almajed H (2020) Smart and agile manufacturing framework, a case study for automotive industry. Energies 13:5766. https://doi.org/10.3390/en13215766

Jan F, Grudowski P (2023) Quality 5.0: A Paradigm Shift Towards Proactive Quality Control in Industry 5.0. Int J Bus Adm 14(2):51–56. https://doi.org/10.5430/ijba.v14n2p51

Jasiulewicz-Kaczmarek M, Antosz K, Zhang C, Ivanov V (2023) Industry 4.0 technologies for sustainable asset life cycle management. Sustainability 15(7):5833. https://doi.org/10.3390/su15075833

Kalman O, Husár J, Lazorík P (2024) Using laser profilometry to investigation FDM printing parameters for outer-perimeter analysis and surface quality improvement. Appl Sci 14:9405. https://doi.org/10.3390/app14209405

Kumar R, Singh K, Jain S (2019) Development of a framework for agile manufacturing. World J Sci Technol Sustain Dev. https://doi.org/10.1108/WJSTSD-05-2019-0022

Laitinen K, Luhtala M, Örmä M, Vaismaa K (2024) Productivity development enablers in the infrastructure sector: capability maturity model integration approach. Built Environ Project Asset Manage 14:201–227. https://doi.org/10.1108/BEPAM-07-2022-0095

Li L, You J, Xu T (2025) Risk analysis of digital twin project operation based on improved FMEA method. Systems 13:48. https://doi.org/10.3390/systems13010048

Lin T-C, Wang KJ (2021) Project-based maturity assessment model for smart transformation in Taiwanese enterprises. PLoS One 16(7):e0254522. https://doi.org/10.1371/journal.pone.0254522

Lishchenko N, Piteľ J, Larshin V (2022) Online monitoring of surface quality for diagnostic features in 3D printing. Machines 10:541. https://doi.org/10.3390/machines10070541

Muzylyov D, Shramenko N, Ivanov V (2021) Management decision-making for logistics systems using a fuzzy-neural simulation. In: Cagáňová D, Horňáková N, Pusca A, Cunha PF (eds) Advances in industrial internet of things, engineering and management, EAI/springer innovations in communication and computing. Springer, Cham. https://doi.org/10.1007/978-3-030-69705-1_11

Palhau M, Sá JC, Ávila P, Dinis-Carvalho J, Rodrigues C, Santos G (2024) Toyota way—the heart of Toyota production system and its impact on sustainable company growth. Qual Innov Prosper 28(3):23–45. https://doi.org/10.12776/qip.v28i3.2073

Pekarčíková M, Dic M, Duda R, Mizerák M (2023) The potential of using the internet of things instead of manufacturing intelligence systems in modern production. In: Balog M, Iakovets A, Hrehova S (eds) EAI international conference on automation and control in theory and practice. EAI ARTEP 2023, EAI/springer innovations in communication and computing. Springer, Cham. https://doi.org/10.1007/978-3-031-31967-9_8

Qamar A, Hall MA, Chicksand D, Collinson S (2019) Quality and flexibility performance trade-offs between lean and agile manufacturing firms in the automotive industry. Prod Plan Control 31(9):723–738. https://doi.org/10.1080/09537287.2019.1681534

Sun B, Jiang Z, Wang Y, Cai W (2023) Special issue on "Green manufacturing and sustainable supply chain management". Processes 11:3294. https://doi.org/10.3390/pr11123294

Wieczorowski M, Trojanowska J, Sokolov O (2023) Digital and information technologies in metrology 4.0. In: Ivanov V, Trojanowska J, Pavlenko I, Rauch E, Piteľ J (eds) Advances in design, simulation and manufacturing VI. DSMIE 2023, Lecture notes in mechanical engineering. Springer, Cham. https://doi.org/10.1007/978-3-031-32767-4_8

Mathematical Modelling of Spherical Piezoelectric Elements for Enhancing the Accuracy of Control and Measurement Systems in Advanced Manufacturing

Constantine Bazilo

Abstract This chapter presents the results of mathematical modelling of spherical piezoelectric elements, enabling the determination of their operating parameters and material constants based on their resonant frequencies and geometric dimensions. The results obtained aim to achieve the highest precision and reliability in measurement and control processes within manufacturing systems aligned with the "Industry 4.0" strategy. The research introduces analytically derived mathematical models and formulas for calculating the electrical impedance of spherical piezoelectric elements and analyzing the relationship between impedance amplitude and frequency. These findings provide valuable insights for developing more efficient measurement and control tools. Furthermore, the results are a foundation for creating innovative methodologies that ensure enhanced control and measurement systems accuracy based on piezoelectric elements. Future research will focus on improving the proposed mathematical model to account for external factors, enabling more precise predictions of element lifespan under varying operating conditions. This approach is expected to significantly enhance the design and application of piezoelectric technologies across various high-tech manufacturing sectors.

Keywords Piezoelectric transducer · Mathematical simulation · Spherical shell

1 Introduction

In modern manufacturing, piezoelectric elements play a key role in control and measurement systems in various industries, including aerospace, automotive, medical and energy (as components of navigation and control systems, in active safety and diagnostic systems for industrial complexes and vehicles, as well as in systems

---

C. Bazilo (✉)
Cherkasy State Technological University, Cherkasy, Ukraine
e-mail: k.bazilo@chdtu.edu.ua

V. Ivanov (ed.), *Emerging Technologies in Manufacturing and Materials Engineering*, Springer Tracts in Mechanical Engineering,
https://doi.org/10.1007/978-3-032-14914-5_7

for monitoring the technical condition of complex mechanisms, for timely detection of structural defects and prevention of accidents). In particular, spherical piezoelectric components are widely used due to their unique properties, such as isotropy and the ability to distribute mechanical stresses evenly, which ensures high accuracy and reliability of control and measurement functions in precision manufacturing processes (Wenjie et al. 2021).

In general, according to a study conducted by Allied Market Research (AMR), an analytical company specialising in market research, industry forecasts and competitive analysis (Allied Market Research 2025), the global piezoelectric materials market was valued at \$1.4 billion in 2021, and is projected to reach \$2.5 billion by 2031, growing at a CAGR of 6.2% from 2022 to 2031. This growth is driven by increased demand for high-precision sensor technologies in various industries.

At the same time, the use of spherical piezoelectric elements reduces maintenance costs and increases the efficiency of production processes. In particular, using such elements in high-precision diagnostic devices helps detect defects and deviations in components and products early, which reduces the overall cost of rejects and repairs of equipment and production systems.

At the same time, a deep understanding of their electromechanical properties is required for the optimal use of spherical piezoelectric elements in these systems. To achieve this, it is necessary to apply mathematical modelling methods that consider the interaction of mechanical, electrical, and material characteristics of such elements (Levy and Shapiro 2023). Such mathematical modelling allows for a highly accurate and reliable prediction of the behaviour of these elements under various operating conditions.

This chapter aims to develop and improve mathematical models of spherical piezoelectric elements to determine their optimal operating parameters and material constants, which will help improve the accuracy of control and measurement systems in modern production.

The significance of the study is to reduce measurement errors, optimise sensor design, reduce testing costs, and expand the use of piezoelectric materials in high-precision technologies.

This research tests the following hypotheses:

- the spherical geometry of piezoelectric elements provides a uniform distribution of mechanical stress, which increases the accuracy of sensors;
- mathematical modeling allows for the effective prediction of the behaviour of spherical piezoelectric elements and reduces the cost of experimental research;
- optimisation of sensor parameters improves their sensitivity and stability in dynamic conditions;
- considering nonlinear effects in the models increases the accuracy of predicting the behaviour of sensors under complex loads.

The expected result is an increase in the accuracy of control and measurement systems, reduced errors, improved sensor design optimization, and expanded use in high-tech production processes.

2 Fundamentals and Challenges

The mathematical modelling of piezoelectric elements is an important area of research that contributes to improving the accuracy and efficiency of control and measurement systems in industrial sectors, which many scientists from both Ukraine and abroad have studied. Let us consider the key publications of leading Ukrainian and foreign authors who have studied this topic.

For example, in (Aladwan et al. 2022), the authors study the peculiarities of using piezoelectric transducers for critical application systems. The primary focus is on developing mathematical models for optimising the characteristics of these elements.

Paper (Xu et al. 2022) investigates a method for simulating coupled electromagnetic and mechanical vibrations on arbitrarily shaped piezoelectric structures. The method is particularly relevant to the design of piezoelectric antennas, resonators, and waveguides exploiting either bulk or surface-acoustic waves.

In (Derraz et al. 2024), the mathematical modelling of piezoelectric elements in production conditions was carried out. The paper focuses on mathematical modelling that propels the advancement of efficient piezoelectric materials for renewable energy applications.

Among other works of scientists, it is worth highlighting the work of the team of authors led by Chuang-Sheng Walter Yang (Chuang-Sheng et al. 2017), where the authors investigate using the finite element method for modelling piezoelectric sensors. The work emphasises the importance of accurate modelling to ensure measurement reliability.

In the study by Bowen Zhong and his colleagues (Zhong et al. 2023), the authors propose analytical models of piezoelectric actuators used for precision control in microelectromechanical systems. The paper emphasises the need to consider nonlinear effects in the models.

Jung Heum Yeon, in his co-authored article (Ledia et al. 2024), investigates the numerical modelling of piezoelectric energy harvesters used in industrial applications. The work emphasises the importance of design optimisation to improve energy harvesting efficiency.

These publications cover various approaches to the mathematical modelling of piezoelectric elements, including finite element methods, analytical models, and numerical simulations. Among the controversial issues in this area are the choice of optimal modelling methods for specific applications, the consideration of nonlinear material properties, and the influence of external factors on model accuracy.

In addition, a detailed introduction to the problems of mathematical modelling of piezoelectric elements in manufacturing industries is provided in the Journal of Intelligent Material Systems and Structures (Zhang et al. 2019). This journal publishes articles on research in intelligent materials and structures, including piezoelectric elements and their modelling.

However, after analysing specialised sources, the author has established that mathematical modelling of spherical piezoelectric elements is an issue of little study for the dynamic field of research and remains relevant to this day, and requires further study to ensure the efficiency and reliability of modern production systems.

Thus, spherical piezoelectric elements are key components of modern control and measurement systems, ensuring their high accuracy and reliability. However, as the efficiency of such elements largely depends on the accurate prediction of their behaviour under various operating conditions, mathematical modelling becomes a critical tool for their optimisation. Further development of such modelling methods will increase measurement accuracy, improve the adaptation of sensors to dynamic loads, and expand the possibilities of their use in high-tech manufacturing processes.

3 Ideas and Solutions

Here, we consider the radial oscillations of a hollow closed sphere of piezoceramics polarized through the thickness.

This situation is a model situation based on its factual content. The object's computational scheme is shown in Fig. 1. The surface of the sphere is covered with a thin metal layer; that is, they have an electrode coating (Medianyk et al. 2018). The wall perforation (Fig. 1), through which the conductor is introduced inside the sphere, has a negligibly small (compared to the wall thickness) size and, for this reason, will be omitted in our subsequent considerations. In other words, we will assume that the discontinuity of the wall due to the perforation introduces insignificant perturbations into the stress-strain state of the sphere, which we can safely ignore. Piezoceramic material of the spherical wall is polarized through thickness, directed along the radial axis r in the spherical coordinate system (r, ϕ, ϑ).

When an electric potential difference is supplied to the electroded surfaces of the sphere $U_0 e^{i\omega t}$ (U_0 is the amplitude; $i = \sqrt{-1}$; ω is the angular frequency of sign inversion; t is time) a radially oriented electric field arises within its wall. The electric field is significant to the total electric field since it represents the algebraic sum of the internal (polarization) electric field and the field induced by the electric potential difference generator. The vector of the total electric field strength is entirely defined by the radial component E_r, which only depends on the radial coordinate r values and is independent from the azimuth angle ϕ and polar angle ϑ values, thus, it has central symmetry. A centrally symmetric electric field generates a centrally symmetric stress-strain state in a piezoelectric material, which is characterized by compression-tensile strains $\varepsilon_{rr} e^{i\omega t}$, $\varepsilon_{\phi\phi} e^{i\omega t}$, $\varepsilon_{\vartheta\vartheta} e^{i\omega t}$ and stress resultants $\sigma_{rr} e^{i\omega t}$, $\sigma_{\phi\phi} e^{i\omega t}$ and $\sigma_{\vartheta\vartheta} e^{i\omega t}$. Herewith, the amplitude values of strain tensor components are $\varepsilon_{rr} = \partial u_r / \partial r$ and $\varepsilon_{\phi\phi} = \varepsilon_{\vartheta\vartheta} = u_r / r$, where u_r is the amplitude value of the radial component of the centrally symmetric and harmonically varying in time displacement vector of the piezoceramic material particles.

The radial component must satisfy the equation of steady harmonic oscillations (Newton's second law in its differential form), which, in the case of a centrally symmetric stress-strain state, is written as follows:

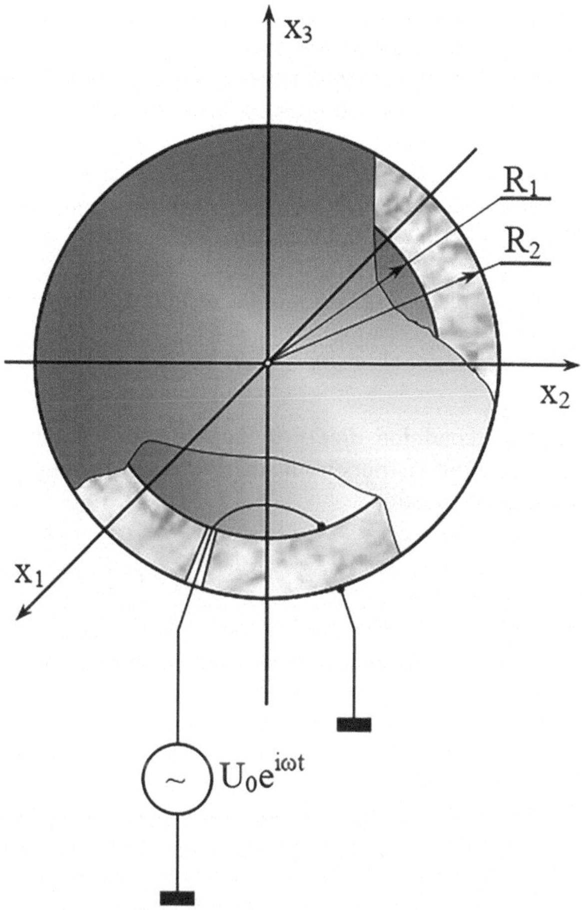

Fig. 1 Computational scheme of a hollow sphere polarized through the thickness

$$\frac{\partial \sigma_{rr}}{\partial r} + \frac{1}{r}\left(2\sigma_{rr} - \sigma_{\phi\phi} - \sigma_{\vartheta\vartheta}\right) + \rho_0 \omega^2 u_r = 0 \forall r \in \left[R_1, R_2\right], \tag{1}$$

where ρ_0 is the density of the piezoceramics.

Amplitude values of the mechanical stress resultants included in expression (1) are defined as follows:

$$\sigma_{rr} = c_{11}^E \varepsilon_{rr} + c_{12}^E \varepsilon_{\phi\phi} + c_{12}^E \varepsilon_{\vartheta\vartheta} - e_{11} E_r, \tag{2}$$

$$\sigma_{\phi\phi} = c_{12}^E \varepsilon_{rr} + c_{22}^E \varepsilon_{\phi\phi} + c_{12}^E \varepsilon_{\vartheta\vartheta} - e_{12} E_r, \tag{3}$$

$$\sigma_{\vartheta\vartheta} = c_{12}^E \varepsilon_{rr} + c_{12}^E \varepsilon_{\phi\phi} + c_{22}^E \varepsilon_{\vartheta\vartheta} - e_{12} E_r. \tag{4}$$

The amplitude value of the radial component E_r of the total electric field strength vector will be determined by the same method that was used when solving the radial vibrations problem of a hollow piezoceramic cylinder. It is appropriate at this point to recall that the method in question was described in the monograph (Petrishchev 2019).

The electrical state of the strained piezoceramics is determined by the electric displacement vector $\vec{D}(r,\phi,\vartheta)e^{i\omega t}$, which, under the framework of the problem being solved, has central symmetry and is entirely determined by the radial component $D_r(r)e^{i\omega t}$. The spatially developed amplitude D_r of the radial component of the electric displacement vector of the resulting electric state is defined as follows

$$D_r = e_{11}\varepsilon_{rr} + e_{12}\left(\varepsilon_{\phi\phi} + \varepsilon_{\vartheta\vartheta}\right) + \chi_{11}^{\varepsilon}E_r. \tag{5}$$

Since the condition $div\vec{D} = 0$ must be satisfied, which prohibits the existence of free electricity carriers in the dielectric, the radial component D_r must satisfy the differential equation

$$\frac{1}{r^2}\left[\frac{\partial}{\partial r}\left(r^2 D_r\right)\right] = 0. \tag{6}$$

Consequently, the solution to the differential equation (6) is the following expression

$$D_r = \frac{C_1}{r^2}, \tag{7}$$

where C_1 is the constant to be determined when solving the problem.

Substituting expression (5) into definition (7), we obtain

$$e_{11}\varepsilon_{rr} + e_{12}\left(\varepsilon_{\phi\phi} + \varepsilon_{\vartheta\vartheta}\right) + \chi_{11}^{\varepsilon}E_r = \frac{C_1}{r^2},$$

whence we derive the expression for calculating the total electric field strength

$$E_r = \frac{C_1}{\chi_{11}^{\varepsilon}r^2} - \frac{e_{11}}{\chi_{11}^{\varepsilon}}\varepsilon_{rr} - \frac{e_{12}}{\chi_{11}^{\varepsilon}}\left(\varepsilon_{\phi\phi} + \varepsilon_{\vartheta\vartheta}\right). \tag{8}$$

By substituting expression (8) into relations (2)–(4), we obtain the following results:

$$\sigma_{rr} = c_{11}^{D}\varepsilon_{rr} + c_{12}^{D}\varepsilon_{\phi\phi} + c_{12}^{D}\varepsilon_{\vartheta\vartheta} - \frac{e_{11}C_1}{\chi_{11}^{\varepsilon}r^2}, \tag{9}$$

$$\sigma_{\phi\phi} = c_{12}^{D}\varepsilon_{rr} + c_{22}^{D}\varepsilon_{\phi\phi} + c_{13}^{D}\varepsilon_{\vartheta\vartheta} - \frac{e_{12}C_1}{\chi_{11}^{\varepsilon}r^2}, \tag{10}$$

$$\sigma_{\vartheta\vartheta} = c_{12}^D \varepsilon_{rr} + c_{13}^D \varepsilon_{\phi\phi} + c_{22}^D \varepsilon_{\vartheta\vartheta} - \frac{e_{12} C_1}{\chi_{11}^\varepsilon r^2}, \tag{11}$$

where $c_{11}^D = c_{11}^E + e_{11}^2 / \chi_{11}^\varepsilon$, $c_{12}^D = c_{12}^E + e_{11} e_{12} / \chi_{11}^\varepsilon$, $c_{22}^D = c_{22}^E + e_{12}^2 / \chi_{11}^\varepsilon$ and $c_{13}^D = c_{12}^E + e_{12}^2 / \chi_{11}^\varepsilon$.

Substituting expressions (9)–(11) into equation (1), and noting the components of the strain tensor in explicit form, we obtain the following differential equation:

$$c_{11}^D \frac{\partial^2 u_r}{\partial r^2} + c_{11}^D \frac{2}{r} \frac{\partial u_r}{\partial r} - \frac{2}{r^2} \left(c_{22}^D + c_{13}^D \right) u_r + \rho_0 \omega^2 u_r = -\frac{2 e_{12}}{\chi_{11}^\varepsilon r^3} C_1,$$

or

$$\frac{\partial^2 u_r}{\partial r^2} + \frac{2}{r} \frac{\partial u_r}{\partial r} - \frac{\mu^2}{r^2} u_r + k^2 u_r = -\frac{2 e_{12}}{\chi_{11}^\varepsilon c_{11}^D r^3} C_1, \tag{12}$$

where $\mu^2 = 2\left(c_{22}^D + c_{13}^D \right) / c_{11}^D$ is a dimensionless parameter, $k^2 = \rho_0 \omega^2 / c_{11}^D$ is the squared wavenumber of centrally symmetric radial vibrations of a hollow piezoceramic sphere.

We note equation (12) in a form that would be more convenient for further calculations:

$$r^2 \frac{\partial^2 u_r}{\partial r^2} + 2r \frac{\partial u_r}{\partial r} + \left[(kr)^2 - \mu^2 \right] u_r = -\frac{2 e_{12}}{\chi_{11}^\varepsilon c_{11}^D r} C_1.$$

In the final entry, we proceed to the dimensionless variable $x = kr$:

$$x^2 \frac{\partial^2 u_r}{\partial x^2} + 2x \frac{\partial u_r}{\partial x} + \left[x^2 - \mu^2 \right] u_r = -\frac{2 e_{12} k}{\chi_{11}^\varepsilon c_{11}^D x} C_1. \tag{13}$$

Equation (13) differs from the standard Bessel equation by number two in the addend. To reduce equation (13) to the standard Bessel equation, we represent the desired solution u_r in the following form

$$u_r = x^{-1/2} U(x), \tag{14}$$

where $U(x)$ is yet unknown function.

It is evident that

$$\frac{\partial u_r}{\partial x} = -\frac{1}{2} x^{-3/2} U(x) + x^{-1/2} \frac{\partial U(x)}{\partial x},$$

$$\frac{\partial^2 u_r}{\partial x^2} = \frac{3}{4} x^{-5/2} U(x) - x^{-3/2} \frac{\partial U(x)}{\partial x} + x^{-1/2} \frac{\partial^2 U(x)}{\partial x^2}.$$

Substituting the derivatives into equation (13) and reducing the fraction by factor $x^{-1/2}$, we obtain

$$x^2 \frac{\partial^2 U(x)}{\partial x^2} + x \frac{\partial U(x)}{\partial x} + \left[x^2 - \nu^2\right] U(x) = -\frac{2 e_{12} k}{\chi_{11}^{\varepsilon} c_{11}^{D} x^{1/2}} C_1, \quad (15)$$

where $\nu^2 = \mu^2 + 1/4$ is a dimensionless parameter of the differential equation (for PZT-19 piezoceramics this parameter will be $\nu = 1, 6869$).

Let us seek a solution of equation (15) in the form

$$U(x) = \left[A + A(x)\right] J_\nu(x) + \left[B + B(x)\right] N_\nu(x), \quad (16)$$

where A and B are constants, $A(x)$ and $B(x)$ are variable constants, $J_\nu(x)$ and $N_\nu(x)$ are order ν Bessel and Neumann functions.

Variable constants form a particular solution of equation (15) and therefore allow a certain arbitrariness on the part of the computing engineer. We require that these constants satisfy the following condition:

$$A'(x) J_\nu(x) + B'(x) N_\nu(x) = 0, \quad (17)$$

where dashes denote the first derivatives of variable constant with respect to variable x.

Substituting the proposed solution (16) into equation (15) and performing differentiation with reference to condition (17), we obtain

$$x^2 A'(x)\left[J_{\nu-1}(x) - \frac{\nu}{x} J_\nu(x)\right] + x^2 B'(x)\left[N_{\nu-1}(x) - \frac{\nu}{x} N_\nu(x)\right] = -\frac{2 e_{12} k}{\chi_{11}^{\varepsilon} c_{11}^{D}} C_1 x^{-1/2}. \quad (18)$$

Condition (17) and relation (18) form a system of algebraic equations, which is uniquely solved with reference to the unknown derivatives of variable constants:

$$A'(x) = \frac{\pi e_{12} k}{\chi_{11}^{\varepsilon} c_{11}^{D}} C_1 x^{-3/2} N_\nu(x),$$

$$B'(x) = \frac{\pi e_{12} k}{\chi_{11}^{\varepsilon} c_{11}^{D}} C_1 x^{-3/2} J_\nu(x).$$

By integrating the resulting expressions with respect to the variable x, we obtain

$$A(x)=\frac{\pi e_{12}k}{\chi_{11}^{\varepsilon}c_{11}^{D}}C_1\int_{x_1}^{x}\xi^{-3/2}N_\nu(\xi)d\xi,$$
$$B(x)=-\frac{\pi e_{12}k}{\chi_{11}^{\varepsilon}c_{11}^{D}}C_1\int_{x_1}^{x}\xi^{-3/2}J_\nu(\xi)d\xi. \tag{19}$$

where $x_1 = kR_1$.

By analogy with the Bessel and Neumann integral functions, we define the integrals in relations (19) as Bessel and Neumann spherical integral functions $Ji_\nu^s(x)$ and $Ni_\nu^s(x)$, consequently

$$Ji_\nu^s(x)=\int_{x}^{\infty}\xi^{-3/2}J_\nu(\xi)d\xi,$$
$$Ni_\nu^s(x)=\int_{x}^{\infty}\xi^{-3/2}N_\nu(\xi)d\xi. \tag{20}$$

The graphs illustrating the Bessel and Neumann integral spherical functions we are demonstrated in Fig. 1. This and further modelling are carried out in MATLAB (Fig. 2).

Definitions (20) of spherical integral functions included, expressions (19) can be cast as follows

$$A(x)=\frac{\pi e_{12}k}{\chi_{11}^{\varepsilon}c_{11}^{D}}C_1\left[Ni_\nu^s(kR_1)-Ni_\nu^s(x)\right],$$
$$B(x)=-\frac{\pi e_{12}k}{\chi_{11}^{\varepsilon}c_{11}^{D}}C_1\left[Ji_\nu^s(kR_1)-Ji_\nu^s(x)\right]. \tag{21}$$

Thus, the radial component u_r of the displacement vector of piezoceramic material particles is to be determined by the following expression

$$u_r=x^{-1/2}\left\{\left[A+A(x)\right]J_\nu(x)+\left[B+B(x)\right]N_\nu(x)\right\}. \tag{22}$$

Constants A and B must enforce Newton's third law on spherical surfaces $r = R_1, R_2$. Generally assuming that the hollow sphere is placed in vacuum, we must ensure that the following boundary conditions are satisfied:

$$\sigma_{rr}\big|_{r=R_1,R_2}=0. \tag{23}$$

By substituting expression (22) into the normal stress σ_{rr} definition (9), and subsequently substituting the obtained result into condition (23), we obtain the system of algebraic equations including the required constants A and B:

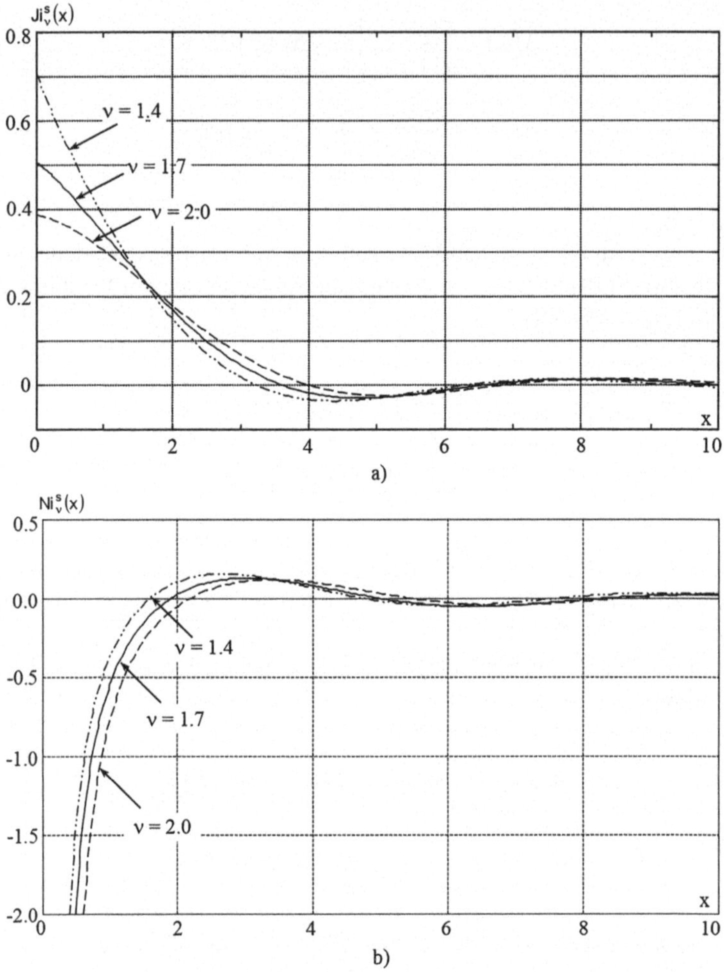

Fig. 2 Integral spherical functions of Bessel (**a**) and Neumann (**b**)

$$AJ\left(kR_1\right)+BN\left(kR_1\right)=\frac{e_{11}kC_1}{\chi_{11}^{\varepsilon}c_{11}^{D}}\left(kR_1\right)^{-3/2}, \tag{24}$$

$$AJ\left(kR_2\right)+BN\left(kR_2\right)=\frac{e_{11}kC_1}{\chi_{11}^{\varepsilon}c_{11}^{D}}\left(kR_1\right)^{-3/2}F\left(kR_2\right), \tag{25}$$

where

$$J\left(kR_j\right)=J_{\nu-1}\left(kR_j\right)-\left(\nu+1/2-2c_{12}^{D}/c_{11}^{D}\right)J_\nu\left(kR_j\right)/x,$$

$$N(kR_j) = N_{\nu-1}(kR_j) - (\nu + 1/2 - 2c_{12}^D / c_{11}^D) N_\nu(kR_j)/x, \quad j = 1, 2,$$

$$F(kR_2) = (R_1/R_2)^{3/2}$$
$$-(kR_1)^{3/2}\left\{J(kR_2)\left[Ni_\nu^s(kR_1) - Ni_\nu^s(kR_2)\right] - N(kR_2)\left[Ji_\nu^s(kR_1) - Ji_\nu^s(kR_2)\right]\right\}.$$

Evidently, the solution of the system of equations (24) and (25) reads

$$A = \frac{e_{11}C_1}{\chi_{11}^\varepsilon c_{11}^D R_1} \alpha(k, R_1, R_2),$$
$$B = \frac{e_{11}C_1}{\chi_{11}^\varepsilon c_{11}^D R_1} \beta(k, R_1, R_2). \tag{26}$$

where

$$\alpha(k, R_1, R_2) = \frac{N(kR_2) - F(kR_2)N(kR_1)}{(kR_1)^{1/2} D(kR_1, R_2)},$$

$$\beta(k, R_1, R_2) = \frac{F(kR_2)J(kR_1) - J(kR_2)}{(kR_1)^{1/2} D(kR_1, R_2)},$$

$$D(kR_1, R_2) = J(kR_1)N(kR_2) - J(kR_2)N(kR_1).$$

The determinant $D(kR_1, R_2)$ of the system of equations (24) and (25) defines the resonant frequencies of an oscillating hollow sphere.

Figure 3 demonstrates the function $1/D(kR_1, R_2)$ module plots calculated at different values of R_2/R_1 ratio. The absolute values of the function $1/D(kR_1, R_2)$ modulus are plotted along the ordinate axis, and the dimensionless wave number kR_1 is plotted along the abscissa axis. The calculation was performed under the assumption that the hollow sphere is made of PZT-19 piezoceramics with a quality factor $Q = 100$. The first resonance (for a sphere with wall thickness $R_2/R_1 = 1, 1$ this resonance is single within the range of dimensionless wave numbers from zero to 30) is determined by the curvature (curvature radius) of the sphere's inner surface, and the second resonance (for thick walls it will be the third resonance, etc.) is determined by the wall's thickness.

After determining constants A and B we can rewrite the expression for calculating the radial component of the material particles displacement vector in a hollow sphere wall as

$$u_r = \frac{e_{11}C_1}{\chi_{11}^\varepsilon c_{11}^D R_1} U_\nu(kR_1, x), \tag{27}$$

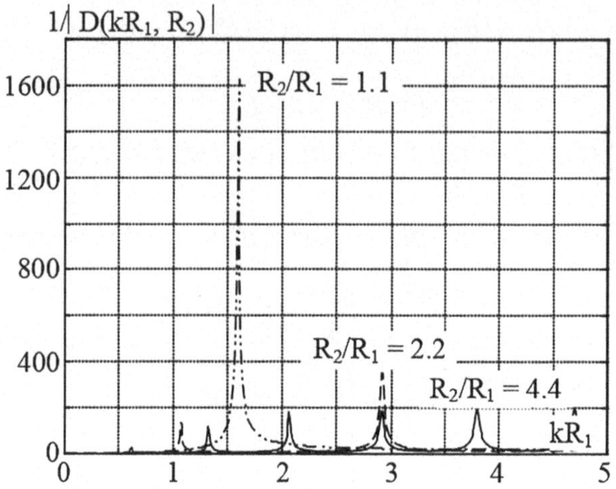

Fig. 3 Resonant properties of a hollow piezoceramic sphere with different wall thicknesses

where

$$U_\nu\left(kR_1,x\right)=x^{-1/2}\left[A\left(k,R_1,R_2\right)J_\nu\left(x\right)+B\left(k,R_1,R_2\right)N_\nu\left(x\right)-\Psi_\nu\left(x\right)\right],$$

$$A\left(k,R_1,R_2\right)=\alpha\left(k,R_1,R_2\right)+\pi\left(kR_1\right)Ni_\nu^s\left(kR_1\right),$$

$$B\left(k,R_1,R_2\right)=\beta\left(k,R_1,R_2\right)-\pi\left(kR_1\right)Ji_\nu^s\left(kR_1\right),$$

$$\Psi_\nu\left(x\right)=Ni_\nu^s\left(x\right)J_\nu\left(x\right)-Ji_\nu^s\left(x\right)N_\nu\left(x\right).$$

The procedure for determining constant C_1 completes the solution of the problem.

Since the radial component $E_r = -\partial\Phi(r)/\partial r$, where $\Phi(r)$ denotes the total electric field potential, it follows from relation (8) that

$$\frac{\partial\Phi\left(r\right)}{\partial r}=-\frac{C_1}{\chi_{11}^{\varepsilon}r^2}+\frac{e_{11}}{\chi_{11}^{\varepsilon}}\frac{\partial u_r}{\partial r}+\frac{2e_{12}}{\chi_{11}^{\varepsilon}}\frac{u_r}{r}. \tag{28}$$

After integrating the left and the right parts of relation (28) over the radial variable we obtain

$$\Phi\left(r\right)=\frac{C_1}{\chi_{11}^{\varepsilon}r}+\frac{e_{11}}{\chi_{11}^{\varepsilon}}u_r\left(r\right)+\frac{2e_{12}}{\chi_{11}^{\varepsilon}}\int\frac{u_r\left(r\right)}{r}dr-C_2, \tag{29}$$

where C_2 is a constant.

Constants C_1 and C_2 are determined from the obvious (Fig. 1) conditions for the electric potential $\Phi(r)$ on the surfaces with electrode coating $r = R_1$ and $r = R_2$:

$$\begin{aligned}\Phi(r)\big|_{r=R_1} &= U_0,\\ \Phi(r)\big|_{r=R_2} &= 0.\end{aligned} \tag{30}$$

By substituting expression (29) into electrical boundary conditions (30), we obtain a system of algebraic equations, which is solved with respect to the desired constants C_1 and C_2 in a unique way. In particular, a constant C_1 is defined by the following expression

$$C_1 = \frac{U_0 \chi_{11}^{\varepsilon} R_1}{(1 - R_1/R_2)\,\Xi(k,R_1,R_2)}, \tag{31}$$

where

$$\Xi(k,R_1,R_2) = 1 - \frac{1}{(1 - R_1/R_2)} \frac{K_r^2}{(1 + K_r^2)}$$
$$\left[A(k,R_1,R_2)\Xi_1(k,R_1,R_2) + B(k,R_1,R_2)\Xi_2(k,R_1,R_2) + \Xi_3(k,R_1,R_2)\right],$$

$K_r^2 = e_{11}^2 / \left(c_{11}^E \chi_{11}^{\varepsilon}\right)$ is the squared electromechanical coupling factor of the piezoceramics in the mode of centrally symmetric radial vibrations of a hollow sphere;

$$\begin{aligned}\Xi_1(k,R_1,R_2) &= (kR_2)^{-1/2} J_\nu(kR_2) - (kR_1)^{-1/2} J_\nu(kR_1)\\ &+ \frac{2e_{12}}{e_{11}}\left[Ji_\nu^s(kR_1) - Ji_\nu^s(kR_2)\right],\end{aligned}$$

$$\begin{aligned}\Xi_2(k,R_1,R_2) &= (kR_2)^{-1/2} N_\nu(kR_2) - (kR_1)^{-1/2} N_\nu(kR_1)\\ &+ \frac{2e_{12}}{e_{11}}\left[Ni_\nu^s(kR_1) - Ni_\nu^s(kR_2)\right],\end{aligned}$$

$$\begin{aligned}\Xi_3(k,R_1,R_2) &= -(kR_2)^{-1/2}\left[Ni_\nu^s(kR_2)J_\nu(kR_2) - Ji_\nu^s(kR_2)N_\nu(kR_2)\right]\\ &+ (kR_1)^{-1/2}\left[Ni_\nu^s(kR_1)J_\nu(kR_1) - Ji_\nu^s(kR_1)N_\nu(kR_1)\right]\\ &+ \frac{2e_{12}}{e_{11}}\left[Ni_\nu^s(kR_1)Ji_\nu^s(kR_2) - Ni_\nu^s(kR_2)Ji_\nu^s(kR_1)\right].\end{aligned}$$

When writing the expression for calculating the function $\Xi(k, R_1, R_2)$, we considered that

$$\int_{kR_1}^{kR_2} \left[Ni_\nu^s(x)J_\nu(x) - Ji_\nu^s(x)N_\nu(x)\right]\frac{dx}{x^{3/2}} = Ni_\nu^s(kR_1)Ji_\nu^s(kR_2) - Ni_\nu^s(kR_2)Ji_\nu^s(kR_1).$$

Substituting expression (31) into the displacement vector radial component definition (27), we obtain a solution to the problem of centrally symmetric radial vibrations of a hollow sphere made of piezoelectric ceramics, which is cast as follows

$$u_r(r) = \frac{U_0 \chi_{11}^{\varepsilon}}{e_{11}} \frac{K_r^2}{\left(1+K_r^2\right)\left(1-R_1/R_2\right)\Xi\left(k,R_1,R_2\right)} U_v\left(kR_1,x\right). \tag{32}$$

The electrical impedance $Z_{el}(\omega)$ of an oscillating sphere is determined, as is usually the case, from Ohm's law for a section of an electrical circuit, i.e.

$$Z_{el}(\omega) = \frac{-U_0}{I}, \tag{33}$$

where I is the alternating electric current amplitude in the conductors that are connected to the electroded surfaces of a hollow sphere (Fig. 1). The minus is indicated because the potential decreases with increasing radial coordinate r.

The current strength is $I = -i\omega 4\pi R_1^2 D_r\left(R_1\right)$ or

$$I = -\frac{i\omega C_0^{\varepsilon} U_0}{\Xi\left(k,R_1,R_2\right)}, \tag{34}$$

where $C_0^{\varepsilon} = 4\pi R_1 \chi_{11}^{\varepsilon} / \left(1-R_1/R_2\right)$ is the dynamic electric capacitance of a hollow piezoceramic sphere.

Substituting expression (34) into the definition (33) of electrical impedance, we obtain the calculation formula

$$Z_{el}(\omega) = \frac{1}{i\omega C_0^{\varepsilon}} \Xi\left(k,R_1,R_2\right). \tag{35}$$

If the dielectric between the spherical electrodes does not possess piezoelectric properties, i.e. $K_r^2 = 0$, then function $\Xi(k, R_1, R_2) = 1$ and expression (35) determine the reactance of a capacitor with electric capacitance C_0^{ε}. Differently, function $\Xi(k, R_1, R_2)$ becomes frequency-dependent, forming local extrema at some frequencies. At the frequencies of the local minima of the $\Xi(k, R_1, R_2)$ function, electromechanical resonances will be observed, and the frequencies of the electromechanical antiresonances will obviously correspond to the frequencies of the local maxima of the function $\Xi(k, R_1, R_2)$.

4 Conclusions

Mathematical models are analysed and different approaches to the description of radial vibrations of a spherical piezoceramic element are compared, which allowed considering the influence of electric field, mechanical stresses, and geometrical parameters of the sphere on its resonant properties.

The problem of constructing a mathematical model of radial vibrations of piezoelectric spherical elements based on the equations of harmonic vibrations and considering the central symmetry of the stress-strain state is proposed, which allows obtaining accurate calculation formulas for predicting the behaviour of such structures.

Analytical dependences for calculating the electric field, mechanical stress, and displacement of material particles in a spherical piezoelectric element are obtained, which allows assessing the influence of design parameters, such as wall thickness and dielectric properties of the material, on the efficiency of the elements.

The dependence between the geometrical parameters of the piezoelectric element and its resonance characteristics is established, in particular, it is shown that the first resonance (for a sphere with wall thickness $R_2/R_1 = 1, 1$ this resonance is single within the range of dimensionless wave numbers kR_1 from zero to 30) is determined by the curvature of the inner surface of the sphere, and the subsequent ones by the wall thickness, which makes it possible to control the spectrum of resonant frequencies when designing such elements.

The effectiveness of mathematical models was evaluated on the example of piezoceramic material PZT-19 with a quality factor $Q = 100$, where it was found that the obtained models allow predicting the frequencies of electromechanical resonances with high accuracy, which is critical for optimising production processes and improving the accuracy of control and measurement systems.

Acknowledgments The section presents the data obtained as a result of the experimental scientific and technical work "Development of an automated ultrasonic system for the extraction of plant materials in the manufacture of multinutrient functional beverages for the rehabilitation and prevention of post-traumatic stress disorders" (state registration number: 0124U000713, 2024–2025), carried out at Cherkasy State Technological University.

The research was partially supported by the International Association for Technological Development and Innovations.

References

Aladwan IM, Bazilo C, Faure E (2022) Modelling and development of multisectional disk piezoelectric transducers for critical application systems. Jordan J Mech Indus Eng 16(2):275–282. https://jjmie.hu.edu.jo/vol16-2/09-48-21.pdf, last accessed 31 Jan. 2025

Allied Market Research, Piezoelectric materials market size, share, competitive landscape and trend analysis report, by material type, by application, by end-use: global opportunity analysis and industry forecast, 2021–2031, https://www.alliedmarketresearch.com/piezoelectric-materials-market-A21105, last accessed 31 Jan. 2025

Chuang-Sheng WY, Yong-An L, Jin-Yeon K (2017) A piezoelectric brace for passive suppression of structural vibration and energy harvesting. Smart Mater Struct 26:085005. https://doi.org/10.1088/1361-665X/aa6e98

Derraz M et al (2024) Mathematical modeling for predicting electrical energy harvested using piezoelectric composite materials for smart system applications. Micro Nano Eng 23:100253. https://doi.org/10.1016/j.mne.2024.100253

Ledia S, Heum JY, Yooseob S (2024) Piezoelectric energy harvesting for civil engineering applications. Energies 17(19):4935. https://doi.org/10.3390/en17194935

Levy DA, Shapiro A (2023) System identification and mathematical modeling of a piezoelectric actuator through a practical three-stage mechanism. Micromachines 14(1):88. https://doi.org/10.3390/mi14010088

Medianyk VV, Bondarenko YY, Bondarenko MO (2018) Research of current-conducting electrodes of elements from piezoelectric ceramics modified by the low-energy ribbon-shaped electron stream. J Nano- Electron Phys 10(6):06012. https://doi.org/10.21272/jnep.10(6).06012

Petrishchev ON (2019) Principles and methods of mathematical modeling of oscillating piezoelectric elements. Hordiienko Publ, Cherkasy

Wenjie W, Yi J, Thomas PJ (2021) Structural design and physical mechanism of axial and radial sandwich resonators with piezoelectric ceramics: a review. Sensors (Basel) 21(4):1112–1127. https://doi.org/10.3390/s21041112

Xu X, Newns V, Oxborrow M (2022) Simulating the magnetic fields generated by piezoelectric devices using FEM software: beyond the quasistatic approximation. J Appl Phys 132:205107. https://doi.org/10.1063/5.0120630

Zhang S-Q, Zhao G-Z, Rao MN, Schmidt R, Yu Y-J (2019) A review on modeling techniques of piezoelectric integrated plates and shells. J Intell Mater Syst Struct 30(8):1133–1147. https://doi.org/10.1177/1045389X19836169

Zhong B, Liao Z, Zhang X, Jin Z, Sun L (2023) Modeling of rapid response characteristics of piezoelectric actuators for ultra-precision machining. Materials 16(6):2272. https://doi.org/10.3390/ma16062272

Analysis of the Quality Indicators of Transient Processes of the Hydraulic Drive for Turning the Container During Loading Municipal Solid Waste into a Garbage Truck

Oleh Bereziuk and Oleksandr Petrov

Abstract This study focuses on establishing analytical relationships that characterize the quality parameters of transient processes in a hydraulic drive, ensuring the container's overturning during the loading of municipal solid waste into the garbage truck. The obtained dependencies were used to determine the hydraulic drive parameters, which contribute to improving the quality of transient processes, ensuring increased durability of garbage trucks. Among the leading quality indicators, characteristics such as relative overshoot, the number of overshoots, and the duration of regulation were considered. As a transient function, the change in the working fluid pressure in the hydraulic drive of the container overturning during the technological operation of loading municipal solid waste was studied. The volume of pipelines in the section between the hydraulic pump and the hydraulic cylinder piston cavity was determined, which ensures an acceptable quality of transient processes. In addition, key quality indicators of these processes were calculated, considering the influence of various system parameters. Response surfaces were constructed, demonstrating the dependence of the quality indicators of transient processes on the main parameters of influence, particularly the area of the hydraulic cylinder piston cavity and the pipeline volume between the hydraulic pump and the piston cavity. Based on the results obtained, it is proposed that the methodology for the hydraulic drive design of garbage trucks be improved.

Keywords Transient quality indicators · Overshoot · Quality area · Mathematical model · Hydraulic drive · Garbage truck · Turning · Container · Manipulator · Municipal solid waste

O. Bereziuk · O. Petrov (✉)
Vinnytsia National Technical University, Vinnytsia, Ukraine
e-mail: petrovov@vntu.edu.ua

V. Ivanov (ed.), *Emerging Technologies in Manufacturing and Materials Engineering*, Springer Tracts in Mechanical Engineering,
https://doi.org/10.1007/978-3-032-14914-5_8

1 Introduction

Specialized vehicles known as garbage trucks are widely used for collecting and transporting municipal solid waste. These vehicles perform an essential function in ensuring environmental cleanliness, as they allow for the efficient movement of waste from its accumulation points to disposal or recycling points. In modern practice, most garbage trucks are equipped with hydraulic drives that ensure the operation of the main functional elements. The use of hydraulic systems ensures high reliability, productivity, and accuracy of operation of such mechanisms, which makes them indispensable in the field of waste management. When starting a hydraulic drive, significant pressure surges in the working fluid may occur, which can be especially dangerous in the loss of system stability or in the event of poor-quality transient processes. Such phenomena can lead to serious consequences, in particular, rupture of high-pressure pipelines, damage or failure of equipment, and loss of working fluid (Fesenko et al. 2018), which is usually a high-cost mineral lubricant. This increases the cost of repairs and maintenance and can also cause equipment operation failures. Studying the characteristics and transient processes quality (Berladir et al. 2022) during the hydraulic drive (HD) start-up is particularly important. The study of these processes makes it possible to determine the functioning rational parameters of the working bodies of garbage trucks. This, in turn, contributes to the increase in durability and reliability of garbage trucks in general, which is of great importance for ensuring the efficiency and safety of the waste management system and achieving sustainability goals (Jasiulewicz-Kaczmarek et al. 2023). Pursuant to the Resolution of the Cabinet of Ministers of Ukraine No. 265, ensuring the introduction of highly efficient modern garbage trucks in municipal services is an urgent task with both scientific and technical significance. This issue is critical in the context of the constant development of waste management infrastructure. When developing hydraulic drives for new lifting machines, such as garbage trucks, and when modernizing existing equipment, one of the key aspects is ensuring rational transient processes in the operation of HD elements. The efficiency and reliability of the machines depend on the correct execution of these processes. Rational design of hydraulic drive systems (Petrov et al. 2019; Polishchuk et al. 2019) reduces the risk of emergencies, increases the durability of equipment, and reduces operating costs, an essential factor for the effective functioning of municipal services. In particular, developing analytical dependencies that describe the indicators and the quality range of transient processes in the HD of the container turning mechanism during loading municipal solid waste (MSW) into a garbage truck is relevant and practically significant. Solving this problem aims to improve transient processes' characteristics, which allows them to achieve greater stability and predictability. Such measures will contribute to improving the overall quality of the hydraulic drives, which, in turn, will become a guarantee of increasing the durability and reliability of garbage trucks as a whole. Rationalization of container turning processes also contributes to reducing the load on individual elements of the system, which allows for extending their service life and reducing the cost of equipment maintenance.

The study aims to determine analytical dependencies that characterize the indicators and quality areas of transient processes of the HD of the container turning mechanism during the loading of MSW into a garbage truck. The implementation of this goal aims to improve the qualitative characteristics of transient processes, which will reduce load on the key elements of the HD system and increase their reliability and efficiency of operation. Achieving the set goal will contribute to extending the service life of garbage trucks, reducing maintenance costs, and ensuring an overall increase in their performance in modern municipal services.

2 Fundamentals and Challenges

During the HD operation, transient processes inevitably occur, which significantly impact the stability and efficiency of their operation. These processes arise due to various factors, in particular, the opening or closing of control elements, the start of a positive displacement pump, and changes in the load on the actuator (Lobov et al. 2019). Such phenomena are an integral part of the hydraulic system's operation and require careful analysis to minimize the negative impact on the drive's operation and the equipment's overall reliability. Of particular interest are transient processes that occur when the load on the actuator, in particular the manipulator, changes. These processes are essential for ensuring the stable operation of the HD and increasing the entire system's reliability. Mathematical modeling represents the most effective approach to studying such processes. Thanks to this approach, obtaining the necessary data with high accuracy for analyzing and predicting the system's behavior is possible. At the same time, mathematical modeling allows you to avoid significant material costs associated with experimental research and ensure efficiency in improving and developing HD.

The scientific article (Tarasik and Savitskiy 2016) presents the experimental study results of a mechatronic control system for an automatic transmission and the results of transient characteristics modeling of this system in the control process. During the study, the influence of various control system parameters on the transient processes' qualitative indicators was analyzed, which allowed us to identify key dependencies and features of the system's dynamic behavior. The paper also presents graphs of transient characteristics that clearly demonstrate the processes' main patterns and significant features. These results are essential for further optimization of the control system parameters, increasing its efficiency, and ensuring stable operation of the automatic transmission.

The transient processes analysis for three types of HD regulators, performed in (Protopopov et al. 2020), allowed us to obtain their dynamic characteristics and assess the impact of different approaches to regulation on the quality of transient processes. Studies have shown that introducing additional pressure feedback significantly reduces the transient process duration and the amplitude of its oscillations.

However, despite the positive impact on dynamic stability, this approach has the disadvantage of increasing the dynamic error of the system. Such a compromise between the speed of system adaptation and its operation accuracy is an important aspect that should be considered when designing hydraulic drives, especially for mechanisms with high requirements for reliability and stability. In the materials of the scientific article (Godorozha et al. 2020), the authors proposed an algorithm for determining the leading indicators of the transient process quality to optimize the control system's components and increase the efficiency of its regulation. The study considered key indicators of the transient process quality, such as relative overshoot, the number of overshoots, and the duration of regulation tp. For a visual presentation of the obtained results, a graphical interpretation of these indicators is given, which illustrates the main characteristics of the control process (Fig. 1). The developed algorithm is an essential tool for improving control systems, since it allows taking into account the influence of each component on the quality of the dynamic operation of the system.

The regulation duration t_{reg} is determined based on the minimum deviation condition of the transition function—the working fluid pressure $p(t)$ from its nominal value p_{nom}. To ensure high-quality transient processes, this deviation remains within 5% of the nominal value (Markov et al. 2022). The number of overshoots N during the regulation time ($t = 0 \ldots t_{reg}$) is defined as the local maxima number of the transition function $p(t)$ during a given time period. For the transient process to meet quality requirements, the number of such overshoots should not exceed 2 … 3. These criteria are the main ones for assessing the stability and efficiency of the hydraulic

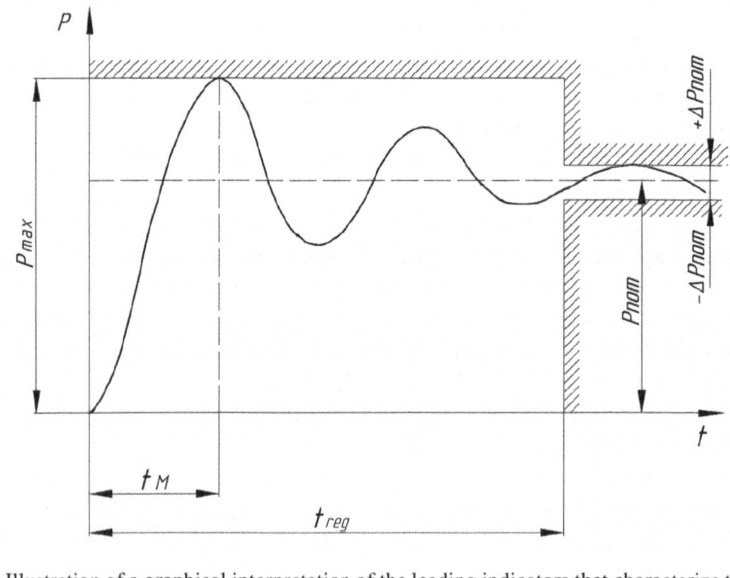

Fig. 1 Illustration of a graphical interpretation of the leading indicators that characterize the quality of transient processes in hydraulic systems

system, since they ensure a reduction in fluctuations and an increase in the accuracy of regulation.

The parameter deviation in the transient process is described by the relative overshoot σ, which is an essential indicator of the quality of regulation. Relative overshoot is defined as the ratio of the regulated quantity's maximum deviation, particularly the working fluid pressure, to its steady value. This parameter reflects the degree of exceeding the set value during the transient process and is a key criterion for assessing the stability of the hydraulic system (Markov et al. 2022).

$$\sigma = \frac{p_{\max} - p_{nom}}{p_{nom}} 100\% = \left(\frac{p_{\max}}{p_{nom}} - 1 \right) 100\%, \tag{1}$$

where p_{nom}, p_{max} are the nominal and maximum pressures of the working fluid, respectively, Pa.

The relative overshoot mustn't exceed 30% to ensure high-quality transient processes.

The scientific article (Polishchuk et al. 2020) presents a mathematical model that is a powerful tool for analyzing the dynamic characteristics of walking robots. This model allows for detailed calculations of forces and moments that arise during the operation of such robots. The main advantage of the model is that it makes it possible to evaluate these dynamic parameters in the context of their compliance with permissible technological loads. The model considered the interaction between the robot and the surface of its movement, which includes an analysis of adhesion forces. Thanks to this, it is possible to determine the limit values of the loads that a walking robot can execute without losing traction with the surface.

In the study described in (Zubov et al. 2015), the lateral mechanized loading process of MSW into the garbage truck body was simulated. A feature of this modeling was considering the influence of free vibrations of the vehicle frame on the system's dynamic behavior during the technological loading operations. The modeling results allowed us to establish the new gripper's optimal design and connection parameters. This minimized the moments and stresses in the "grip-container-grip" system during operation. In particular, such dimensions and parameters were determined to reduce the impact of dynamic loads on the system, which, in turn, contributes to increasing its reliability and durability. Such results are essential for improving the structural elements of garbage trucks, as they allow for more efficient and safe operation of loading mechanisms, reducing component wear and increasing the overall efficiency of municipal equipment operation.

The article (Hoskins and Stoianov 2014) presents an innovative approach to continuously analyzing and monitoring dynamic hydraulic conditions in water distribution and transmission systems. This approach is based on using computationally efficient embedded algorithms that provide adaptive thresholding for identifying hydraulic transients.

The study (Semenov et al. 2020) considers transients arising from homogeneous mechanical stresses and ultrasonic vibrations in silicon wafers. The study aims to analyze the material's dynamic behavior in the direction of its thickness under the influence of ultrasonic waves. Particular emphasis is placed on studying the conditions under which vibrations and stresses remain homogeneous, which is critically important for ensuring the stability of the wafers' mechanical characteristics.

In the materials of the article (Wang et al. 2014), complex dynamic and hydraulic models were created that describe the operation of the electrohydraulic flow matching control system in a hydraulic excavator. These models provide a deep understanding of the interaction between the system components, allowing for the study of their behavior in different operating modes. In particular, detailed modeling of dynamic characteristics was carried out, which made it possible to estimate the key parameters that affect the performance and stability of the system. The dynamic processes studied in the HD made it possible to determine the features of the system operation when changing load modes and assess the efficiency of flow control in the hydraulic circuit.

The study of jump-like phenomena in hydraulic systems caused by transient processes, taking into account the real topological characteristics of the system, was carried out in (Agostinho et al. 2018). The authors used 14 key hydraulic indicators, allowing them to conduct a detailed analysis of the sensitivity of such phenomena and assess their potential impact on the system's stability. The study's main objective was to identify and assess the effects of the relevance of inertia and fluid compressibility, which play an essential role in forming dynamic phenomena in fluid transportation systems. The results allowed us to determine the critical conditions under which undesirable transient phenomena occur, which can cause an increase or a sharp decrease in pressure. In addition, the work proposes protection strategies to weaken the impact of transients and minimize the risks of damage to system elements, particularly pipelines. A key focus is placed on developing measures that help avoid excess pressure or water hammer, which are the leading causes of pipe breakage.

In the materials of the article (Nikolenko and Ryzhakov 2013), a thorough study of transient processes that occur in HD based on axial-piston hydraulic machines with discrete regulation and an inclined cylinder block was conducted. The main attention was paid to the behavior analysis of the HD during its start-up and when changing operating modes, which are critical for ensuring the stability and efficiency of its functioning.

In the paper (Bereziuk et al. 2023), analytical models of the quality indicators of HD transient processes for turning the manipulator lever of a garbage truck were proposed, which made it possible to determine the drive parameters that allow improving the quality of the transient processes to increase the durability of garbage trucks in general.

In the paper (Berezyuk and Savulyak 2017), approximate analytical dependences of the pressure in the hydraulic cylinder pressure line, the angular velocity, and the overturning angle of the container on time were obtained based on the proposed linearized mathematical model of the HD for overturning the container during the technological operation of loading MSW into a garbage truck. An approximate dependence of the duration of container turning on the main parameters of the hydraulic drive was also found.

However, the authors did not find specific multifactorial mathematical dependencies of the quality indicators of the HD transient processes of container turning during the loading of municipal solid waste into a garbage truck, which requires further study.

3 Ideas and Solutions

Several methods were used to conduct the study, covering various aspects of analyzing and modeling hydraulic drive systems. In particular, the method of analyzing literary sources was used to obtain the necessary theoretical basis and summarize existing scientific achievements in this field.

One of the primary methods is the mathematical modeling of processes occurring in hydraulic drives. For this purpose, the fundamental laws of solid mechanics and the principles of the theory of hydraulic drives and hydropneumatic automation were used, which allowed the description of dynamic processes and the interaction of system components. In addition, the study used approaches from the modeling and systems analysis theory, which contributed to a comprehensive consideration of systems and processes in hydraulic drives.

For the numerical study of mathematical models, numerical methods allow for the accurate and effective solving of differential and algebraic equations.

Thanks to computer information processing, large volumes of data were automated, significantly accelerating the research process and obtaining more accurate results.

The MathCAD application package was actively used to solve the problems presented in the study.

In the study presented in (Agostinho et al. 2018), a linearized mathematical model of an HD was developed, which is used to turn a container over during the loading of municipal solid waste into a garbage truck. As an analytical solution result of the constructed linearized mathematical model using the Laplace transform method, it was possible to obtain several important analytical dependencies that characterize the HD operation. Among them, the approximate analytical dependency that describes the change in the working fluid pressure in the hydraulic cylinder piston (pressure) cavity is of particular importance

$$p_1(t) \approx \frac{GR\cos(\delta-\gamma)}{S_{C1}l_p \sin\left[\dfrac{Q_P}{2S_{C1}l_p\sin\left(\bar{\phi}_1+\psi\right)}t+\alpha\right]} + \frac{Q_P}{S_{C1}l_p\sin\left(\bar{\phi}_1+\dfrac{\alpha+\psi}{2}\right)}$$
$$\times\sqrt{\frac{J}{2KW_{12}}}e^{-\frac{\sigma_n}{2KW_{12}}t}\sin\left\{S_{C1}l_p\sin\left(\bar{\phi}_1+\frac{\alpha+\psi}{2}\right)\sqrt{\frac{2}{KW_{12}J}}t\right.$$
$$\left.-\arcsin\left[\frac{GR\cos(\delta-\gamma)}{Q_P\sin\alpha}\sin\left(\bar{\phi}_1+\frac{\alpha+\psi}{2}\right)\sqrt{\frac{2KW_{12}}{J}}\right]\right\}, \quad (2)$$

where Q_P—actual hydraulic pump flow, m^3/sec; G—weight of moving elements, N; J—inertia moment of moving elements, kg·m^2; l_P—distance between the rotation centers of the gripper and the hydraulic cylinder rod, m; R—turning radius of moving elements, m; S_{C1}—piston cavity area of the hydraulic cylinder, m^2; α—angle between the axes of the lever and the hydraulic cylinder arm; ψ—angle between the axis of the hydraulic cylinder arm and the axis passing between the rotation centers of the gripper and the hydraulic cylinder; γ—angle that takes into account the deviation of the center of mass position; δ—angle between the gripping arm and the horizontal; $\bar{\phi}_1$—averaged in the first approximation the value of the container overturning angle; K—working fluid compressibility coefficient, Pa^{-1}; σ_n—working fluid flow rate, m^5/(N·sec); W_{12}—pipelines volume between the hydraulic pump and the hydraulic cylinder piston cavity, m^3 (Fig. 2).

Figure 2 also includes the following symbols: Cn—container, Gp—gripper, L—lever, HD—hydraulic distributor, HC—hydraulic cylinder, HP—hydraulic pump, SV—safety valve, T—tank with working fluid, F—filter.

To determine the nonlinear analytical dependences of the quality indicators of the transient processes, mathematical transformations were used using expressions corresponding to the extreme and working fluid pressure nominal values in the HC piston (pressure) cavity. To calculate the acceptable values of the quality indicators of transient processes using nonlinear and transcendental analytical dependences of the form $x = f(x)$ the iteration method (method of successive approximations) was used, which makes it possible to obtain a sequence of approximate values that converge to the exact solution of the equation.

A detailed analysis of the approximate analytical expression (2) used to determine the time t_m at which the pressure p reaches its maximum yielded the following equation:

$$\sin\left\{S_{C1}l_p\sin\left(\bar{\phi}_1+\frac{\alpha+\psi}{2}\right)\sqrt{\frac{2}{KW_{12}J}}t\right.$$
$$\left.-\arcsin\left[\frac{GR\cos(\delta-\gamma)}{Q_P\sin\alpha}\sin\left(\bar{\phi}_1+\frac{\alpha+\psi}{2}\right)\sqrt{\frac{2KW_{12}}{J}}\right]\right\}=1, \quad (3)$$

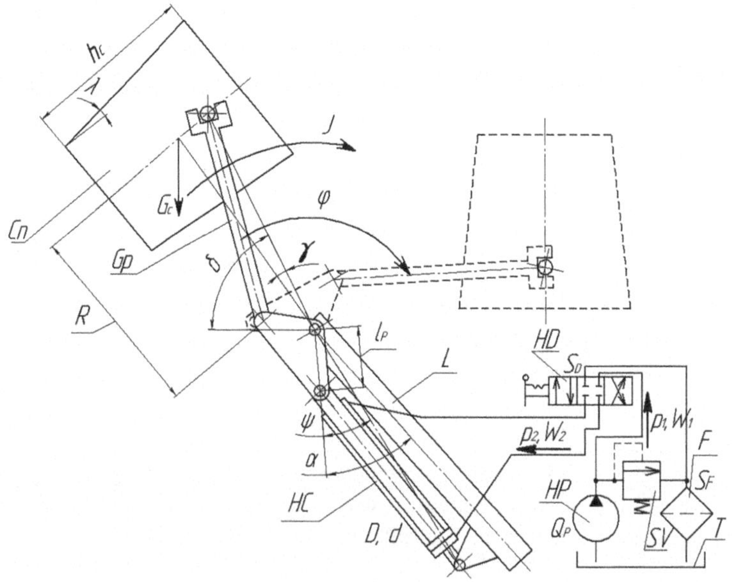

Fig. 2 Calculation scheme corresponding to the linearized mathematical model of the HD operation, which ensures the overturning of the container when loading municipal solid waste into a garbage truck

where

$$t_m = \frac{\dfrac{4\pi N - 3\pi}{2} + \arcsin\left[\dfrac{GR\cos(\delta-\gamma)}{Q_P \sin\alpha}\sin\left(\bar{\phi}_1 + \dfrac{\alpha+\psi}{2}\right)\sqrt{\dfrac{2KW_{12}}{J}}\right]}{S_{C1} l_p \sin\left(\bar{\phi}_1 + \dfrac{\alpha+\psi}{2}\right)\sqrt{\dfrac{2}{KW_{12}J}}}, \quad (4)$$

where N is the number of overadjustments of the working fluid pressure in the piston cavity of the container turning HC.

After substituting dependence (4) into equation (2), it becomes possible to determine the working fluid pressure p maximum value that occurs in the piston (pressure) cavity of the HC. This maximum pressure value is expressed by the following formula

$$p_{\max} = \frac{GR\cos(\delta-\gamma)}{S_{C1}l_p \sin\left\{Q_P \frac{\frac{4\pi N-3\pi}{2}+\arcsin\left[\frac{GR\cos(\delta-\gamma)}{Q_p\sin\alpha}\sin\left(\bar{\phi}_1+\frac{\alpha+\psi}{2}\right)\sqrt{\frac{2KW_{12}}{J}}\right]}{2S_{C1}^2 l_p^2 \sin(\bar{\phi}_1+\psi)\sin\left(\bar{\phi}_1+\frac{\alpha+\psi}{2}\right)\sqrt{\frac{2}{KW_{12}J}}}+\alpha\right\}}$$
$$+\frac{Q_P}{S_{C1}l_p\sin\left(\bar{\phi}_1+\frac{\alpha+\psi}{2}\right)}\sqrt{\frac{J}{2KW_{12}}}e^{-\frac{\sigma_n}{2}\sqrt{\frac{J}{2KW_{12}}}\frac{\frac{4\pi N-3\pi}{2}+\arcsin\left[\frac{GR\cos(\delta-\gamma)}{Q_p\sin\alpha}\sin\left(\bar{\phi}_1+\frac{\alpha+\psi}{2}\right)\sqrt{\frac{2KW_{12}}{J}}\right]}{S_{C1}l_p\sin\left(\bar{\phi}_1+\frac{\alpha+\psi}{2}\right)}}. \quad (5)$$

Based on the analysis of formula (5), which allows for evaluating the key parameters of the hydraulic drive operation, it was established that the nominal pressure value p can be determined using the following mathematical expression

$$p_{nom} = \frac{GR\cos(\delta-\gamma)}{S_{C1}l_p\sin\alpha}. \quad (6)$$

After substituting dependencies (5) and (6) into equation (1), we can obtain an extended formula that allows us to calculate the relative overshoot of the working fluid pressure in the HC piston (pressure) cavity.

$$\sigma = \frac{\sin\alpha}{\sin\left\{Q_{=} \frac{\frac{4\pi N-3\pi}{2}+\arcsin\left[\frac{GR\cos(\delta-\gamma)}{Q_p\sin\alpha}\sin\left(\bar{\phi}_1+\frac{\alpha+\psi}{2}\right)\sqrt{\frac{2KW_{12}}{J}}\right]}{2S_{C1}^2 l_p^2 \sin(\bar{\phi}_1+\psi)\sin\left(\bar{\phi}_1+\frac{\alpha+\psi}{2}\right)\sqrt{\frac{2}{KW_{12}J}}}+\alpha\right\}}$$
$$+\frac{Q_P}{S_{C1}l_p\sin\left(\bar{\phi}_1+\frac{\alpha+\psi}{2}\right)}\sqrt{\frac{J}{2KW_{12}}}e^{-\frac{\sigma_n}{2}\sqrt{\frac{J}{2KW_{12}}}\frac{\frac{4\pi N-3\pi}{2}+\arcsin\left[\frac{GR\cos(\delta-\gamma)}{Q_p\sin\alpha}\sin\left(\bar{\phi}_1+\frac{\alpha+\psi}{2}\right)\sqrt{\frac{2KW_{12}}{J}}\right]}{S_{C1}l_p\sin\left(\bar{\phi}_1+\frac{\alpha+\psi}{2}\right)}} - 1100\%. \quad (7)$$

Provided that $N = 1$, i.e., for the first override, the formula for determining the relative override of the working fluid pressure in the HC piston cavity takes the following form

$$\sigma = \frac{\sin\alpha}{\sin\left\{ Q_{=} \dfrac{\dfrac{\pi}{2} + \arcsin\left[\dfrac{GR\cos(\delta-\gamma)}{Q_P \sin\alpha} \sin\left(\bar{\phi}_1 + \dfrac{\alpha+\psi}{2}\right)\sqrt{\dfrac{2KW_{12}}{J}} \right]}{2S_{C1}^2 l_p^2 \sin\left(\bar{\phi}_1 + \psi\right) \sin\left(\bar{\phi}_1 + \dfrac{\alpha+\psi}{2}\right)\sqrt{\dfrac{2}{KW_{12}J}}} + \alpha \right\}}$$

$$+ \frac{Q_P}{S_{C1} l_p \sin\left(\bar{\phi}_1 + \dfrac{\alpha+\psi}{2}\right)} \sqrt{\frac{J}{2KW_{12}}}\, e^{-\frac{\sigma_n}{2}\sqrt{\frac{J}{2KW_{12}}} \frac{\frac{\pi}{2} + \arcsin\left[\frac{GR\cos(\delta-\gamma)}{Q_P \sin\alpha} \sin\left(\bar{\phi}_1 + \frac{\alpha+\psi}{2}\right)\sqrt{\frac{2KW_{12}}{J}}\right]}{S_{C1} l_p \sin\left(\bar{\phi}_1 + \frac{\alpha+\psi}{2}\right)}} - 1100\%. \quad (8)$$

To visualize the dependence (8) in the response surface form of the objective function—relative overshoot of the working fluid pressure in the HC piston cavity, we construct a three-dimensional diagram, shown in Fig. 3. This surface demonstrates the dependence of the relative overshoot on the main parameters of influence.

Based on the mathematical dependence (8), the region depicted in Fig. 4 was constructed, which characterizes the quality of transient processes in the HC piston

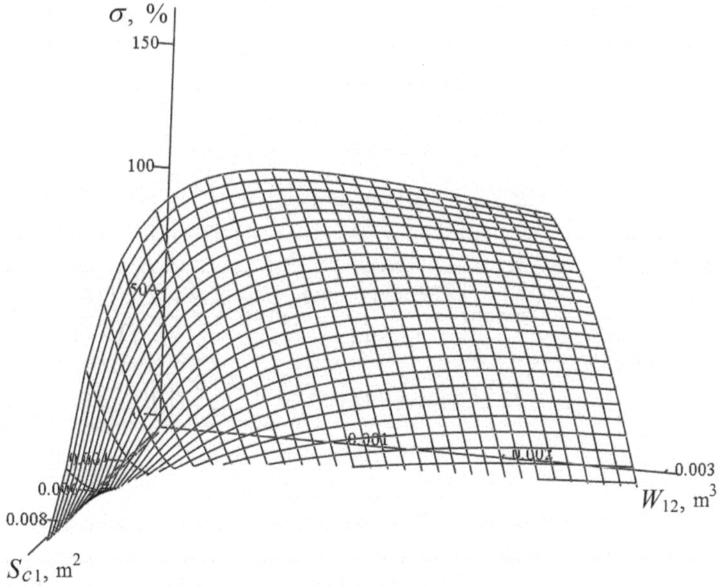

Fig. 3 The objective function response surface, which characterizes the pressure relative overregulation of the working fluid in the piston cavity of the container turning HC during MSW loading into a garbage truck in the coordinates of the main influence parameters

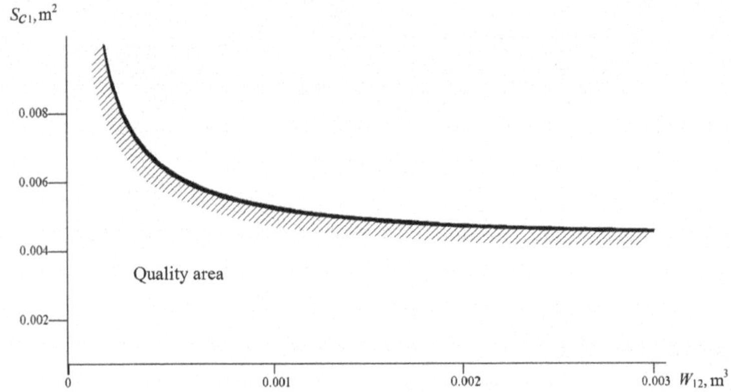

Fig. 4 The quality area of transient processes occurring in the HC piston cavity responsible for turning the container during MSW loading into the garbage truck body

cavity designed to turn the container over during the MSW loading into the body of a garbage truck.

By setting the left-hand side of mathematical expression (8) equal to the normalized relative overshoot value of 30%, we can determine the iterative formula for calculating the volume of pipelines connecting the HP and the HC piston cavity.

$$W_{12} \le \frac{J\sigma_n^2 \frac{\pi}{2} + \arcsin\left[\frac{GR\cos(\delta-\gamma)}{Q_P \sin\alpha}\sin\left(\bar{\phi}_1+\frac{\alpha+\psi}{2}\right)\sqrt{\frac{2KW_{12}}{J}}\right]}{8KS_{C1}^2 l_p^2 \sin^2\left(\bar{\phi}_1+\frac{\alpha+\psi}{2}\right)}$$

$$\div \ln^2 \frac{\dfrac{Q_P \sin\alpha}{GR\cos(\delta-\gamma)\sin\left(\bar{\phi}_1+\frac{\alpha+\psi}{2}\right)}\sqrt{\dfrac{2}{KW_{12}J}}}{1{,}3-\dfrac{\sin\alpha}{\sin\left\{Q_P\dfrac{\frac{\pi}{2}+\arcsin\left[\frac{GR\cos(\delta-\gamma)}{Q_= \sin\alpha}\sin\left(\bar{\phi}_1+\frac{\alpha+\psi}{2}\right)\sqrt{\frac{2KW_{12}}{J}}\right]}{2S_{C1}^2 l_p^2 \sin(\bar{\phi}_1+\psi)\sin\left(\bar{\phi}_1+\frac{\alpha+\psi}{2}\right)\sqrt{\frac{2}{KW_{12}J}}}+\alpha\right\}}}. \quad (9)$$

Applying the iterative dependence (9) taking into account the initial value of the pipeline volume $W_{12.0} = 2\cdot10^{-3}$ m³ on the 13th iteration, a refined volume value was obtained, which should not exceed the pipelines volume in the section of the hydraulic pump—piston cavity of the container turning HC over during the MSW loading into the garbage truck body $W_{12} \le 1.21\cdot10^{-3}$ m³ to ensure the transient processes quality. The results obtained are based on parameters that correspond to the technical characteristics of a real serial model of the garbage truck KO-436, which is manufactured by the "Turbiv Machine-Building Plant" (Ukraine): $Q_P = 55$ l/min; $K = 10^{-9}$ Pa^{-1}; $\sigma_n = 9.24 \cdot 10^{-11}$ m^5/(H·sec); $S_{C1} = 5.027 \cdot 10^{-3}$ m^2; $\alpha = 40°$; $\psi = 30°$; $\gamma = 20°$; $\delta = 65°$; $\bar{\phi}_1 = 15.5°$; $l_P = 55$ mm; $J = 112.6$ kg m^2; $GR = 1254$ H m.

By equating the left side of formula (8) to the permissible error value, the value of which should not exceed 5%, we managed to obtain a formula for calculating the number of overadjustments of the working fluid pressure in the area between the HP and the HC piston cavity, where *Int* is a mathematical function that determines the integer part of a number.

$$N = Int \frac{S_{C1} l_p}{\pi \sigma_n} \sin\left(\bar{\phi}_1 + \frac{\alpha + \psi}{2}\right)\sqrt{\frac{2KW_{12}}{J}}$$

$$\times \ln \frac{\dfrac{Q_P \sin\alpha}{GR\cos(\delta - \gamma)\sin\left(\bar{\phi}_1 + \dfrac{\alpha+\psi}{2}\right)}\sqrt{\dfrac{2}{KW_{12}J}}}{1{,}05 - \dfrac{\sin\alpha}{\sin\left\{Q_= \dfrac{\dfrac{\pi}{2} + \arcsin\left[\dfrac{GR\cos(\delta-\gamma)}{Q_P \sin\alpha}\sin\left(\bar{\phi}_1 + \dfrac{\alpha+\psi}{2}\right)\sqrt{\dfrac{2KW_{12}}{J}}\right]}{2S_{C1}^2 l_p^2 \sin(\bar{\phi}_1 + \psi)\sin\left(\bar{\phi}_1 + \dfrac{\alpha+\psi}{2}\right)\sqrt{\dfrac{2}{KW_{12}J}}} + \alpha\right\}}}$$

$$- \frac{1}{2\pi}\arcsin\left[\frac{GR\cos(\delta-\gamma)}{Q_P \sin\alpha}\sin\left(\bar{\phi}_1 + \frac{\alpha+\psi}{2}\right)\sqrt{\frac{2KW_{12}}{J}}\right] + \frac{3}{4}, \quad (10)$$

Using dependence (10), we will construct the objective function response surface, which is shown in Fig. 5 and characterizes the number of overadjustments of the working fluid pressure in the HC piston (pressure) cavity of the container turning system during loading of municipal solid waste into a garbage truck.

After substituting the mathematical dependence (10) into equation (4), it becomes possible to obtain an expression for calculating the pressure regulation duration of the working fluid, which is located in the HC piston cavity, ensuring the overturning of the container during the loading of municipal solid waste into the garbage truck.

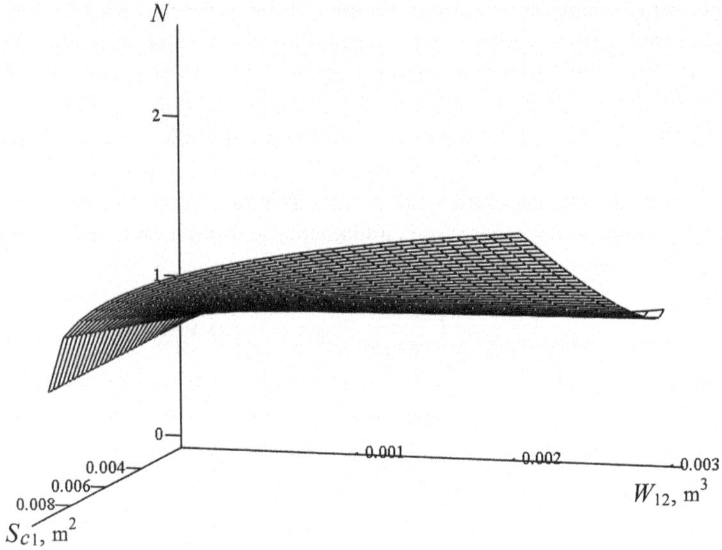

Fig. 5 The objective function response surface, which determines the number of overadjustments of the working fluid pressure in the piston cavity of the container turning HC during MSW loading into a garbage truck, in the plane of the influence of main parameters

$$t_{reg} = \frac{2KW_{12}}{\sigma_n} \times \ln \frac{\dfrac{Q_P \sin\alpha}{GR\cos(\delta-\gamma)\sin\left(\bar{\phi}_1 + \dfrac{\alpha+\psi}{2}\right)}\sqrt{\dfrac{2}{KW_{12}J}}}{1{,}05 - \dfrac{\sin\alpha}{\sin\left\{Q_P \dfrac{\dfrac{\pi}{2} + \arcsin\left[\dfrac{GR\cos(\delta-\gamma)}{Q_P \sin\alpha}\sin\left(\bar{\phi}_1 + \dfrac{\alpha+\psi}{2}\right)\sqrt{\dfrac{2KW_{12}}{J}}\right]}{2S_{C1}^2 l_p^2 \sin(\bar{\phi}_1+\psi)\sin\left(\bar{\phi}_1 + \dfrac{\alpha+\psi}{2}\right)\sqrt{\dfrac{2}{KW_{12}J}}} + \alpha\right\}}}. \tag{11}$$

Using the mathematical dependence (11), we will construct the objective function response surface—the adjustment duration of the pressure of the working fluid in the HC piston cavity for turning the container during MSW loading into a garbage truck in the plane of the main influence parameters, shown in Fig. 6.

Based on the obtained mathematical dependencies (8), (10) and (11), using the previously determined volume value of pipelines in the section of the HP—HC piston cavity $W_{12} = 1.21 \cdot 10^{-3}$ m^3, which provides the transient processes required quality, we determine the key indicators values of the operation quality of the

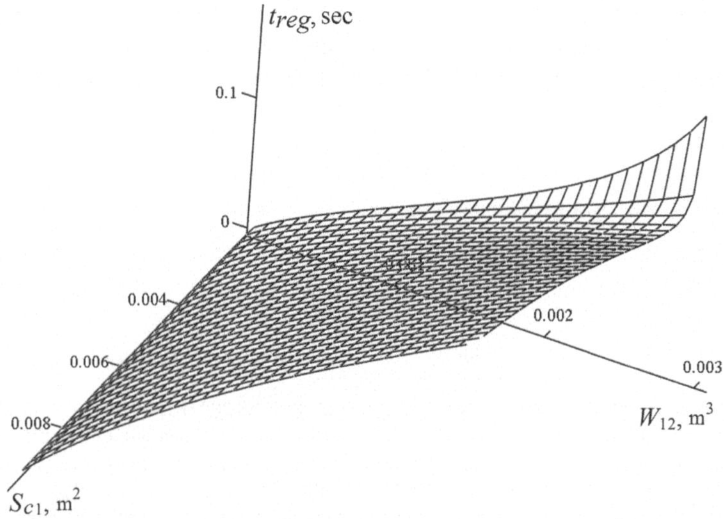

Fig. 6 The objective function response surface, which reflects the adjustment duration of the working fluid pressure in the piston cavity of the container turning HC during MSW loading into a garbage truck in the plane of the main influence parameters

hydraulic drive for turning the container during loading municipal solid waste into a garbage truck: relative overshoot $\sigma = 30\%$, number of overshoots $N = 1$, regulation duration $t_{reg} = 0.0888$ s.

4 Conclusions

Analytical dependencies have been determined that describe the quality indicators of the hydraulic drive transient processes, which ensure the container overturning during the loading of municipal solid waste into the garbage truck, and are necessary for determining the drive parameters that allow improving the quality of the transient processes to increase the efficiency of the garbage trucks as a whole. The value of the pipeline volume in the hydraulic pump section—the hydraulic cylinder piston cavity has been determined $W_{12} = 1.21 \cdot 10^{-3}$ m^3, which provides acceptable transient quality with the following values of leading transient quality indicators: regulation duration $t_{reg} = 0.0888$ sec, relative overshoot $\sigma = 30\%$, overshoot number $N = 1$. The response surfaces of the indicators and the quality area of the hydraulic drive transient processes for turning the container over during the loading of municipal solid waste into a garbage truck have been constructed in the plane of the key influence parameters: the area of the hydraulic cylinder piston cavity, and the volume between the hydraulic pump and the hydraulic cylinder piston cavity. The obtained analytical dependencies of the indicators and the quality area of the hydraulic drive transient processes for turning the container over during the loading of municipal solid waste into a garbage truck suggest improving the corresponding design calculation methodology. Quality studies of the hydraulic drive's transient processes in other garbage truck technological operations require further research.

Acknowledgments The research was partially supported by the International Association for Technological Development and Innovations.

References

Agostinho MSP, Fernandes CVS, Jung BS (2018) Assessment of hydraulic transient indicators in water supply network. WDSA/CCWI Joint Conf Proc 1:074

Bereziuk O, Petrov O, Lemeshev M, Slabkyi A, Sukhorukov S (2023) Transient processes quality indicators of the rotation lever hydraulic drive for the dust-cart manipulator. In: Advances in design, simulation and manufacturing VI. DSMIE 2023, Lecture notes in mechanical engineering, vol 2, pp 3–12. https://doi.org/10.1007/978-3-031-32774-2_1

Berezyuk O, Savulyak V (2017) Approximated mathematical model of hydraulic drive of container upturning during loading of solid domestic wastes into a dustcart. Tech Sci 20(3):259–273. https://doi.org/10.31648/ts.5426

Berladir K, Trojanowska J, Ivanov V, Pavlenko I (2022) Materials selection in product development: challenges and quality management tools. In: Advances in manufacturing III. MANUFACTURING 2022, Lecture notes in mechanical engineering, pp 72–86. https://doi.org/10.1007/978-3-031-00218-2_7

Fesenko A, Yevsiukova F, Basova Y, Ivanova M, Ivanov V (2018) Prospects of using hydrodynamic cavitation for enhancement of efficiency of fluid working medium preparation technologies. Period Polytech Mech Eng 62(4):269–276. https://doi.org/10.3311/PPme.11877

Godorozha S, Protopopov A, Ryabinin M (2020) Development of an algorithm for determining the main quality indicators of the transition process on the example of a LUDV hydraulic system. IOP Conf Series Mater Sci Eng 963(1):012034. https://doi.org/10.1088/1757-899X/963/1/012034

Hoskins A, Stoianov I (2014) Infrasense: a distributed system for the continuous analysis of hydraulic transients. Procedia Eng 70:823–832. https://doi.org/10.1016/j.proeng.2014.02.090

Jasiulewicz-Kaczmarek M, Antosz K, Zhang C, Ivanov V (2023) Industry 4.0 technologies for sustainable asset life cycle management. Sustainability 15:5833. https://doi.org/10.3390/su15075833

Lobov NV, Maltsev DV, Genson EM (2019) Improving the process of transport of solid municipal waste by automobile transport. IOP Conf Series Mater Sci Eng 632:012033. https://doi.org/10.1088/1757-899X/632/1/012033

Markov VA, Barchenko FB, Neverov VA, Savastenko AA, Lotfullin SR (2022) Computational analysis of the dynamic qualities of a diesel engine and a tracked vehicle. BMSTU J Mech Eng 744(3):31–52. https://doi.org/10.18698/0536-1044-2022-3-31-52

Nikolenko I, Ryzhakov A (2013) Modelling of operational cycle of hydraulic drive of lifting mechanism on the basis of axial piston hydraulic machines with discrete control. Teka Komisji Motoryzacji i Energetyki Rolnictwa 13(1). https://doi.org/10.7463/0915.0812337

Petrov O, Kozlov L, Lozinskiy D, Piontkevych O (2019) Improvement of the hydraulic units design based on CFD modeling. In: Advances in design, simulation and manufacturing II. DSMIE 2019, Lecture notes in mechanical engineering, pp 653–660. https://doi.org/10.1007/978-3-030-22365-6_65

Polishchuk L, Kozlov L, Burennikov Y, Strutinskiy V, Kravchuk V (2019) Application of hydraulic automation equipment for the efficiency enhancement of the operation elements of the

mobile machinery. Informatyka, Automatyka, Pomiary w Gospodarce i Ochronie Środowiska 9(2):72–78. https://doi.org/10.5604/01.3001.0013.2553
Polishchuk M, Suyazov M, Opashnyansky M (2020) Study on numerical analysis of dynamic parameters of mobile walking robot. J Mech Eng Sci 14(1):6380–6392. https://doi.org/10.15282/jmes.14.1.2020.14.0499
Protopopov A, Gontarev A, Manenak Y (2020) Mathematical modeling and analysis of the influence of various regulators on the quality of the transient hydraulic drive with throttle control. IOP Conf Series Mater Sci Eng 963(1):012028. https://doi.org/10.1088/1757-899X/963/1/012028
Semenov A, Baraban S, Baraban M, Zhahlovska O, Tsyrulnyk S, Rudyk A (2020) Development and research of models and processes of formation in silicon plates p-n junctions and hidden layers under the influence of ultrasonic vibrations and mechanical stresses. Key Eng Mater 844:155–167. https://doi.org/10.4028/www.scientific.net/KEM.844.155
Tarasik VP, Savitskiy VS (2016) Experimental research of electro-hydraulic proportional valve of mechatronic control system of automatic gearbox. Tract Agri Mach 83(9):10–18. https://doi.org/10.17816/0321-4443-66189
Wang X, Tong S, Ge J, Zhang J (2014) Study on dynamic performance of EFMC system in hydraulic excavator. ZhongguoJixieGongcheng (China Mech Eng) 25(15):2030–2037. https://doi.org/10.3969/j.issn.1004-132X.2014.15.009
Zubov VV, Domnitskiy AA, Kargin RV (2015) Calculation and choice of grip parameters for garbage truck manipulator. Procedia Eng 129:896–902. https://doi.org/10.1016/j.proeng.2015.12.124

Ensuring Joint Quality with Self-Cutting Fastening Components

Liudmyla Danylova and Serhii Lapkovskyi

Abstract Self-Drilling Thread-Forming Screws (SDTFs) are widely utilized in industry and construction due to their ability to eliminate the need for separate drilling and thread-forming operations in joined parts. They also ensure precise screw positioning in the pilot hole during thread formation. Threads formed through plastic deformation create tight, vibration-resistant connections and offer higher static durability than cut threads, although cyclic durability remains challenging. This study proposes installing SDTFs with a controlled axial load, offering a compromise integrating drilling and thread-forming into a single process. Applying axial force ensures proper helical motion during thread formation, enhancing the cyclic durability of the thread. This paper presents experimental research and empirical models to determine the optimal axial force and pilot hole diameter for thread forming, based on the theory of wedge indentation in ductile materials. The short drilling section of SDTFs has high strength and stiffness, and its one-time use significantly alters drilling parameters compared to traditional methods. This research establishes a balanced and efficient assembly process for SDTFs by matching thread-forming requirements with drilling constraints. The proposed methods improve the reliability and quality of threaded connections. The findings can improve the reliability and quality of threaded connections. This methodology facilitates the development of automated assembly systems and introduces a progressive approach to industrial and construction assembly.

Keywords Assembly operation · Thread-forming · Plastic deformation

L. Danylova (✉) · S. Lapkovskyi
National Technical University of Ukraine «Igor Sikorsky Kyiv Polytechnic Institute», Kyiv, Ukraine
e-mail: danylova.liudmyla@lll.kpi.ua

V. Ivanov (ed.), *Emerging Technologies in Manufacturing and Materials Engineering*, Springer Tracts in Mechanical Engineering,
https://doi.org/10.1007/978-3-032-14914-5_9

1 Introduction

Threaded connections are the most commonly used mechanical connections, accounting for over half of all connections in devices and machinery. However, traditional screw-based connections are labor-intensive and challenging to automate, requiring significant effort during assembly operations. The labor intensity of assembly operations with threaded parts depends on the type of connection (bolt, screw, pin), which determines the number of parts that are fastened together, the number of technological operations, and the executive devices. Among these, screw connections are preferred for automation, typically requiring no more than 10 executive devices for automatic installation (He et al. 2022).

Several factors, including productivity, operational efficiency, labor intensity, and the quality of the resulting connections, determine the efficiency of threaded fastenings. One effective way to enhance these indicators is through the use of Self-Drilling Thread-Forming Screws (SDTFs). These screws eliminate the need for pre-drilling, tapping threads in the hole, and precise locating of the screw and hole during threading, forming, and simplifying the assembly process. Thread-Forming Screws, also known as thread-rolling screws, form their own threads during assembly without requiring tapping in the pilot hole. This makes them ideal for use in industries such as electronics manufacturing, where chip formation must be minimized due to strict technical requirements. Their widespread introduction into production began with the assembly technology of the so–called "smooth-threaded connections" and self-tapping screws. Threads formed through plastic deformation create tight, vibration-resistant connections as the material is not weakened by material removal. The degree of tightness (the ratio of the thread unwinding torque to the thread driving torque) ranges from 0.7 to 1.1, depending on the material. Magnesium alloys tend to have lower values, while aluminum alloys exhibit higher values. SDTFs also offer several advantages, including rapid insertion because of the absence of unnecessary nuts and access from only one side of the joint. The close fit between the screw threads and the mating threads ensures tight connections even under vibrational loads (Drumheller 2021).

The installation of SDTFs is particularly advantageous because it minimizes the required actuators. Assembly with SDTFs consists of the sequential execution of two diverse operations—first drilling and then rolling the thread. The technological process consists of the sequential execution of two traditional methods achieved using combined tools that integrate these functions within a single screw body.

The primary criterion for the efficiency of threaded fastenings is their strength. Thread-forming screws are particularly suitable for applications involving alternating loads, such as in the automotive, aerospace, instrumentation, and shipbuilding industries, as well as in building structures exposed to severe weather conditions. However, most failures in threaded fastenings result from fatigue, driven by the variable stresses these connections endure. Smooth-threaded fastenings are obtained by self-tightening, and interference fits range from 30% to 100%, yielding static strength that is 1.5 to 2 times higher than that of cut fasteners (Kenny and Patterson

1989). However, their cyclic strength often falls short. An analysis of the factors influencing the performance of threaded elements under cyclic loading reveals the following contributions to failures: structural defects, process defects, material defects, and defects arising during use. The thread forming process, also called form tapping, shares similarities with rolling out internal threads using rollers. However, thread-forming elements have a distinct purpose and operate under different conditions. These elements serve a dual function as both tools and fasteners, imposing higher operational requirements on them despite their relatively lower cross-sectional strength than rollers. Failures in threaded parts are predominantly caused by fatigue due to alternating stresses, with stress levels from external loads often lower than the permissible limits for static strength.

The research objectives of this study provide a detailed examination of the drilling and thread-forming processes and their interrelation within a unified technological process, striking a balance between drilling and thread-forming requirements to ensure reliable performance of a high-quality joint and compatibility with automated assembly machines. The quality of a threaded fastening is primarily determined by the quality of the thread formation, making this the critical criterion for optimizing the assembly parameters of self-drilling thread-forming screws. The study hypothesizes that axial thread-forming force correlates with joint strength and that SDTFs can be installed with axial force within a unified process.

Threaded connections operate under alternating loads, and their cyclic durability is essential. While threads formed by plastic deformation exhibit high static strength, their cyclic durability often falls below that of cut threads due to the residual stresses introduced during the process. This highlights the importance of improving the quality of smooth-threaded connections. Enhancing the durability and reliability of such connections under cyclic loads is a central focus of this work.

2 Fundamentals and Challenges

Proper organization of the cold plastic deformation process in thread formation significantly enhances the cyclic durability of threaded fastenings. This improvement is achieved by generating compressive stresses at the thread's root, which counteract tensile stresses and increase fatigue resistance. However, cold plastic deformation inherently produces non-uniform deformations and stresses, leading to an inhomogeneous stress-strain state in the material. Residual stresses formed during cold working significantly impact the material's mechanical properties. One of the primary factors reducing the strength of structural materials is the presence of surface defects, such as folds and cracks. These defects often arise during manufacturing and become sites of localized high stress concentrations, which can ultimately lead to failure. Defects at the thread root are particularly problematic, as the root is a critical area of stress concentration, directly contributing to decreased fatigue resistance.

The primary causes of surface defects in threads are related to the geometry and kinematics of the thread-forming process, which can result in axial displacement of the metal's surface layers. Thus, the factor influencing fatigue resistance becomes the process factor, which does not lead to axial displacement of the surface layers of the metal. Key conditions are as follows: 1. The formation of the thread profile should adhere to the "unfilled contour" scheme; 2. The kinematics of the thread-forming process should promote symmetrical deformation, ensuring that axial displacements of the metal are avoided. Figure 1 shows the profile of a metric thread produced by a thread-forming screw in accordance with the unfilled contour and filled contour.

When a screw is installed into a workpiece, the entry taper profile deforms the material forming the thread root. The displaced metal flows along the flank surfaces of the thread profile and accumulates on the sides of the root, forming ridges. These ridges grow in height as material is progressively displaced, eventually shaping the crests of the final thread form. During this plastic deformation, contact stresses develop at the interface between the screw and the workpiece, influencing both the deformation process and the resulting thread quality (Tao et al. 2024). Understanding and controlling these stresses is critical for ensuring a robust, fatigue-resistant thread profile.

The entry taper of the screw carries out the primary work in forming the thread profile. Threads formed through plastic deformation often exhibit a distinct split crest, referred to as a "crater," at the top of the thread (Fig. 1a). The formation and size of this crater depend on the dimensions of the pilot hole for thread formation. To prevent axial displacement of the surface layers of the material—a critical factor in ensuring fatigue resistance—the thread profile must be formed according to the "unfilled contour" scheme. This approach minimizes the risk of defects and ensures the metal's symmetrical deformation. This study also focuses on determining the optimal diameter of the pilot hole for thread formation. The pilot hole should be as large as possible to facilitate efficient material deformation while remaining within the allowable tolerances to maintain the structural integrity of the thread. The balance between these parameters is crucial for achieving high-quality thread profiles without compromising the fastening performance.

Research studies (Wan et al. 2023; Chen et al. 2021; Hou et al. 2024) emphasize that the pilot hole size is a critical factor in determining the strength of a threaded

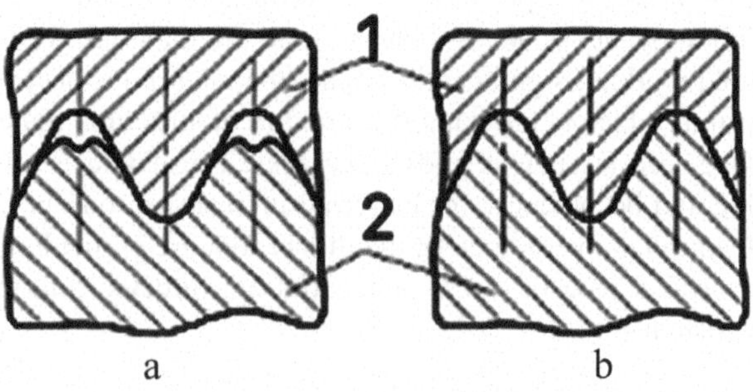

Fig. 1 Profile of a metric thread formed by extrusion: (**a**) unfilled contour; (**b**) filled contour. 1—profile of the formed thread, 2—screw profile

connection. It directly affects the magnitude of residual stresses, which influence the durability and performance of the thread. In (Chen et al. 2021), a formula for calculating the diameter of the pilot hole during internal fine-thread cold extrusion was derived based on the theory of equal volume in the plastic deformation of metals. Another study (Stéphan et al. 2012) calculated the pilot hole diameter for thread-forming screws with a trilobed core using the slip line method. This approach incorporated a flat model to account for two successively cut threads. Additionally, (Sarafraz et al. 2021) highlighted the need for further investigation into determining the optimal pre-drilling diameter to achieve a balance between the dimensional and mechanical quality of threads. This research used the slip line method to analyze the thread-forming process. This method allows for the determination of the geometric and profile parameters of the thread, as well as the optimal diameter of the pilot hole.

Relatively little is known about the detailed mechanics of the thread-forming process. Most existing models and studies have focused on internal thread forming using extrusion taps (He et al. 2022; Tao et al. 2024; Wan et al. 2023; Hou et al. 2024) and thread-forming screws (Stéphan et al. 2012; Mathurin et al. 2009). Much of the research in this area is experimental (He et al. 2022; Tao et al. 2024; Wan et al. 2023; Chen et al. 2021), with limited theoretical development. Studies have employed mechanical models or 3D finite element modeling (FEM) to analyze the assembly process and the associated loads when using thread-forming screws (Kenny and Patterson 1989; Mathurin et al. 2009). For example, in (Hou et al. 2024), a regression response theoretical model was developed to investigate the interactive effects of pilot hole diameter and friction coefficient on extrusion torque, extrusion temperature, and thread height, and in (Mathurin et al. 2009), a 3D FEM analysis of material flow during the thread-forming process was presented. The study also analyzed parametrically to identify the most influential process parameters. In their studies, Florestan Mathurin, Jean Guillot, and Pierre Stephan used the slip-line method to account for two consecutively formed threads. This approach enabled the calculation of pressure on the thread flank as a function of thread height, ultimately determining the maximum forming torque. Despite these advances, existing models for internal thread forming primarily focus on predicting torque values, often neglecting the role of axial forces in the process.

The study of the axial force of threaded connections has been given attention in the following works. Experiments for two axially consecutive teeth are discussed in the material (Chowdhary et al. 2003). The mechanistic model relates the forces developed during the process to the contact area between the tool and the workpiece through specific forming energies. Additionally, (Zhang et al. 2016) introduced a new method for calculating the stiffness of screw connections based on load distribution across the threads. Several notable contributions have informed this research. G. Fromentin and G. Poulachon explored surface integrity in threads formed by taps. A. R. Yao and J. A. Doranoing analyzed stress in screw threads. At the same time, Feargal Peter Brennan studied fatigue and fracture mechanics in threaded connections. Wentao Chen and Zhifeng Liu analyzed axial load distribution in threads under plastic deformation, and Veliko Kolev Ivanov investigated the rolling process of internal threads. These studies highlight the complex interplay of forces, material flow, and deformation mechanics in thread forming, providing a foundation for further exploration.

Deformation during the thread-forming process is characterized by planar deformation, where movement and deformation occur within a single plane. This specific condition enables graphical methods to analyze and solve related problems. Slip-line field theory is a widely applied approach for modeling plastic deformations in metallic materials under plane strain conditions. Slip-line fields represent the trajectories of maximum shear stresses in plastically deformed bodies. These fields consist of straight lines oriented at 45° to the direction of the applied load. Researchers have extensively utilized slip-line theory to derive analytical equations for understanding plastic deformation behaviors.

The pressing scheme influences the nature of the plastic zone and the corresponding slip-line field, which depends on the geometry of the thread fastener. For instance, determining whether it is a fine or coarse thread fastener, and whether the thread develops on the entry taper or the cylinder, affects the deformation process. A self-drilling screw designed with a coarse thread, where the thread develops on the cylinder for one turn or less, is depicted in Fig. 2.

This configuration defines the interplay between the geometry of the screw and the resulting deformation mechanics.

The slip-line method is a valuable tool for calculating contact stress and deformation forces, providing solutions through graphical techniques, and enabling the determination of the geometric parameters of the thread profile. The slip-line field constructed for a self-drilling screw (Fig. 3) is based on the solution of the problem of plasticity theory about the wedge indentation of ductile materials (Hill et al. 1947).

In this context, the formation of the thread profile occurs within one or fewer turns of the screw. The main parameters of these processes are:

$$\text{The compression ratio of the workpiece material } R = \frac{H_1}{H},$$

where

H and H_1 are the thickness of the workpiece before and after compression,

Fig. 2 Self-drilling screw design

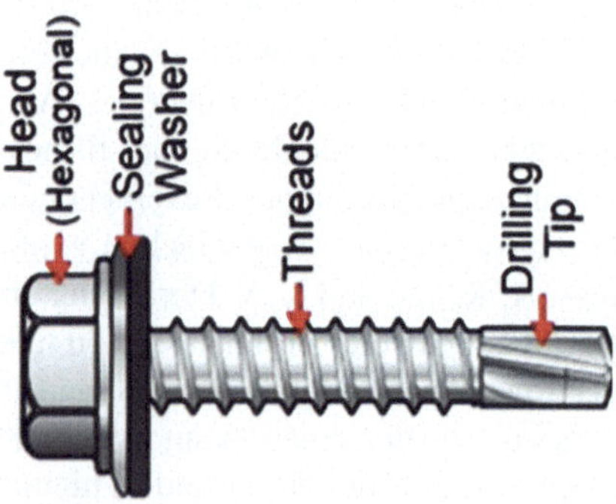

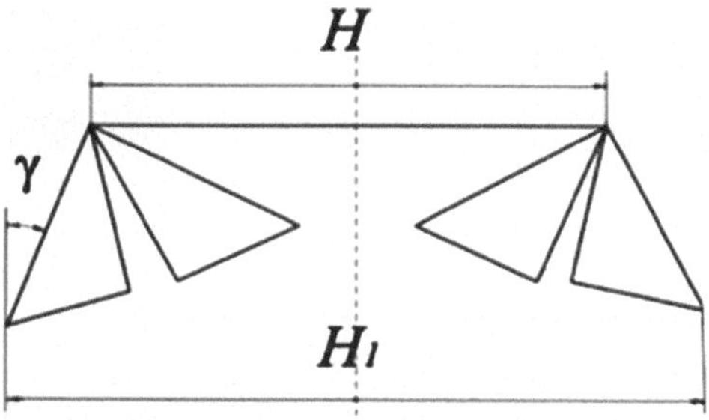

Fig. 3 The description of the slip-line field for the coarse thread screw

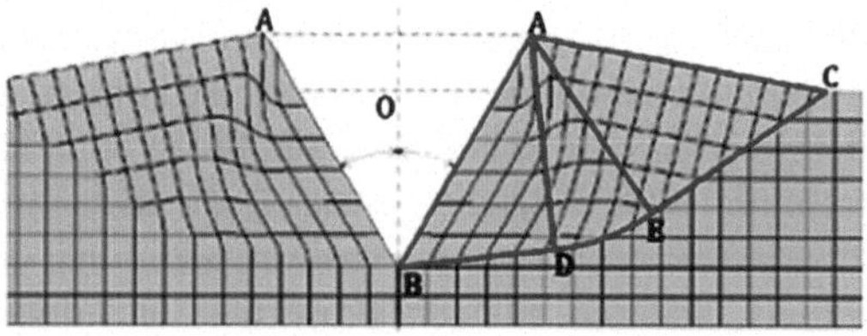

Fig. 4 Deformed square grid after indentation by a lubricated rigid wedge

γ is the semi-wedge angle of the thread profile.

If it is known that the fields of sliding lines cover the entire area of the screw groove, then the following relationship is valid (Danylova et al. 2022):

$$R = \frac{H_1}{H} = \frac{1}{1 + 2\sin\gamma} \tag{1}$$

If R exceeds the value calculated from the limit load condition, then it corresponds to the solution of the problem of plasticity theory about wedge indentation of ductile materials. For a Self-drilling screw $\gamma = 30^{\circ}$, the calculated value of R is 0.5. Parameters of the coarse thread screws are such that the ratio H_1/H is significantly greater than 0.5. As a result, the slip-line field for the plastic region formed by coarse-thread screws corresponds with the case of the slip-line field for wedge indentation (Fig. 4). Pressing a hard wedge (or conical indenter) into a rigid-plastic material provides an example of unsteady flow with geometric similarity.

In this condition, the shape of the plastic region remains unchanged, independent of the penetration depth of the wedge. As the penetration depth increases, the magnitude of the plastic area also increases, but its shape remains constant. This behavior highlights the predictable nature of the deformation process, making it possible to analyze and optimize the geometry of coarse-thread screws based on slip-line theory.

The range of thread-forming fasteners continues to grow, with variations in the ratio of profile height to pitch even within the same standard. Figure 5 illustrates a thread profile formed during an experiment using a self-drilling screw of size M6. The experiment demonstrates the influence of thread-forming parameters on the resulting profile, showcasing how adjustments to process conditions can impact

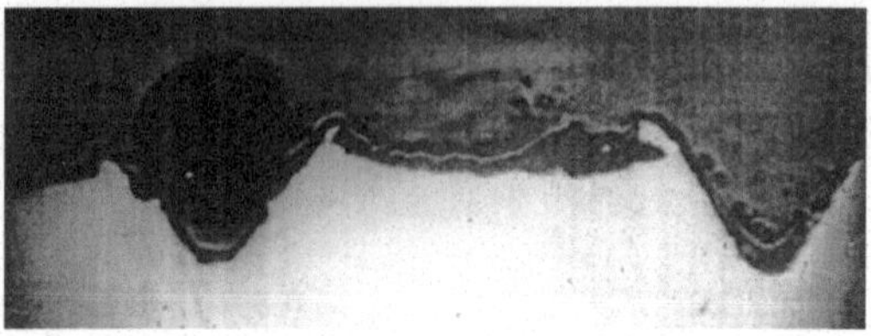

Fig. 5 Profile of the second turn of the formed thread. Original magnification ×100

thread geometry and performance. The parameters of the "crater" depicted in Fig. 5 are significant, and they reveal substantial differences between the thread profiles of the screw and the workpiece. This discrepancy makes it impossible to calculate the diameter for thread formation based solely on the similarity of the profiles, even though the condition of preserving the constancy of metal volume before and after deformation is maintained. Traditionally, the diameter of the pilot hole for thread forming is determined by ensuring the metal volume remains constant before and after plastic deformation. However, additional considerations, such as friction conditions, material strength properties, and dimensional tolerances, must also be accounted for to reduce the risk of overfilling the thread profile.

For metric fine thread profiles, the pilot hole diameter is calculated based on volume constancy and to ensure similarity between the internal and external threads. In contrast, the physical model for forming coarse-thread fasteners aligns with the theory of wedge indentation for ductile materials. Solving this classical problem requires incorporating corrections for actual material properties. The "crater" size and the ability to compress its peaks help mitigate the risk of overfilling the thread profile, allowing for an expanded tolerance range in pilot hole diameter during thread forming. This ensures better adaptability to variations in material properties and process conditions while maintaining thread quality.

3 Ideas and Solutions

To address the problem of wedge indentation in ductile materials, it is essential to construct a field of characteristics, as shown in Fig. 6. Due to symmetry, the construction is performed only for the right side of the smooth wedge, with the angle γ equal to half the angle of the wedge. The characteristics field comprises two regions, 0.0-1.0-1.1 and 2.0-2.1-3.0, characterized by a uniform stress state. These regions are connected by a central field, 1.0-1.1-2.0, where: ξ = represents characteristics drawn at an angle of 45° to the lip b and η = represents characteristics that are arcs of circles centered at 1.1-2.1. Constructing this field requires knowing the length of the contact edge of the wedge and the plastic region. For a smooth wedge, this length equals the length of the inclined sides 2.1-3.0. Due to symmetry, the length of the contact edge and the plastic region equals the length of the lip b. The length b can be determined using the following formula:

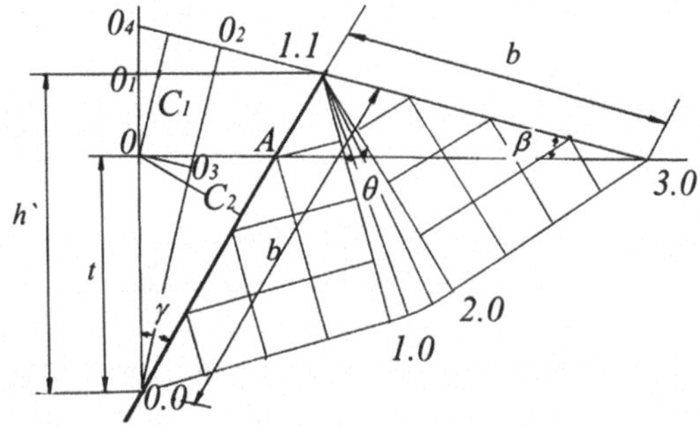

Fig. 6 Slip-line fields for indentation by lubricated rigid wedge

$$b = \frac{h'}{\cos\gamma}, \tag{2}$$

where h' is the height of the formed profile.

If d represents the diameter of the screw, then the diameter of the pilot hole $d_0 = d - 2t$ i.e., is determined by the depth of penetration t of the rigid wedge into the rigid-plastic region. This penetration causes the displaced metal to rise along the flank surfaces of the wedge by a total value b.

For this, we use the conditions:

1. $t = 0.0\text{-}O_1 - O\text{-}O_1$

$$t = b\left(c{>}s\gamma - s\mathbf{V}n\beta\right), \tag{3}$$

2. the equality of the areas of triangles 0.0-O-A and A-1.1,2.1-3.0, which arises due to the incompressibility of the material. The equality of these areas implies the equality of perpendiculars C_1 and C_2 drawn from point O to the sides b:

$$\frac{0_4 - 3.1 \cdot C_1}{2} = \frac{b \cdot C_2}{2} + \frac{0_4 - 1.1 \cdot C_1}{2},$$

$C_2 \cdot b = C_1(0_4\text{-}3.0\text{–}04\text{-}1.1) = C_1 \cdot b$, thus, $C_2 = 0.0\text{-}O_2 - 0.0\text{-}O_3$,

$$ts\mathbf{V}n\gamma = bc{>}s\left(\gamma - \beta\right) - tc{>}s\beta. \tag{4}$$

Considering this $t = b(\cos\gamma - \sin\beta)$, formula (4) transforms into a relationship that determines the angle β:

$$bs\mathbf{V}n\gamma\left(c{>}s\gamma - s\mathbf{V}n\beta\right) = bc{>}s\left(\gamma - \beta\right) - c{>}s\beta b \cdot \left(c{>}s\gamma - s\mathbf{V}n\beta\right),$$

$$c{>}s\left(\gamma - \beta\right) = \left(s\mathbf{V}n\gamma + c{>}s\beta\right)\left(c{>}s\gamma - s\mathbf{V}n\beta\right). \tag{5}$$

From Fig. 8 and triangle A-1.1-3.0, this dependence can be expressed as follows:

$\beta = \gamma - \theta$ and therefore dependence (5) can be written in the form:

$$\cos(2\gamma - \theta) = \frac{\cos\theta}{1+\sin\theta}. \tag{6}$$

The calculations required to determine θ have been completed, with the calculated value $\theta = 17^\circ$. Since θ and β are independent of t, the shape of the field of characteristic remains unchanged, expanding uniformly as the rigid wedge penetrates the material. This property simplifies the determination of the metal lifting height (h'-t) and the insertion depth t depending on the required height of the formed profile h'. However, the volume of metal displaced by the wedge from the region below the line O-3.0 does not match the volume of the formed region above the same line. This discrepancy violates the condition of area equality between triangles O-0.0-A and A-1.1,2.1-3.0. Consequently, the line O-3.0 is shifts from the center to the screw axis, taking the position O_2–3.2. This adjustment is used to determine the corrected pilot hole diameter d_0 (Fig. 7a). Since the trapezoid O-O1-1.1-A is common to the triangle 0.0-01-1.1 and the trapezoid O-O1-1.1-3.1it is convenient to calculate and compare the volumes obtained by rotating these figures around the screw axis. For simplicity, these volumes can be represented as the sum of several simpler shapes: $V_{(O_1-0.0-1.1)} = V_{(O_1-O_2-3.2-3.1)} - V_{(2.1-3.0-3.1)}$. After performing the necessary calculations, the following formula for the pilot hole diameter d_0 is obtained:

$$d_0 = \frac{\frac{1}{3}(r-h)\cos\beta}{4\left(\frac{1}{2}+\frac{2}{3}\cos\beta\right)} + \frac{\sqrt{\left[\frac{1}{2}(r-h)\cos\beta\right]^2 + 4\left(\frac{1}{2}+\frac{2}{3}\cos\beta\right) - \frac{1}{3}\left[\begin{array}{l} r^2\left(\cos\beta - \frac{2}{3}\right) + rh\left(\frac{3}{2} - 2\cos\beta\right) \\ -h^2\left(\frac{1}{2}+\cos\beta\right)\end{array}\right]}}{4\left(\frac{1}{2}+\frac{2}{3}\cos\beta\right)}. \tag{7}$$

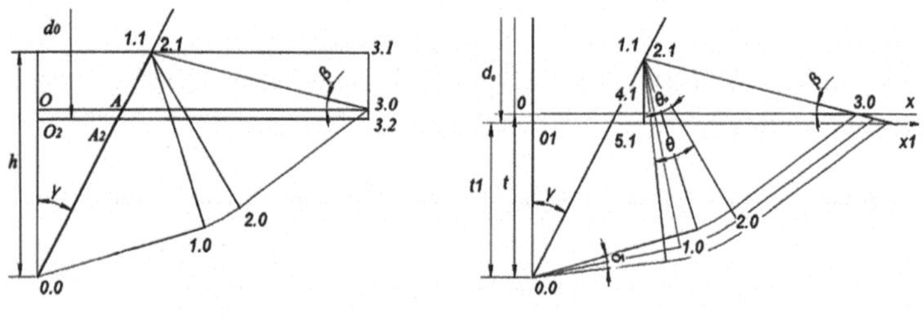

Fig. 7 Scheme for determining the correction of the pilot hole diameter for thread forming: (**a**) cylindricity correction; (**b**) friction correction

In this formula d_0 is pilot hole diameter, *r* is half the **major** Self-Drilling Thread-Forming Screw diameter, h is desired height of the formed profile, β is independent parameter.

The desired height of the formed profile *h* can be defined as:

1. Equal to half the difference between the external thread diameter (maximum major diameter) and the internal thread diameter (minimum minor diameter).
2. Smaller, if there is a risk of overfilling the thread profile.
3. Larger, for scenarios involving significant crater sizes in the thread.

The solution for the case of rough wedge indenter differs from that of smooth wedge indenter due to the effect of friction. In the characteristics diagram, the slip lines approach the contact surface not at 45°, but at an angle of (45° + δ) (Fig. 7b).

For a smooth wedge, the depth of penetration is *t*, and the length of the inclined sides is b. For a rough wedge, however, the depth is reduced to t_1, and the length of the inclined sides increases to b_1. The reduction in the depth of wedge insertion due to friction can be estimated using the following formula (based on Fig. 8):

$$t - t_1 = \frac{b\sin\left(45^\circ + \delta\right)}{\sin 45^\circ}\sin\beta - b\sin\beta = \frac{h^1}{\cos\gamma}\sin\beta\left[\frac{b\sin\left(45^\circ + \delta\right)}{\sin 45^\circ}\right], \quad (8)$$

where *t* is depth of penetration for a smooth wedge, t_1 is reduced depth of penetration for a rough wedge, *b* is length of the inclined lips for a smooth wedge, b_1 is length of the inclined lips for a rough wedge, β is independent parameter, γ is semi-wedge angle of the thread profile, h_1 is height of the formed thread profile, δ is friction angle.

The range of δ is defined as $0 < \delta < 45^\circ$. This corresponds to the coefficient of friction values for different materials: 06–0.08 for copper alloys, 0.07–0.10 for aluminium alloys, 0.12–0.30 for steel.

The previously obtained formula for determining the pilot hole diameter d_0, as described in Formula (8), must be corrected to account for the friction coefficient. This correction is achieved by including the value $2(t\text{-}t_1)$, where *t* represents the penetration depth for a smooth wedge and t_1 *is* the reduced penetration depth for a rough wedge. In Table 1 highlights the parameters required for calculating the pilot hole diameter d_0 for thread-forming screws, as explained in conjunctions with Fig. 7

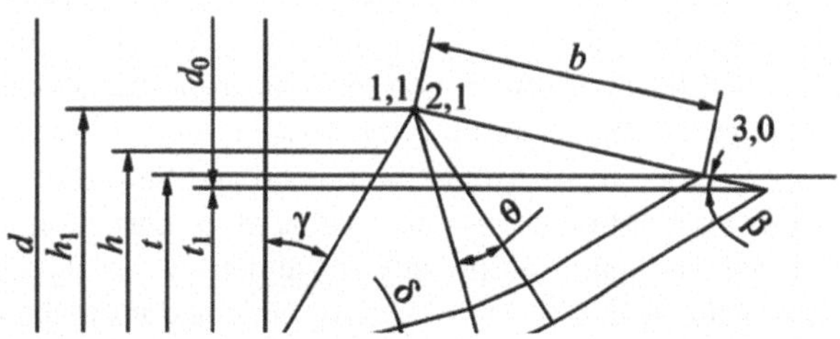

Fig. 8 Scheme for determining the diameter of a pilot hole for cutting threads

Table 1 Parameters for calculating the diameter of the hole for thread-forming screws

d	d_1	h	t	d_0	Δt ($\delta = 15^\circ$)	Δt ($\delta = 30^\circ$)	Δt ($\delta = 45^\circ$)
2.5	1.7	0.4	0.30	1.92	0.0233	0.0814	0.2064
2.5	2	0.25	0.19	2.13	0.0145	0.0175	0.2295
3	2.1	0.45	0.33	2.34	0.0262	0.0316	0.2522
3	2.3	0.35	0.26	2.49	0.0204	0.0246	0.2673
4	2.8	0.6	0.44	3.13	0.0349	0.0422	0.3363
4	2.9	0.55	0.41	3.20	0.0320	0.0387	0.3440
5	3.5	0.75	0.56	3.91	0.0437	0.0527	0.4204
5	3.9	0.55	0.41	4.19	0.0320	0.0387	0.4514
6	4.2	0.9	0.67	4.69	0.0524	0.0633	0.5045
6	4.9	0.55	0.41	5.19	0.0320	0.0387	0.5588
7	3.6	2.2	1.63	4.87	0.1282	0.1548	0.5237
7	6.2	0.9	0.67	6.68	0.0524	0.0633	0.7191

and Fig. 8, that illustrates the geometry used for determining the pilot hole diameter for cutting threads. The table provides specific details such as the external screw thread diameter (d), internal screw thread diameter (d_1), thread profile height (h), penetration depth (t), and corrections (Δt) for friction angles $\delta = 15^\circ$, $\delta = 30^\circ$, and $\delta = 45^\circ$.

The formation of surface defects in threads is primarily influenced by factors related to the geometry and kinematics of the thread-forming process. These factors often cause axial displacement of the surface layers of the metal, leading to imperfections in the thread profile. Fatigue resistance strongly depends on the distribution of residual stresses across the thread's cross-section. For optimal performance, the residual stress distribution should avoid zones of increased deformation and residual tensile stresses, especially in areas of greatest deformation. Such stresses can significantly reduce the fatigue resistance and overall durability of the threads. The conditions under which the thread profile is formed are key contributors to surface defects. These conditions are closely tied to the kinematics of the thread-forming process, which involves the redistribution of elementary volumes of metal displaced by the screw profile. During thread formation, as the screw profile is pressed into the material, it encounters axial resistance. This resistance is overcome either through the self-feed force of the screw or by the application of additional external force.

The screw serves a dual purpose as both a tool and a copy screw that facilitates movement. The second component in this kinematic pair is the thread being formed, which is made from a softer material compared to the screw. However, when excessive forces are applied to the screw, it may deviate from the helical motion law, resulting in defects in the thread-forming process. Figure 9 illustrates the microstructure of the thread root area between the second and third turns of the thread, formed by self-feed force. Microdefects are visible at the transition point between the radial section and the flank. Additionally, asymmetry is observed in the profile of the formed thread. These defects are indicative of deviations in the kinematics of the thread-forming process. Such observations highlight the importance of

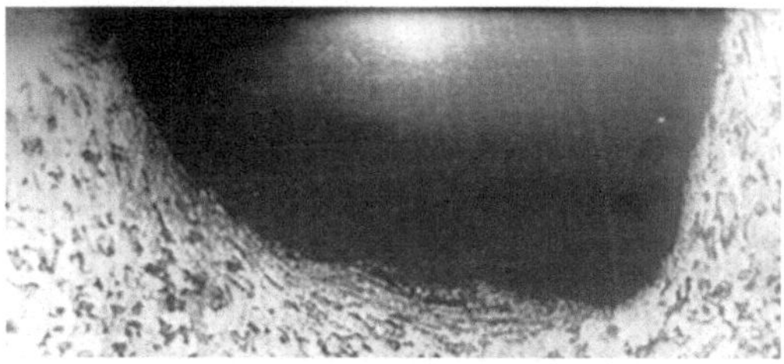

Fig. 9 Profile of the thread root which screw by self-feed force. Original magnification ×200

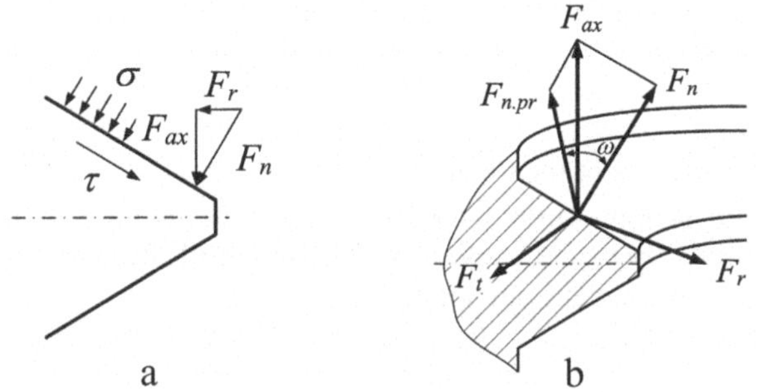

Fig. 10 Stresses of material resistance during deformation and the deformation forces: (**a**) action of normal and tangential stresses; (**b**) the projection of the F_n on the screw axis

maintaining optimal force levels and controlling the thread-forming kinematics to prevent defects, ensure profile symmetry, and improve the overall quality of the threads. The presence of an uncompensated axial force during thread formation increases the defect density in the hardened layer and introduces tensile stresses that counteract the beneficial compressive stresses at the thread root. This leads to the relaxation of useful compressive stresses, negating the positive effects of plastic deformation. As a result, the number of internal defects increases, and the overall quality of threads formed by plastic deformation is lower compared to cut threads.

The reverse movement of material during thread formation neutralizes the benefits of plastic deformation, causing surface defects and reducing the cyclic durability of the thread compared to cut threads. The primary cause of this reduction in cyclic durability is the uncompensated axial force acting during thread formation. It is proposed that thread forming be performed with an axial compensating force to address this issue. This approach ensures symmetrical deformation, preventing axial displacements of the material and reducing surface defects. The value of this compensating force is determined by calculating the radial thread-forming force, which is the product of the specific radial force and the contact area. The specific radial force is derived from the plasticity theory about the task of wedge indentation of ductile materials. Determining the axial uncompensated thread-forming force requires analyzing the kinematics of the process and the forces acting on the thread during formation. Figure 10a illustrates the normal σ and tangential τ stresses of material resistance during deformation and the deformation forces. The normal

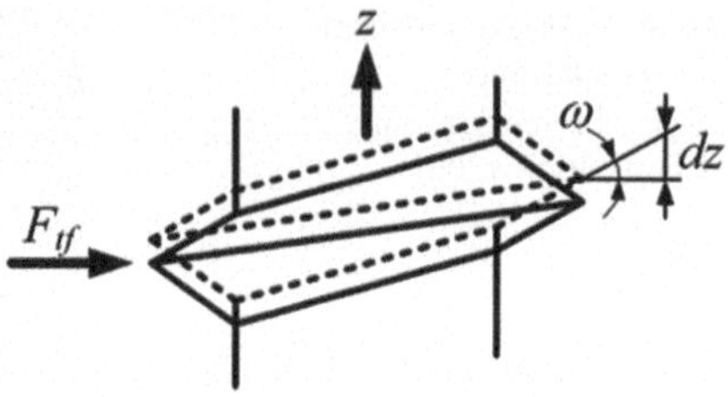

Fig. 11 A deformation diagram for the thread profile, highlighting the relationship between axial and radial displacements due to the applied forces

deformation force F_n consists of the radial force F_r and the axial force F_{ax}. From Fig. 10b the projection of the normal pressure force on the screw axis is expressed as:

$$F_{n.pr} = F_n \cdot \cos\omega, \tag{9}$$

where ω—the lead angle of the thread.

The axial displacement of the flank by the value dz and the radial displacement of the thread profile by dr are caused by the rotation of the screw by an angle φ_i (Fig. 11). These displacements are interrelated as follows:

$$\frac{dF_{ax}}{dFF_{tf}} = \tan\omega.$$

The additional axial load dF_{ax} is an uncompensated axial force resulting from the action of the thread-forming force on the thread and can be expressed as:

$$F_{ax} = F_n \sin\gamma \cdot \cos\omega \cdot \tan\omega = F_n \sin\gamma \cdot \sin\omega, \tag{10}$$

where F_n is the normal deformation force, γ is the semi-wedge angle of the thread profile.

Thread-forming forces result from the contact stresses exerted by the metal's resistance on the thread flank. Therefore, determining the calculated dependencies of thread-forming forces requires knowledge of the level of contact stresses and the surface area over which they act. As previously noted, the slip-line method enables the calculation of contact stresses and deforming forces. This method also allows for the complete determination of stress and velocity fields in the cross-sections of the deforming metal. By leveraging the slip-line field (as shown in Fig. 8), the total force per unit length of the wedge in the normal direction to the contact surface, including the effect of friction, is determined using the following equation.

$$F = 2kb\left[\frac{\pi}{2} - \gamma + \delta + \frac{\cos\delta}{\cos\gamma}\cos(\gamma - \delta)\right], \tag{11}$$

where $k = n \cdot n_v \cdot n_H \cdot \sigma_T$—plastic constant, γ is semi-wedge angle, δ is friction angle, n is deformation type coefficient, equal to 0.57 for planar deformation, n_v is deformation rate coefficient equal to 1 for low deformation speeds, n_H is strain hardening coefficient, (calculated using Eq. 12) (Maksymenko and Izmailova 2016):

$$n_H = \frac{\sigma_u}{\sigma_s} \cdot \frac{1}{1-\psi_{\%}} \left(\frac{\psi_{\%}}{\psi} \right)^{\frac{\psi_{\%}}{1-\psi}}, \tag{12}$$

where σ_u is ultimate tensile strength of the material, σ_s is yield strength of the material, ψ is relative narrowing of the material during tensile testing (ductility parameter), $\psi_{\%}$ is degree of deformation (for threads, typically $\psi_{\%} = 0.43$), b is length of the contact area of the deforming lip.

When using a compensating axial force F_{ax}, the kinematics of the thread formation process ensures symmetrical deformation, preventing axial displacements of the material. This axial force, created by the wedge, is determined as the product of the specific thread-forming force and the contact area. When establishing thread-forming parameters, it is essential to ensure that the compressive residual stresses are maximized and distributed consistently across the entire thickness of the surface layer, extending to the depth of the non-propagating defects. Since the maximum compressive residual stress occurs at a depth proportional to the applied force, forming the thread within one turn or less is optimal. Figure 12a illustrates the geometry of the deforming turn of the screw, where the variable contact length is represented by the straight-line segment RP. The total contact area S of the deforming turn with the part is the sum of all segments RP as the radius vector OP rotates around a circle. In the figure: φ—angle of the deforming sector, ρ—the radius vector, expressed as a function of φ based on the equation of the curve.

$$S = \frac{1}{2} \int_0^{\varphi} \left(\frac{r - r_1}{\varphi \cos \gamma} \right)^2 \varphi^2 d\varphi + \int_0^{\varphi} \left(\frac{r - r_1}{\varphi \cos \gamma} \right) \varphi \cdot r_1 d\varphi. \tag{13}$$

The uncompensated axial thread forming force, the occurrence of which causes the appearance of surface defects and relaxation of useful compressive residual stresses, considering (10), (11) and (13), can be calculated:

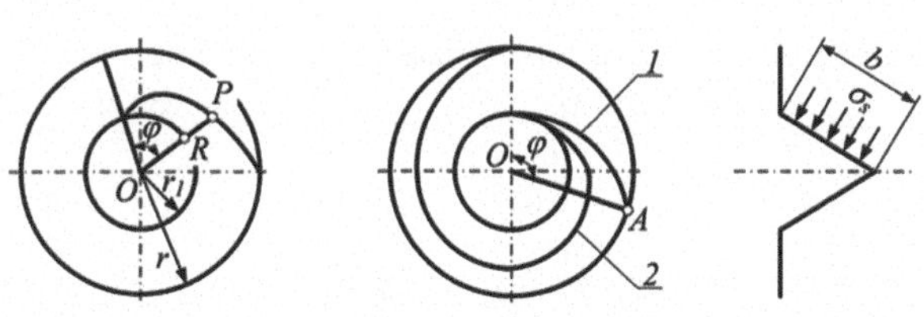

a b

Fig. 12 Forming wedge of the forming screw: (**a**) the determination of the contact area; (**b**) forming wedges that develop threads over one full turn (1) or less than one turn (2)

$$F_{ax} = 2k\left[\frac{1}{2}\int_0^{\varphi}(\mathrm{r}-\mathrm{r}_1)^2\,d\varphi + \int_0^{\varphi}(\mathrm{r}-\mathrm{r}_1)\mathrm{r}_1 d\varphi\right]\left(\frac{\pi}{2}-\gamma+\delta+\frac{\cos\delta}{\cos\gamma}\cos(\gamma-\sigma)\right)tg\gamma\cdot\cos\omega, \quad (14)$$

where F_{ax} is axial thread-forming force, k is plastic constant, as defined earlier, r is external screw thread radius, r_1 is internal screw thread radius, γ is semi-wedge angle, δ is friction angle, ω is lead angle of the thread, φ is angle of the deforming sector.

The calculation was performed for specific angles of φ, including 90°, 180°, 360°.

Figure 12b illustrates the geometry of a forming wedge in the section OA, where the contact length b is variable. The deforming wedge 1 forms a thread over one full turn, while the deforming wedge 2 forms a thread over less than one turn.

Table 2 provides the calculated axial thread-forming forces for self-drilling screws with a diameter of 6 mm. The calculations take into account the coefficient of friction and the strength characteristics of the material being deformed. The materials commonly used for connections involving self-drilling screws typically exhibit a tensile yield strength σ_y ranging from 50 *MPa* to 250 *MPa.* The thread-forming parameters presented in Table 2 were used in a study to investigate the effects of an uncompensated axial force. Evidence of reverse material movement is observed in the asymmetry of the formed thread profile. The study was conducted on steel samples with the following material properties: ultimate tensile strength σ_u = *410 MPa,* yield strength σ_s = *245 MPa.* For the experiment, the most commonly used screw size, M6, was selected. Thread formation was performed sequentially under a constant axial load of 100 *N*, 150 *N*, 200 *N*, 250 *N*, 300 *N*, 350 *N* and 400 *N*.

Table 2 Results of calculation of axial compensating force

S (φ = 90°) 1.802 mm²			S (φ = 180°) 3.860 mm²			S (φ = 360°) 7.2 mm²		
σ_y50 *MPa*		F_{ax}, *N*	σ_y50 *MPa*		F_{ax}, *N*	σ_y50 *MPa*		F_{ax}, *N*
δ	0	47.15	δ	0	94.301	δ	0	188.62
	0.087	54.27		0.087	108,540		0.087	217.08
	0.122	56.95		0.122	113.911		0.122	227.62
	0.175	60.78		0.175	121.570		0.175	243.14
	0.262	66.58		0.262	133.159		0.262	266.32
	0.524	78.83		0.524	157.765		0.524	315.31
	0.785	82.78		0.785	165.567		0.785	331.13
S (φ = 90°) 1.802 mm²			S (φ = 180°) 3.860 mm²			S (φ = 360°) 7.2 mm²		
σ_y250 *MPa*		F_{ax}, *N*	σ_y250 *MPa*		F_{ax}, *N*	σ_y250 *MPa*		F_{ax}, *N*
δ	0	110.018	δ	0	220.03	δ	0	440.07
	0.087	126.630		0.087	253.26		0.087	506.52
	0.122	132.896		0.122	265.80		0.122	531.58
	0.175	141.831		0.175	283.66		0.175	567.33
	0.262	155.352		0.262	310.71		0.262	621.41
	0.524	183.931		0.524	367.86		0.524	735.72
	0.785	193.161		0.785	386.32		0.785	772.64

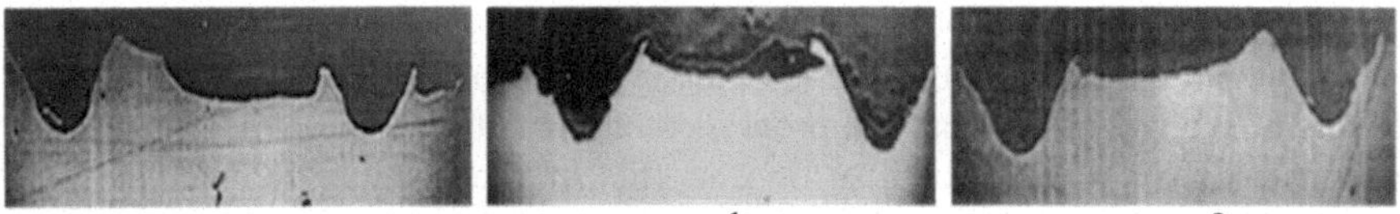

Fig. 13 Profile of the second turn of the formed thread: (**a**) load 250 *N* (original magnification ×100); (**b**) load 300 *N* (original magnification ×100); (**c**) load 350 *N* (original magnification ×100)

The experiment results are documented in cross-sectional photographs of the second turn of the formed thread, presented in Fig. 13 (original magnification: ×100). The photographs in Fig. 13 are arranged in ascending order of axial load, illustrating the changes in the profile parameters of the formed thread under insufficient or excessive loading. The most symmetrical thread profile was achieved with an axial load of 300 *N*.

The assembly process using self-drilling screws involves the sequential execution of two distinct operations: drilling and thread forming. Consequently, the technological process for installing such screws must either adopt two traditional, separate methods or implement a compromise approach. By applying an axial force during installing the Self-drilling screws, it is possible to achieve compromise prose for the installation of self-drilling screws. In this approach, the compromise is made on the thread forming side of the operation, as the overall quality of the threaded connection ultimately depends on the thread's quality.

Self-drilling screws are commonly manufactured from carbon steel, stainless steel, or bi-metal materials. Carbon steel screws often undergo heat treatment, surface treatment, and coatings (such as zinc) to enhance their strength and resistance to corrosion. The unique feature of self-drilling screws is their drill point, which functions as a drill bit (Fig. 3). These screws are subjected to the same forces as high-speed steel drill bits; however, their durability is significantly lower. This limitation necessitates a specialized approach for determining the axial force during the drilling process. The short drilling section of self-drilling screws provides high rigidity, and their single-use nature allows for significant alterations in cutting parameters compared to traditional processes.

Failures at the cutting edges of the drill can generally be classified into three categories:

1. Local failures without operational disruption. These failures remain within the critical wear limits and do not affect the operation.
2. Local failures requiring regrinding.
3. Critical failures, these failures render the drill unusable and prevent further operation.

When determining drilling parameters for the drill point of self-drilling screws, it is essential to adhere to the strength conditions of the cutting edges, which correspond to the second category of failures. Existing empirical formulas for determining axial force during drilling are typically designed for drills made from cutting tool materials, such as high-speed steel. However, these formulas are unsuitable for self-drilling screws due to differences in geometry, material, and operational constraints. Instead, an analytical-empirical approach is recommended for calculating drilling parameters using the specific cutting force (Fischer et al. 2010). This technique isolates the effects of the workpiece material properties on the cutting force. Each material is characterized by a specific cutting force coefficient, representing the force required to cut a chip area of 1 mm^2 with a thickness of 1 mm and when a rake angle of 0°. The actual cutting conditions for self-drilling screws introduce additional considerations:

1. The tool life of the drill corresponds to a single cutting cycle.
2. The drill geometry is non-standard.
3. Material properties of the self-drilling screws.

The cutting force varies exponentially with chip thickness h, following the relationship h^{1-y}, where y is a material-dependent coefficient. The axial force during drilling can be determined using the following relationship:

$$F_{ax} = D \cdot h^{1-y} \cdot K_{v.1.1}. \tag{15}$$

The analytical-experimental determination of the optimal axial drilling force for the drilling portion of self-drilling screws is conducted in the following sequence:

1. Experimental determination of axial force range: A series of axial force values during drilling are experimentally determined. These values range from the maximum force at which the cutting edges are destroyed to the minimum force at which drilling becomes impossible. This step is performed for one of the standard sizes of self-drilling screws.
2. Experimental determination of feed rates: The feed rates corresponding to the range of axial forces identified in step 1 are experimentally determined.
3. Material coefficients selection: The coefficients $K_{v.1.1.}$ dependent on the workpiece material are selected from reference tables.
4. Calculation of average cutting condition coefficient: The average value of $(1 - y)$, which reflects the actual cutting conditions, is calculated for the given screw diameter.
5. Calculation of axial forces: Axial forces corresponding to different diameters and feed rates are calculated using the derived coefficients and cutting conditions.
6. Comparison with experimental range: The calculated axial force values are compared to the experimentally determined range of axial forces during drilling to ensure they fall within this range.

Table 3 presents the calculated axial drilling forces for self-drilling screws. For the M6 screw, the calculated values of the force during drilling and thread formation have close values, as can be seen from Tables 2 and 3.

The experimental device used for the study is illustrated in Fig. 14. The device consists of the components:

1. Housing (1): Equipped with grooves (2) to house the drill bush halves.

Table 3 Results of calculation of axial drilling force

D, mm	2.5	3	4	5	6	8
f, mm/rev	0.02	0.03	0.03	0.04	0.05	0.07
F_{ax}, *N* (experimental)					300	
1-y	0.84	0.84	0.84	0.84	0.837	0.84
F_{ax}, *N* (estimated)	57.1	79.9	135.7	204.6	300.0	485.8

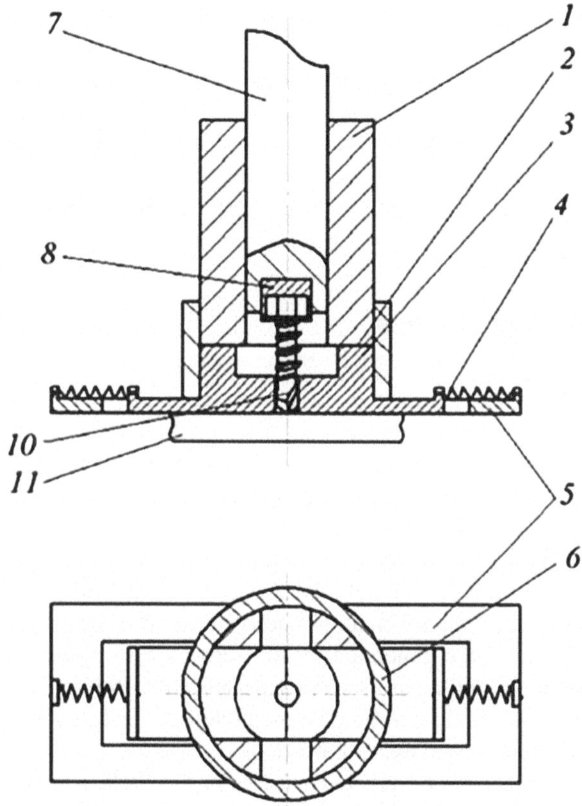

Fig. 14 Experimental device

2. Drill Bush Halves (3): Positioned within the grooves (2) with the ability to move radially.
3. Springs (4): Mounted on a template (5) and connected to the drill bush halves (3).
4. Ring Retainer (6): Installed on the outer surface of the housing (1) with the capability of axial movement.
5. Spindle (7): Mounted in the housing (1) and contains an insert (8) for fixing the head of the screw (10). The spindle (7) has a larger diameter place for installing weights (not shown in the figure).
6. Part (11): Mounted under the template (5) to install the screw (10) in it.

This setup ensures that the drill remains securely located, prevents displacement during operation, and facilitates proper alignment of the threaded part along the drilled hole. The design also supports the passage of the screw head, ensuring accurate and efficient installation.

The study was conducted on steel samples with the following material properties: $\sigma_u = 410\ MPa$, $\sigma_s = 245\ MPa$, screw size M6 was selected. Drilling was performed sequentially under a constant axial load of 100 N, 150 N, 200 N, 250 N, 300 N, 350 N and 400 N with fixing the required number of revolutions. As a result of the experiment, it was found that the force 100 N, 150 N in is insufficient to perform drilling. The force of 400 N is excessive and leads to the critical failures of the cutting edge, these failures render the drill unusable and prevent further operation. Drilling at 250 N requires a lot of RPM. Further investigation consisted of analysing the geometry of the formed thread in this range of axial load. The results were discussed earlier in this article. This advancement will facilitate the automation of assembly processes and the development of assembly devices.

4 Conclusions

Devices for automated SDTF installation use two primary thread-forming methods: (1) forced feed rate method (based on thread pitch)—this approach is complex due to tool geometry and the need to synchronize feed movement with tool rotation; (2) self-feed method—this method can cause deviations from the expected helical motion, leading to thread-forming defects.

This study proposes a technological process for installing self-drilling screws with a calculated axial load. Applying axial load provides a balanced compromise between thread formation and drilling. The thread-forming force aligns with the drilling force generated by the screw's drilling tip, keeping it within acceptable limits. Manufacturers specify permissible drilling force values for various materials. This study introduces a methodology for calculating the required drilling force to optimize and expand this range.

The proposed process ensures proper helical motion during thread formation, preventing reverse material movement, surface defects, and the relaxation of beneficial compressive stresses. These improvements enhance the cyclic durability of the thread. Additionally, forming a symmetrical thread profile increases the connection's strength and prevents the fastener from loosening.

Using the theory about of wedge indentation of ductile materials, analytical dependencies have been derived to calculate the technological parameters of thread formation and the pilot hole diameter d_0 for thread-cutting screws. The influence of axial thread-forming force on the thread profile parameters has been established by analyzing the rigid-plastic zone in the region of maximum deformation. This approach enables modelling of any combination of thread-forming conditions, including friction, tightness, and connection accuracy, for various thread profile heights and thread pitches. The calculated d_0 is smaller than traditionally accepted values, while the diameter tolerance for the thread can be increased due to compensation by the "crater".

The results of this study will improve the quality of connections using self-drilling screws through the application of thread-forming processes with compensating axial forminfg loads.

Acknowledgments This research was partially supported by the International Association for Technological Development and Innovations.

References

Chen X, Zhao Y, He Y, Wan g C (2021) The influence of the bottom hole of cold extruding internal thread on thread quality. Mechanika 27(4):335–341. https://doi.org/10.5755/j02.mech.27435

Chowdhary S, Kapoor SG, DeVor RI (2003) Modeling and analysis of internal thread forming. J Manuf Sci Eng 125(4):681–688. https://doi.org/10.1115/1.1619178

Danylova L, Lapkovskyi S, Prykhodko V (2022) Features of calculating the diameter of the hole for installing a thread-forming part. Int J Mech Adv Technol 6(2):151–160. https://doi.org/10.20535/2521-1943.2022.6.2.264828

Drumheller J (2021) Fundamentals of torque-tension and coefficient of friction testing. PCB Piezotronics, North Carolina. https://www.pcb.com/Contentstore/mktgcontent/WhitePapers/WPL_21_Fund_Torque-Tension.pdf

Fischer U, Heinzler M, Naher F, Oesterle S (2010) Mechanical and metal trades handbook, 2nd edn. Europa Lehrmittel Nourney, Vollmer GmbH & Co., Germany. https://www.academia.edu/27033995/Mechanical_and_Metal_Trades_Handbook

He Q, Jiang Y, Jing X, Jiang Y, Zhou H, Fu B (2022) Research and optimization of process parameters for internal thread forming based on numerical simulation and experimental analysis. Materials 15(9):3160. https://doi.org/10.3390/ma15093160

Hill R, Lee EH, Tupper SJ (1947) The theory of wedge indentation of ductile materials. Proc R Soc Lond A Math Phys Eng Sci 118(1013):273–290. https://doi.org/10.1098/rspa.1947.0009

Hou, H., Mei, Y., Chen, X., Zhou, J.: Multi-objective collaborative optimization of internal thread cold extrusion process based on response surface method. In: Allwood, J., Tekkaya, J Advanced manufacturing technology, vol. 135(1–2), pp. 695–707. Springer (2024). https://doi.org/10.1007/s00170-024-14447-1

Kenny B, Patterson E (1989) The distribution of load and stress in the threads of fasteners. J Mech Behav Mater 2(1–2):87–106. https://doi.org/10.1515/JMBM.1989.2.1-2.87

Maksymenko O, Izmailova M (2016) Theoretical foundations of processes. Part two. DDTU, Kamianske. https://www.dstu.dp.ua/Portal/Data/5/26/5-26-kl66.pdf

Mathurin F, Guillot J, Stéphan P, Daidié A (2009) 3D finite element modeling of an assembly process with thread forming screw. J Manuf Sci Eng 131(4):041015. https://doi.org/10.1115/1.3160377

Sarafraz Y, Koch A, Walther F (2021) Influence of pre-drilling diameter on fatigue behavior of internal threads manufactured by friction drilling and thread forming in AZ91 alloy. In: 4th conference & exhibition on light materials at virtual conference, 02-04. https://www.researchgate.net/publication/362416371

Stéphan P, Mathurin F, Guillot J (2012) Experimental study of forming and tightening processes with thread forming screws. J Mater Process Technol 212(4):766–775. https://doi.org/10.1016/j.jmatprotec.2011.10.029

Tao Y, Wan N, Xu D, He Q (2024) Research and optimization of extrusion tap structure based on numerical simulation and experimental analysis. Materials 17(8):1891. https://doi.org/10.3390/ma17081891

Wan, N., He, Q., Jiang, Y., Fu, B.: Numerical and experimental investigation of the influence of the machining parameters on residual stress distribution of internal thread cold extrusion. CIRP J Manuf Sci Technol, vol. 40(2), pp. 142–154. Elsevier Inc. (2023). https://doi.org/10.1016/j.cirpj.2022.10.005

Zhang D, Gao S, Xu X (2016) A new computational method for threaded connection stiffness. Adv Mech Eng 8(12). https://doi.org/10.1177/1687814016682653

Advanced Materials and Surface Engineering

Laser Surface Texturing for Metal Products

Gennadii Khavin and Borys A. Aleksenko

Abstract The principles of metal products laser processing and the features of the radiation effect on solid media are considered. The physical processes occurring on the surface of solids during laser heating and technological applications of laser action are presented. A classification of laser processing types is given. Their technological features, advantages, and disadvantages are described. The physical effects of light absorption, light energy conversion into heat, and the formation of thin-film technologies are considered. The research on the 3D surface structure formation during laser processing of various metals and alloys is systematized. The observed changes significantly distinguish it from the surface structure typical for traditional processing methods. This indicates an increase in the change process in the surface layer, its periodicity at the micro- and macro levels, depending on the processing conditions. The influence of technological modes of AISI 321 steel surface treatment by nanosecond laser radiation on the nature and magnitude of the treated surface roughness is determined. In this case, the main attention is paid to the process of periodic and possible 3D structures during surface modification by line-by-line laser radiation movement.

Keywords Laser treatment · Roughness · Surface structures formation · Technological radiation modes

1 Introduction

Controlling surface microrelief is a critical task in applied mechanical engineering, as topography directly determines the functional properties of components, such as wear resistance, light absorption, and adsorption capacity. In this context,

G. Khavin (✉) · B. A. Aleksenko
National Technical University "Kharkiv Polytechnic Institute", Kharkiv, Ukraine
e-mail: Borys.Aleksenko@khpi.edu.ua

V. Ivanov (ed.), *Emerging Technologies in Manufacturing and Materials Engineering*, Springer Tracts in Mechanical Engineering,
https://doi.org/10.1007/978-3-032-14914-5_10

nanosecond laser radiation represents an up-and-coming tool for surface modification due to its versatility and controllability. This experimental study aimed to establish relationships between the regimes of laser processing of austenitic steel, the resulting 3D surface topography, and the ensuing optical and mechanical characteristics. The results confirmed that varying the radiation power enables the controllable creation of a full spectrum of structures: from stochastic ones with a polishing effect at low processing intensity to strictly periodic ones at high intensity. This is reflected in the systematic change of the Ssk and Sku parameters. An important finding is the established anisotropy of the resulting surface roughness, manifested in higher roughness parameter values along the beam scanning direction. From an optical perspective, it was shown that initial laser processing effectively reduces reflectivity; however, after completing the deep relief formation phase with vertical walls, further high-power processing does not significantly enhance absorption. Furthermore, phase transformations occur in the surface layer upon reaching a critical radiation intensity, forming martensite and dispersed carbides, leading to a significant increase in hardness. Thus, this work demonstrates that nanosecond laser processing is a comprehensive technological tool for the targeted synthesis of surfaces with predetermined topographic and functional properties.

2 Fundamentals and Challenges

2.1 Lasers and Basic Laser Application Technologies

Over the past few decades, various types of lasers have been developed, each with its own application in metalworking. Lasers have become essential tools in industry and engineering, allowing high precision and efficiency in processing metallic materials. The development of laser technology has led to significant advances in metalworking and the expansion of its capabilities.

Unlike mechanical processing technology, laser achieves high precision in cutting metal sheets. Laser cutting significantly accelerates and simplifies the process of creating metal products, while facilitating the work of process engineers. At present, two types of laser equipment are used in industry: solid-state, which is characterized by low power (up to 6 kW) and small wavelength (1 μm), and gas, which has high power (up to 15 kW) and uses helium, argon, nitrogen, and carbon dioxide to create a laser beam (Tdtam.com 2024). The most popular types of lasers used in the metalworking industry are:

1. CO_2 lasers. Tools in the field of cutting metal materials that provide high precision and purity reduce the amount of internal stress in the material being processed. Due to their ability to combine high power and processing accuracy, CO_2 lasers are also widely used for marking and welding metals.
2. Fiber lasers are highly efficient and reliable. Their specific design allows for the transmission of laser radiation via optical fiber, which reduces energy loss and

ensures process stability. Fiber lasers are used in cutting and welding metals, especially in industries that require high productivity.
3. Diode lasers are compact and efficient devices characterized by low energy costs. Diode lasers are used for processing operations such as laser soldering, engraving, and marking. They are widely used in the medical and electronic industries where precise and soft processing is required.
4. The use of ytterbium lasers has opened a new stage in the development of laser metalworking. They operate at a wavelength of about 1 μm, which ensures high precision and variability. These lasers are especially effective when working with thin metal sheets and products requiring high detail. Ytterbium lasers have high stability and reliability, which makes them an excellent choice for industrial metalworking tasks and additive manufacturing processes (Machinemfg.com 2024).

According to the pulse duration, lasers are classified as nanosecond, picosecond, and femtosecond. This determines the pulse power and plays a role in choosing the areas of application and the possibility of using these lasers. Nanosecond lasers were first introduced into the industry in the late 1990s as diode-pumped solid-state (DPSS) lasers. Most of these lasers have a pulse duration of tens to hundreds of nanoseconds. A picosecond laser has a short pulse duration, high energy, adjustable repetition rate, and is ideal for use in biomedicine, optical parametric oscillators (Rp-photonics.com 2024), and microbiology. A femtosecond laser is a laser with ultrashort, high-intensity pulses. The transition to femtosecond lasers represents a paradigm shift in the mechanisms of laser-material interaction. While nanosecond lasers rely primarily on photothermal processes, femtosecond lasers utilize photoablation and photochemical interactions. This increases processing precision, reduces thermal effects, and provides a broader range of materials, including traditionally difficult-to-process substances such as transparent materials and superhard alloys (Facebook.com/machmfg 2024).

The main physical processes of laser energy impact on a material are considered to be ablation and/or modification of the material surface. Ablation is understood as the complete or partial removal of the material. Modification is a physical impact that changes the properties and/or appearance of the material (Ulsinc.com 2024). In this case, during the modification process at the micro level, a certain amount of material is removed (partial ablation).

2.2 Main Characteristics of Laser Processing Equipment

The technical characteristics of laser devices are primarily determined by the properties of the laser as a radiation source and are divided into (Archive.org 2024):

1. Spectral (frequency) characteristics of laser radiation, which describe the distribution of energy (power) of radiation by wavelengths or frequencies. These include: the frequency of laser radiation; wavelength of laser radiation; width of

the laser radiation spectrum line; width of the envelope of the spectrum of laser radiation; width of the active part of the envelope of the spectrum; and number of spectral lines of radiation. Several inverted transitions can be realized at once in the active medium of a laser. Suppose they are located (on the frequency scale) close enough to each other. In that case, the laser radiation co-occurs at several frequencies (wavelengths), such as in an argon laser. If the transitions are located far enough from each other, the radiation occurs, as a rule, at the frequency of one selected transition, as in a helium-neon laser. Finally, if the composition of the active medium changes smoothly, then the frequency of the radiation also changes within a particular spectral interval. In this case, the frequency (wavelength) of laser radiation is determined by the nature of the active medium.

The spectral characteristics that determine the possibilities of practical application of lasers include: the visibility of the laser beam; conditions of passage of radiation through optical elements and the radiation propagation medium; the magnitude of radiation divergence and the size of the focused spot; radiation receivers that can be used to detect the signal; the magnitude of the background received along with the signal.

2. Spatial parameters and characteristics describe the distribution of the laser radiation field intensity in space. These include: near and far zones; beam diameter; divergence; radiation pattern; and radiation pattern axis.

   An essential place in the mathematical modeling of physical laser action is occupied by the concept of a Gaussian beam, in which the power (energy) density in the beam cross-section is well approximated by the Gaussian function. In models describing the energy interaction of a laser beam with a surface (Researchgate.net 2024), an exponential dependence often specifies the energy distribution. The temporal shape of a single pulse in the models is often specified by a Gaussian.

3. Energy parameters and spatial parameters determine photometric relationships in a laser beam: radiation energy, radiation pulse energy, radiation power, average radiation power, maximum radiation pulse power, energy density and radiation power, average energy density and power (averaged over the beam cross-section).

   Continuous radiation is characterized by its magnitude. The power and radiation in the form of pulses are characterized either by power or energy, and these quantities are related to each other through the pulse duration. The average radiation power characterizes the mode of a regular pulse sequence.

4. Laser operating modes (radiation modes): continuous generation of laser radiation (continuous mode); pulsed generation of laser radiation (pulsed mode); free generation of laser radiation (free mode); mono-pulse mode; resonator Q-modulation mode.

5. The stability of the parameters and characteristics of laser radiation depends on: the pumping system; resonator misalignment; changes in the inverse population over the volume of the active element, etc.

   Laser equipment enables processing in which the material's temperature reaches the evaporation temperature (ablation). Depending on the temperature achieved, three groups of technological operations are distinguished: heat treatment, thermal shock at a temperature below the melting point of the material being processed; cutting, drilling, fitting, milling, and adjustment of material at a temperature above its melting point; welding, remelting, surface alloying at a temperature above the melting point but below the evaporation point of the material.

2.3 *Physical Processes of Laser Technologies in the Surface Processing of Workpieces*

The main feature of laser action is the intense increase in the internal energy of the electron gas. The energy transfer to the lattice in the surface layer of the target occurs relatively slowly due to the significant difference in the masses of electrons and ions. Since the time it takes to establish equilibrium in the electron gas is significantly shorter than it takes to establish equilibrium between the electrons and the lattice, two subsystems arise in the metal: electron and phonon, which have different temperatures. Electrons in the metal, which make a significant contribution to the energy transfer to the lattice, move at speeds of the order of the Fermi speed of ≈ 10^8 cm/s, which significantly exceeds the speed of sound in the metal, equal to approximately $3 \cdot 10^5$ cm/s.

Energy losses by an electron are calculated as the work of the force with which the electron acts on the lattice, considered an elastic continuum. In the temperature range 10^2 … 10^4 °K, the primary energy transfer mechanism is electron thermal conductivity. Radiative energy transfer plays a noticeable role at temperatures greater than 10^4 °K. With rapid, intense energy action, the equilibrium between the electrons and the lattice is disturbed, and a noticeable difference arises between the temperatures of the lattice and the electrons.

The destruction of the surface layer begins under the condition that the value $q\sqrt{t_0}$ (where q—is the flux density t_0—is the pulse duration) exceeds a specific value depending on the properties of the metal (Anisimov and Baerbel 1997). The energy density of the laser action at which material removal begins is called the ablation threshold. For stainless steel, this value is 0.30 J/cm^2. Thus, the guided energy density distribution over the treated surface allows for the controlled removal of material, forming the required surface structures.

In general, removing materials under the influence of laser radiation includes three phases: reflection and absorption with insignificant radiation penetration into the material. Then, the radiation is converted into heat in the surface layer, which is accompanied by an increase in temperature and heating of deeper zones of the product to phase transformations (liquid—steam). In the third phase, the melting and evaporation zone moves from the surface into the depth of the product with continuous ejection of material.

With a considerable thickness of the metal, the proportion of the liquid phase in the destruction products does not decrease, since the bulk of the liquid phase in the destruction products appears as a result of the melting of the crater walls under the action of the flow of escaping steam and, partly, under the radiation scattered inside the crater. As a result of such "erosion" of the crater walls, the increase in its depth is also accompanied by an increase in the diameter, as a result of which the shape of the crater differs significantly from cylindrical. The proportion of liquid metal in the destruction products turns out to be little dependent on the radiation flux density and in a wide range of q values represents the bulk of the destroyed mass.

The structured surfaces obtained by the specified method of controlled cratering (Dobrotvorskiy et al. 2024) have several useful properties in terms of reducing friction (Conradi et al. 2018), hydrophobicity (Wu et al. 2009), and scattering of a narrow beam (Khadka et al. 2024). Studies (Aleksenko et al. 2023) confirm that such periodic structures have the property of a reflective diffraction grating.

Another physical result of the laser's interaction with the material is the laser-induced self-organized periodic structuring of the surface (LIPSS) (Birnbaum 1965). Studies have confirmed a stable correlation between the resulting LIPSS and the laser polarization, with the periodic structures being located perpendicular to the direction of the electrical component (Zhang et al. 2020).

Various hypotheses have been put forward regarding the formation of LIPSS, and this process is described using various models. There is a theory that the formation of LIPSS is due to the interference effect between light scattered by the surface (Siegman and Fauchet 1986; Sipe et al. 1983; Young et al. 1984; Jost et al. 1986), and the electromagnetic theory is currently used (Guosheng Zhou et al. 1982; Dauscher et al. 1996; Seehafer 1994; Fuentes-Edfuf et al. 2019). In (Bonch-Bruevich 1992), surface electromagnetic waves (SEW) in the optical range were described for the first time.

Unique properties of light excitation at the boundary of a condensing medium and their participation in forming periodic micro- and nanostructures were also demonstrated. In (Sergey Dobrotvorskiy et al. 2024), a theory was developed based on the assumption that the periodicity of LIPSS is explained not by a coherent superposition of waves scattered by surface inhomogeneities (Tsibidis et al. 2022), but by the oscillatory effect on charged particles in the plasma cloud in the polarization direction, which is due to the Rayleigh effect. The resulting plasma periodic structure forms a complex periodic power distribution on the target surface. This is also confirmed by a stable correlation between the type of polarization initiating characteristic wave processes in the plasma cloud and the type of the resulting structure (Hnatovsky et al. 2013; Stoian et al. 2013). The spatial period is close to the wavelength of the laser radiation and is also oriented perpendicular to the linear polarization of the laser beam.

Alternative models based on the reorganization of the surface material have also been proposed (Bonse et al. 2017a; Sokolowski-Tinten et al. 2010; Garcia-Lechuga et al. 2016). In this case, potential causes of LIPSS formation include: material instability; self-organization, such as defect development; diffusion or erosion effects; hydrodynamic effects of temporarily melted surfaces, and thermodynamic phase shifts.

Depending on the laser energy density and material characteristics, the laser-induced process can lead to the formation of unusual microstructures in the subsurface region and the complex hydrodynamic motion and redistribution of material on the irradiated surface.

The papers (Bonse et al. 2017a; Varlamova et al. 2010; Bonse et al. 2017b) provide evidence that the laser electric field controls the excited electrons, and the electrons move fastest in the direction of the laser electric field. In contrast, the electrons move slowly when parallel to the electric field. These works overwhelmingly support the concept of self-organization.

Using a large-scale MD-TTM model, including the molecular dynamics (MD) model and the two-temperature model (TTM), the mechanism of single-pulse LIPSS generation under intense ablation was studied and presented in (Shugaev et al. 2017). This mechanism explains the phase transition caused by an ultrashort pulse laser, and the TTM model describes the electron-phonon equilibrium.

Currently, the process of LIPSS structure formation is being intensively studied. Research in this area continues, new theories are put forward, and mathematical models are created to describe them.

It was also noted that in the $10^{-3}\ldots10^{-2}$ cm layer, the metal temperature changes from the melting point to room temperature, suggesting the occurrence of a large temperature gradient of ~$10^{5}\ldots10^{6}$ deg./cm by the end of the pulse, due to which rapid cooling and hardening of the heated layer occurs. The zone adjacent to the crater walls consists mainly of untempered martensite. With distance from the wall, the microhardness decreases and reaches a minimum at the point where annealing of the work-hardened material occurs.

2.4 Methods of Surface Laser Processing and some Features of the Impact on the Treated Surface

When exposed to a laser beam, due to localized heating, high solidification rate, and the ability to melt and evaporate virtually any metal, the effect of thermal deformation (Dzyubyk et al. 2024; Berladir et al. 2024) is reduced. Laser exposure is the most effective way to change the structure of the surface layer, allowing, for example, ultra-fast hardening of a very thin (tens of micrometers deep) layer, including from a liquid state. In this case, changes in the structure occur, such as amorphization, grain refinement, and the appearance of quasi-periodic or multi-zone structures. The primary methods of laser surface treatment include: laser heat treatment (hardening), melting, creating surface coatings, vacuum laser deposition, and the impact action of laser radiation.

Laser heat treatment is similar to traditional heat treatment methods, except that laser radiation carries out heating. High heating and cooling rates lead to the formation of hardening structures in alloys with high hardness. If the thickness of the part is commensurate with the size of the laser impact zone and heat removal is insufficient, then laser annealing takes place. In fact, this is laser heating to a temperature below the critical one, at which residual thermal stresses are removed.

Changing the state of the treated surface in one way or another is accompanied by changes in surface roughness. This is the main characteristic of the surface state in mechanical engineering after treatment. The type and magnitude of roughness are also responsible not only for strength, tribological, corrosion, heat resistance (Loskutova et al. 2025), and other mechanical characteristics, but also for the reflective and absorbent properties of the surface. Laser destruction of the surface layer material due to the appearance of oriented roughness changes the absorbent capacity of the surface layer. A formed structured layer determines the surface roughness. This leads to new surface properties with reduced or increased absorbent properties. The latter circumstance makes laser surface treatment with predicted roughness, which allows a sharp increase in absorbent capacity, very promising in many industrial applications.

2.5 Formation of a Structured Surface Layer and Change in Absorption Capacity During Laser Heating

The structuring of the surface layer is associated with the formation of surface periodic structures (SPS), which represent a system of ordered linear protrusions and valleys of the surface relief (Horn 2022). The appearance of these structures is associated with the occurrence of surface electromagnetic waves (SEW) or polaritons, which strongly affect the material through heating and destruction. A sign of SEW generation is the structuring of the surface layer and vice versa.

Periodic structures have a period of the order of the wavelength of light. They are predominantly oriented with their lines perpendicular to the projection of the electric vector of the light wave onto the surface (with linear radiation polarization). The formation of SPS is a consequence of the force action of laser radiation, partially converted into SEW. This conversion occurs on resonant periodic gratings, which exist in a real surface's set of random irregularities. The interference of two waves running in different directions occurs in their propagation. As a result, a resulting "stationary" distribution of the acting radiation arises, modulated in space with the same period as the resonant grating. Ultimately, after the surface cools, the induced relief remains in the form of SPS. Specific processes of their formation can include evaporation, surface melting, and displacement of the melt by excess vapor pressure, thermo-capillary phenomena, thermochemical reactions, thermal deformations, etc.

The formation of SPS is observed under the influence of laser radiation in almost the entire range of wavelengths, from infrared to ultraviolet spectrum regions. The range of pulses is from femtoseconds to continuous exposure. In this case, the relief can increase during a series of light pulses.

Qualitatively and quantitatively, the period of the structures depends on the angle of incidence and the state of polarization of the radiation. For the SPS obtaining, it is necessary to heat the material with a laser beam to a temperature approximately equal to the melting point, but not higher than the temperature of developed evaporation, which usually refers to the range of moderate luminous flux densities (0.1–100) • 10^6 W/cm^2, depending on the pulse duration.

The appearance and growth of SPS are directly related to such a phenomenon as a change in the specular reflection coefficient (SRK). Moreover, the growth of SPS leads to a decrease in SRK. This leads to a change in the absorption capacity of the surface of the processed material, which can increase significantly up to 1. In addition, an effect appears in which the absorption capacity depends on the direction and path length of the SPS. This allows you to control the degree of thermal impact of radiation by changing the orientation of the electric vector of the light wave relative to any other selected direction.

The described influence of SPS on the force action of radiation can directly influence the control of the absorption capacity and the nature of the distribution of light intensity on the surface, which is of great practical importance.

2.6 Estimation of Parameters of Roughness and Periodicity of Structures Formed Using Laser Radiation by Considering Them as a Diffraction Grating

Suppose the surface after laser treatment acquires an ordered microstructure. In that case, promising methods allow for the estimation of the geometry of the microrelief of this surface, considering it to be a diffraction grating. We have built a mathematical model of the reflection of a sliding beam from a surface that has been laser-treated and on which a periodic structure has been formed. The mathematical apparatus of ray optics was used when building the mathematical model.

A diffraction grating covers a wide range of applications. This device usually has periodic structures that can scatter incident light in different directions. A characteristic of a reflection grating, which indicates the basic architecture of gratings, is that they consist of many periodically distributed grating lines. For monochromatic light, the grating can create an array of evenly spaced rays in different directions.

A unidirectional diffraction grating is an optical component with a surface containing a series of parallel lines or grooves. When light strikes these lines, it is scattered across the surface of the diffraction grating, forming a complex diffraction pattern. The key characteristic of a unidirectional diffraction grating is that the lines are oriented in one direction. This means that light diffracts primarily in one dimension, creating a spectra pattern along the lines.

Unlike a unidirectional diffraction grating, a bidirectional diffraction grating has grooves or lines oriented in two perpendicular directions, which form a grating pattern. This leads to the diffraction of light in two dimensions, creating a more complex pattern of diffraction maxima, which can be observed when the grating is irradiated in two perpendicular directions.

To promptly control the process of forming periodic structures during laser processing, we have developed a technology for controlling surface periodicity by the sliding reflection method of a laser beam, which meets the requirements of simplicity and accessibility (Dobrotvorskiy et al. 2024; Aleksenko et al. 2023), and allows for current and prompt control of the state of the formed surface microstructure with minimal costs and sufficient efficiency.

Since the grating pitch is a crucial parameter of the grating structure for determining the angular gap between the diffraction orders, more precisely, a small period creates a significant angular separation, and the grating pitch is usually designed according to the application requirements and the applied fabrication method. The shape of the grating profile is another significant parameter that will affect the diffraction efficiency and the relative power of each diffraction order.

Since the trajectory of the laser beam during surface processing is ordered and is a set of many parallel linear trajectories, the structure is ordered and uniform. It can be considered that the surface microrelief formed during the laser processing process is a surface cut by parallel, elongated depressions. This organization of the surface relief elements allows it to be considered a diffraction reflection grating.

Since in our case, the width of the recess does not exceed 4e-3 [cm], and the wavelength when using a red laser is about 7e-5 [cm], the diffraction type criterion is <1, which corresponds to the Fraunhofer diffraction type. In this case, a diffraction pattern appears on the observation projection screen, a set of alternating maxima and minima.

The direction of the diffracted rays determines the intensity distribution in the diffraction spectrum of each grating slit, and the diffraction patterns created by each longitudinal recess will be the same.

The grating structure emits secondary rays of arbitrary diffraction order. The transmitted wave of diffraction order m corresponds to the angle θm to the limiting normal. The interference pattern diagram can be used to visualize the bands resulting from the interference of two or more rays on the surface.

When calculating the propagation of rays in space, we obtained the points of projection of diffraction maxima onto the projection plane, the coordinates of which coincide with the diffraction pattern obtained practically with an error of 5 mm. The error may be associated with the inaccuracy of the location of the components, the imperfect curvature of the surfaces, and some randomness of the studied microrelief.

The experimental data that we obtained correlates with the data from atomic force microscopy and the data we obtained when calculating the constructed mathematical model.

Thus, the proposed method of monitoring the surface microstructure by reflecting a sliding laser beam makes it possible to control the state of the surface microrelief without using complex, expensive, and hard-to-reach equipment. Control can be carried out by sequentially comparing experimental data with data obtained after a series of parametric calculations of a computer model and by comparing experimentally obtained data with a previously calculated reference value. In either case, the result can be obtained quickly, efficiently, and without great expense, with the possible additional use of, for example, neural networks to assess the comparability of the actual and calculated diffraction patterns. The proposed control method clearly determines that the relief is two complexes of the same type of rectilinear parallel depressions on the surface, relatively equidistant from each other. In this case, the two specified groups of linear depressions are mutually perpendicular, which at the stages of the experiment forms clearly pronounced bidirectional diffraction patterns. A similar result is observed on crossed gratings, when the grating periods are different and located at right angles, i.e., rotated so that their slits are mutually perpendicular. The different distance between the bright spots of the diffraction maxima observed by us is explained by different periods, and the difference in shape is explained by the difference in the structure of the microrelief, which causes ordered diffuse reflection. Thus, the characteristic of the surface microrelief obtained by us by the sliding reflection method of a laser beam is fully confirmed by atomic force microscopy data.

Thus, we have established that the mathematical model we have constructed gives adequate results and can be used both to confirm the compliance of the existing microstructure with the stated requirements and to estimate the periodicity of the microstructure with an accuracy of 10…20 microns.

The experiment also shows the applicability and effectiveness of the integrated use of the sliding laser beam method and the constructed mathematical model to verify and determine the properties of the microstructure of the reflecting surface. Our proposed method is simple, inexpensive, and does not require complex equipment.

3 Ideas and Solutions

3.1 Study of the Influence of Nanosecond Laser Processing Modes on the Nature and Magnitude of Surface Roughness of AISI 321 Steel Samples

The main objective of the experimental studies was to determine the effect of various technological modes of surface treatment of AISI 321 steel with nanosecond ytterbium pulsed fiber IPG-Photonics laser radiation by line-by-line scanning on the nature and magnitude of the roughness of the treated surface. In this case, the main attention was paid to the formation process of periodic 3D structures during modifying the steel surface with laser radiation. The subject of the experimental studies was also the study of the reflective characteristics of the surface of a part made of AISI 321 steel after laser treatment in various exposure modes.

The scientific and theoretical prerequisites for conducting the experiments were previous studies on laser surface treatment of metals and alloys, published in various literary sources. They showed that laser modification of the surface forms a 3D surface structure with significant differences from the surface structures typical of traditional processing methods. Such structures characterize the intensification of structuring the treated surface and its periodicity at the micro- and macrolevels, depending on the processing conditions.

Another equally important task of the experimental studies was to prove the probability of forming all the main types of 3D structures under laser action on the surface of AISI 321 steel, with the formation of a separate micro-level, nano-level, and their mutual integration. By controlling the technological parameters, this allows creating two-level and multi-level 3D structures.

3.2 *The Technology Used for the Experiments and the Laboratory Equipment*

Laser processing of the surface of AISI 321 steel samples was carried out per the process parameters given in Table 1, for 4 samples measuring 35x75x1 mm, obtained from a sheet of AISI 321 stainless steel. Scanning was carried out in lines along the *X* axis with a line shift along the *Y* axis with a step of 25 μm, and with variable parameters of scanning speed—*v*, m/s, laser pulse frequency—*f*, kHz, and laser power—*w*, W. The surface roughness was studied in the exact directions.

The serial installation "MiniMarker 2" was used as a laser instrument in the experiments. This compact equipment, based on a fiber laser, as operating experience has shown, does not require special maintenance and consumables, is very reliable, and provides long-term uninterrupted operation even in difficult operating conditions.

A high-resolution digital microscope, KEYENCE VHX 7100, was used to perform the roughness measurements. This instrument has an optical shadow effect mode. It combines specially designed high-resolution objectives, a 4 K CMOS camera, and powerful illumination, which allows you to view images in color and analyze changes in illumination from different directions. Smooth transition between magnifications from ×20 to ×6000 using a nosepiece revolving support provides an easy-to-use viewing environment.

Table 1 Values of technological processing parameters

Sample number	Scanning speed, m/s	Laser pulse frequency, kHz	laser power, %
1	100	50	15
2	150	40	20
3	150	40	30
4	100	60	40
5	Base surface		

The Brucker Alicona Portable RL precision system was also used to process the experimental data. This is a contactless, optical, three-dimensional, Focus Variation-based mobile measurement system for quality control of microstructured surfaces. Checking is possible for fields up to 50 × 50 × 25 mm. The extensive vertical measurement range allows for measuring various geometric shapes.

For metallographic and materials science measurements of roughness, a universal digital optical microscope ZEISS AXIO was used, which allows measurements in transmitted and reflected light and fluorescence analysis.

3.3 The Main Results of the Distribution of the Roughness Value of the Treated Experimental Surface

The Brucker Alicona Portable RL precision system was used to process the experimental data in accordance with DIN EN ISO 25178-2:2023-09 (DIN EN 2022). The surface structure of 4 samples was analyzed in the *X* and *Y* directions at a microscope magnification of ×50.

Figure 1 shows the results of the study of the roughness of the treated surface (Dobrotvorsky et al. 2024). Analysis of these figures allows us to draw the following conclusions.

Dark dots on the surface indicate the direction of scanning by the laser beam. The small size of the dots indicates the beginning of surface modification. The data in Fig. 1 shows that the height range for the 1st sample, Fig. 1a, is distributed in the interval ~ +1.0 to −1.0 µm. A more detailed analysis of the profilograms of the processed surface in the *X* and *Y* directions showed the distribution of the height range in the *X* direction within the limits from −0.8 µm to +0.8 µm. Analysis of the profilogram in the transverse direction of processing *Y* showed that the roughness of the processed surface has lower values and corresponds to the range from −0.6 to +0.15 µm. It can be concluded that the surface roughness of sample 1 is uneven and has significant differences. The distribution of roughness as a whole is random.

To evaluate the roughness characteristics of AISI 321 alloy, we will use the ISO parameter (*FLTt*). The ISO parameter (*FLTt*) is not included in ISO 2518 but is widely used in roughness studies. It characterizes the deviation of the measured surface from the ideal and is the distance between two parallel planes obtained by Chebyshev approximation to the surface. This mathematical method uses two parallel planes to "squeeze" the data surface points from the inside and outside, adjusting the angle to minimize the distance between the planes.

The treated surface for sample 2, shown in Fig. 1b, characterizes the beginning of the formation of periodic structures on the surface of the treated steel. The general distribution of surface roughness along the treatment plane is within the range from –0.9 µm to +0.9 µm. To clarify the range of roughness heights of the treated sample, studies of surface profilograms were performed in the longitudinal and transverse directions. They showed that in the *X* direction, the distribution of the

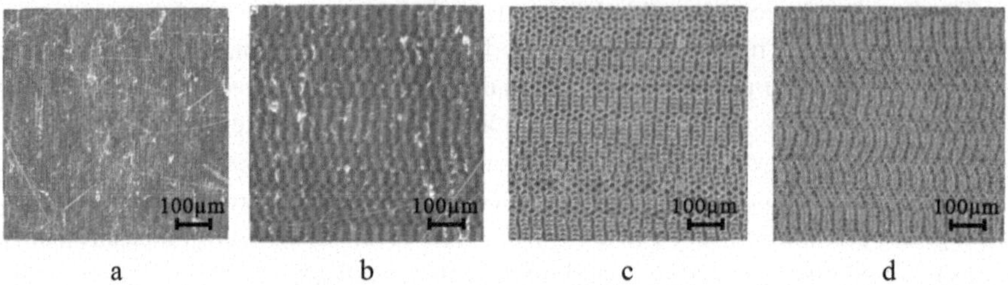

Fig. 1 General view of processed samples: (**a**) sample 1; (**b**) sample 2; (**c**) sample 3; (**d**) sample 4

height range is within the range from −0.5 μm to +0.5 μm, in the transverse direction *Y*, the roughness of the treated surface has higher values in the range from −0.7 μm to +0.75 μm.

In principle, it can be concluded that the treated surface of the 2nd sample is not uniform and has significant deviations, and its character can be described as quasi-periodic.

Studies for sample 3 showed the following. In Fig. 1c, the shape of the craters formed by the laser beam and its scanning lines is clearly visible. The large size of the dots indicates an intensive modification of the surface, which is more significant compared to samples 1 and 2. The roughness range is from −1 μm to +1 μm. The analysis of the profilograms in the longitudinal direction showed the distribution of the height range within the range from −0.93 μm to +0.75 μm. In the transverse direction, the roughness of the processed surface has higher values and corresponds to the range of −1 μm to +0.8 μm. It can be concluded that the surface for sample 3, Fig. 2a, is uneven and has differences. It differs from the previous surfaces by the paucity of minus values and has a pronounced periodic structure, which is clearly observed at the micro level in the *X* direction. Characteristic paired small oscillations at the peaks of the spectrum reflect the presence of roughness at the nanolevel, which is present on the main 3D microstructure, Fig. 2a.

In the *Y* direction, the periodicity of the 3D structures is also present, but there are areas of reduced roughness with implicit periodicity. This behavior is most likely associated with the zone of the spots, where there is no overlap of the laser processing.

In Fig. 2b, one can observe the presence of obvious periodic surface structures resulting from the changed nature of ablation due to the increased laser radiation power. The overall distribution of the surface roughness of the processed sample is within the range from −2 μm to +2.5 μm. To clarify the range of roughness heights of the processed sample, surface profilograms were studied in the longitudinal and transverse directions, from which it can be concluded that the range of heights is distributed in the interval from −2 μm to +2.5 μm. The periodicity of the irregularities by their location on the surface is also traced. From the data analysis, it can be concluded that the surface of sample 4 is not uniform and is characterized by significant differences in heights and roughness valleys. The difference from the previous samples is in the partial absence of large minus values and a pronounced periodic

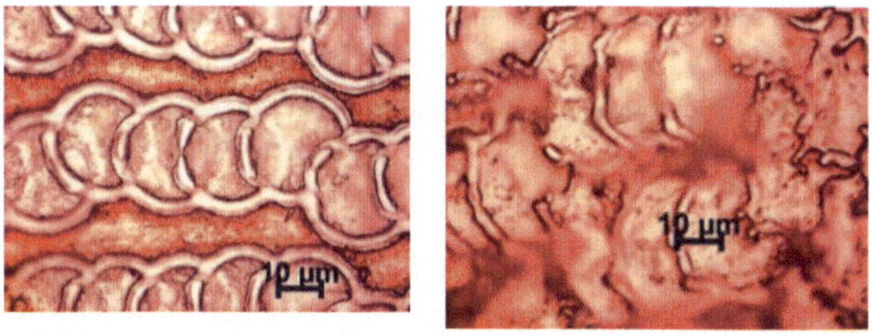

a b

Fig. 2 Surface of samples after treatment with nanosecond laser irradiation at microscope magnification ×100: (**a**) sample 3; (**b**) sample 4

structure. The periodicity is clearly visible at the micro level in the *X* direction, Fig. 2b. The presence of nanostructures on the surface of 3D microstructures can be noted on the roughness diagrams in the *Y* direction. Characteristic paired small oscillations at the peaks of the spectrum reflect the presence of roughness at the nanoscale on the underlying 3D microstructure.

The obtained profilograms for the sample "Base" (the original surface not processed by the laser beam—the base surface) of measuring roughness in the *X* and *Y* directions, allow us to conclude that the surface of the sample "Base" is uneven and has significant deviations. The surface roughness in the *X* direction is greater (−0.8 μm … +0.8 μm) than in the *Y* direction (−0.7 μm … +0.25 μm). This can be explained by the direction of plastic deformation of the steel surface during rolling production.

In general, determining the exact values of the surface roughness parameters after laser modification is a practically unsolvable task. This is due to the ambiguous and complex nature of surface destruction, which is quasi-stable. Therefore, to obtain more reliable information, the roughness parameters for all samples were measured at least 5 times, and the parameters were averaged. An area of 25 cm^2 and a threshold value of 0.05 μm were taken. Such generalized research results are summarized in Table 2.

The surface topography of the AISI 321 alloy before laser processing (sample "Base") and after (samples 1–4) indicates an apparent increase in changes in the surface structure with increasing processing intensity, which is confirmed by the appearance of periodicity of the surface 3D structures. The periodicity of the structure is also reflected in the value of the surface texture parameters *Ssk* and *Sku* (Table 2). The *Ssk* value reflects the degree of displacement of the roughness shape and indicates a deviation below the average plane displacement in sample 4 (*Ssk* = 0.42). In comparison with sample 4, other samples do not show significant deviations. An important issue is the excess of Sku of the selected determination region, indicating that the height distribution for sample 2—*Sku* = 2.8, 3—*Sku* = 2.89 and 4—*Sku* = 2.64 is skewed above the average plane. However, the proximity to the value =3 indicates the coexistence of sharp and depressed parts. In the case of sample 1 and sample "Base" *Sku* > 3, which confirms the spike-like nature of the height distribution.

Table 2 Results of the roughness study of samples 1–4 and the base sample, averaged over five measurements

Parameter	Sample 1	Sample 2	Sample 3	Sample 4	Sample 5 "Base"
Sa, μm	0.24	0.28	0.36	1.02	0.19
Sq, μm	0.31	0.34	0.45	1.24	0.24
Sp, μm	1.38	1.32	1.83	5.05	1.60
Sv, μm	1.40	1.64	1.73	3.10	1.10
Sz, μm	2.78	2.96	3.56	8.15	2.70
S10z, μm	2.54	2.48	3.27	7.55	1.85
Ssk	0.02	−0.06	−0.05	0.42	−0.15
Sku	3.59	2.80	2.89	2.64	3.59
Sdq	0.04	0.04	0.09	0.22	0.04
Sdr, %	0.07	0.08	0.39	2.47	0.07
FLTt, μm	2.78	2.96	3.56	8.15	2.70

The *Sdr* parameter (the percentage of the additional area of the definition region that falls on the texture compared to the flat definition region) has the highest values for the roughness of sample 3 ($Sdr = 0.39\%$) and 4 ($Sdr = 2.47\%$). This confirms the fact that the deviations during creating a pre-flat base surface was small—$Sdr = 0.07\%$.

The distribution of the maximum peak height of the selected region Sp indicates the dominance of the peaks in sample 4 ($Sp = 5.05$). This result may indicate melt accumulations occurring due to the melting of the material or the formation of a significant amount of oxides, especially since the Sv ratio between all samples is more similar to each other than Sp. The *Sz* and *S10z* parameters confirm the above results. The roughness profile parameters *Ra, Rq,* and *Rz* of the treated surface are presented in Table 3. From the analysis of the experimental data, it is evident that the sections of samples 3 and 4 demonstrate a developed profile relief. A significant influence of the measurement direction on the dependence of the parameters *Ra, Rq,* and *Rz* is observed for the surfaces of samples 3 and 4, and for samples 1 and 2 such dependence is expressed much weaker.

It should be noted that higher values of the measured parameters are observed on the surface when measuring the profile in the *X* direction, which corresponds to the scanning direction of the laser beam. This allows us to conclude that surface modification is associated with the intensity of the irradiation process and the scanning direction of the laser beam.

The analysis of the obtained data allowed us to conclude that roughness with a probabilistic (random) distribution of amplitudes and 3D shapes of the created structures is formed on the surface at low laser radiation intensity. The effect of surface polishing is observed for the treated surface of sample 1. This effect is explained by smoothing the maximum protrusions of the profile peaks, as evidenced by a decrease in the *Sp* parameter. In the case of an increase in the laser radiation power, for example, for sample 2, a more intensive restructuring of the surface and the formation of periodic structures begin, which can be characterized as

Table 3 Parameters of the roughness profile of the machined surface for different samples

Parameter	Direction	«Base»	1	2	3	4
Ra	*X*	0.046	0.049	0.080	0.195	0.532
	Y	0.036	0.043	0.027	0.074	0.183
Rq	*X*	0.057	0.063	0.094	0.234	0.639
	Y	0.046	0.053	0.035	0.092	0.231
Rz	*X*	0.234	0.297	0.340	0.874	2.384
	Y	0.188	0.241	0.157	0.396	1.021

quasi-random. A quasi-periodic (sample 3) and a completely periodic profile (sample 4) are formed with a further increase in intensity. This is reflected in the change in the *Ssk* and *Sku* parameters.

It should be noted that laser radiation allows for the shifting degree of shape roughness and excess. These parameters are essential when considering the absorption and scattering processes of electromagnetic waves over the surface. An increase in the surface area (parameter *Sdr*) is a favorable factor when considering adsorption processes. Anisotropy of surface roughness along the scanning lines of the laser beam can be very effective when considering tribology issues. In addition, it should be noted that nanosecond laser radiation is a unique technological tool that allows the creation of random and periodic surface structures.

4 Conclusions

The results of the experiments confirmed the possibility of creating, using nanosecond laser radiation, both periodic 3D structures and 3D structures with a probabilistic distribution of parameters. The created 3D structures are usually multi-level, depending on the technological parameters of laser processing.

It is shown that periodic structures in scattering spectra also create diffraction patterns. The predominance of the mirror component in the scattering spectra characterizes surfaces with a probabilistic distribution of 3D structures.

The main theoretical and practical provisions confirmed by the experimental results in this section can be formulated as follows. Higher values of the roughness parameters occur for the direction corresponding to the scanning direction of the laser beam. This indicates that the development of the surface is associated not only with the intensity of the irradiation process but also with the scanning direction of the laser beam. The cross-section of the roughness profile demonstrates an increase in the structure's periodicity level with the intensification of laser processing. This observation confirms the change in excess and the directional dependence of the roughness profile parameters.

During the processing, we obtained a blackened surface with the index Sk = 1.615 μm. We performed repeated processing, as a result of which the parameter increased by 1.68 times and amounted to 2.717 μm. At the same time, the measurements showed a change in reflectivity by 5%.

Further increase in the depth of the depressions with vertical walls (increase in Sk and arithmetic mean height Sa) does not change the surface properties regarding the reflection of the beam, with the absorption k unchanged. This can be explained by the fact that the depth of the depressions formed as a result of the first treatment is sufficient for its complete absorption (due to a sufficient number of reflections of the beam from the walls in the depression). And further deepening of the relief (Sv) does not lead to a corresponding blackening of the surface. The parameter Sku also does not change due to repeated ablation and repeated deposition of microparticles on the surface.

When changing the intensity of laser processing, it can be assumed that at low intensity, the processes occurring in the surface layer of the processed material correspond to the nature of traditional heating-cooling. However, when the critical irradiation conditions (irradiation power, heating and cooling rate, deformation gradient, etc.) are met, laser strengthening of austenitic steel occurs with an increase in hardness, i.e., hardening. The mechanism for increasing hardness is based on the partial transformation of austenite into martensite and the formation of hard carbides and oxides in the surface layer of the processed metal.

Acknowledgments This work was carried out within the framework of project 0124U00048, funded by the Ministry of Science and Education of Ukraine. This research was partially supported by the International Association for Technological Development and Innovations.

References

Aleksenko BA, Dobrotvorskiy S, Basova Y, Prykhodko V, Kościński M (2023) Computer modeling and comparative analysis of surface microrelief control by the method of reflection of a sliding laser beam. In: Arsenyeva O, Romanova T, Sukhonos M, Biletskyi I, Tsegelnyk Y (eds) Smart technologies in urban engineering. STUE 2023, Lecture notes in networks and systems, vol 808. Springer, Cham, pp 436–446. https://doi.org/10.1007/978-3-031-46877-3_39

Anisimov SI, Baerbel R (1997) Theory of ultrashort laser pulse interaction with a metal. In: Proceedings of SPIE, the international society for optical engineering/Proc. SPIE 3093, Nonresonant Laser-Matter Interaction (NLMI-9). https://doi.org/10.1117/12.271674

Archive.org. Advanced lasers: laser physics and technology for applied and fundamental science. https://archive.org/details/advancedlasersla0193unse, last accessed 14/12/2024

Berladir K, Hovorun T, Trojanowska J, Ivanov V, Iakovets A (2024) Failure analytics of defects in mechanical engineering parts using root cause analysis: case study. In: Advances in manufacturing IV. MANUFACTURING 2024, Lecture notes in mechanical engineering. https://doi.org/10.1007/978-3-031-56444-4_26

Birnbaum M (1965) Semiconductor Surface damage produced by ruby lasers. J Appl Phys 36(11):3688–3689. https://doi.org/10.1063/1.1703071

Bonch-Bruevich AM (1992) Surface electromagnetic waves in optics. Opt Eng 31(4):718. https://doi.org/10.1117/12.56133

Bonse J et al (2017a) Applications of laser-induced periodic surface structures (LIPSS). In: Proceedings of SPIE 10092, Laser-based micro- and nanoprocessing XI, 100920N. https://doi.org/10.1117/12.2250919

Bonse J, Höhm S, Kirner SV, Rosenfeld A, Krüger J (2017b) Laser-induced periodic surface structures—a scientific Evergreen. IEEE J Select Topic Quant Electron 23(3). https://doi.org/10.1109/JSTQE.2016.2614183

Conradi M, Drnovšek A, Gregorčič P (2018) Wettability and friction control of a stainless steel surface by combining nanosecond laser texturing and adsorption of superhydrophobic nanosilica particles. Sci Rep 8(1):7457. https://doi.org/10.1038/s41598-018-25850-6

Dauscher A, Feregotto V, Cordier P, Thomy A (1996) Laser induced periodic surface structures on iron. Appl Surf Sci 96–98:410–414. https://doi.org/10.1016/0169-4332(95)00495-5

DIN EN ISO 25178-2:2023-09 (2022) Geometrical product specifications (GPS) - Surface texture: Areal - Part 2: Terms, definitions and surface texture parameters (ISO 25178-2:2021); German version EN ISO 25178-2

Dobrotvorskiy S, Aleksenko BA, Basova Y, Gnilitskyi IM, Kościński M, Machado J (2024) Light beam scattering from the Metal Surface with a complex mono- and two-periodic microstructure formed with femtosecond laser radiation. Appl Sci 14(19):8662–8662. https://doi.org/10.3390/app14198662

Dobrotvorsky SS, Khavin GL, Basova EV, Aleksenko BO, Prykhodko VO (2024) Surface roughness during laser processing of stainless steel. NTU "KhPI", Kharkiv. https://doi.org/10.20998/978-617-05-0513-2

Dzyubyk A, Kusyi Y, Dzyubyk L, Nazar I, Ivanov V (2024) Influence of the welding cycle on the parameters of material damageability of the high-strength steel connection with an austenitic seam. Vibroengineering Proc 55. https://doi.org/10.21595/vp.2024.24483

Facebook.com/machmfg. Nanosecond vs picosecond vs femtosecond Laser: explained | MachineMFG. Mechanical engineering. https://www.machinemfg.com/ru/nanosecond-vs-picosecond-vs-femtosecond-laser, last accessed 14/12/2024

Fuentes-Edfuf Y, Sánchez-Gil JA, Florian C, Giannini V, Solis J, Siegel J (2019) Surface plasmon polaritons on rough metal surfaces. Role in the formation of laser-induced periodic surface structures. ACS Omega 4(4):6939–6946. https://doi.org/10.1021/acsomega.9b00546

Garcia-Lechuga M, Puerto D, Fuentes-Edfuf Y, Solis J, Siegel J (2016) Ultrafast moving-spot microscopy: birth and growth of laser-induced periodic surface structures. ACS Photon 3(10):1961–1967. https://doi.org/10.1021/acsphotonics.6b00514

Guosheng Zhou, Fauchet PM, Siegman AE (1982) Growth of spontaneous periodic surface structures on solids during laser illumination. Phys Rev B 26(10):5366–5381. https://doi.org/10.1103/physrevb.26.5366

Hnatovsky C, Shvedov VG, Krolikowski W (2013) The role of light-induced nanostructures in femtosecond laser micromachining with vector and scalar pulses. Opt Express 21(10):12651. https://doi.org/10.1364/oe.21.012651

Horn A (2022) The physics of laser radiation–matter interaction. Springer, Cham. https://doi.org/10.1007/978-3-031-15862-9

Jost D, Lüthy W, Weber HP, Salathé RP (1986) Laser pulse width dependent surface ripples on silicon. Appl Phys Lett 49(11):625–627. https://doi.org/10.1063/1.97060

Khadka N, Yang Y, Hoffman A, Rosenberger M, Kinzel EC (2024) Spectral selectivity and blackening through direct-write femtosecond micromachining. In: Kaierle S, Kleine KR (eds) High-power laser materials processing: applications, diagnostics, and systems XIII, vol 12878. SPIE, San Francisco, pp 86–91. https://doi.org/10.1117/12.3003098

Loskutova T, Scheffler M, Ivanov V, Krueger M, Kharchenko N, Taran V, Hatala M, Karpets M, Stelmakh Y, Hasemann G (2025) Influence of diffusion siliconizing and boron–siliconizing in chlorine-containing environments on the structure, composition, and protective properties of TZM molybdenum alloy. Appl Surf Sci Adv 27. https://doi.org/10.1016/j.apsadv.2025.100769

Machinemfg.com. Explore over 10 types of lasers for various applications, https://www.machinemfg.com/types-of-lasers/, last accessed 14/12/2024

Researchgate.net. Guillaume Savriama. tutorial: two temperature model for ultrashort pulse laser with comsol multiphysics, https://doi.org/10.13140/RG.2.2.24211.81449, last accessed 14/12/2024

Rp-photonics.com. Optical parametric oscillators. https://www.rp-photonics.com/optical_parametric_oscillators.html, last accessed 14/12/2024

Seehafer N (1994) Current helicity and the turbulent electromotive force. Europhys Lett 27(5):353–357. https://doi.org/10.1209/0295-5075/27/5/004

Sergey Dobrotvorskiy, Aleksenko BA, Yevheniia Basova, Iaroslav Gnilitskyi, Zawadzki P, Mikołaj Kościński (2024) hypothesis of the lips structures formation mechanism during a femtosecond laser processing. ISSN 15565068. https://doi.org/10.2139/ssrn.4891347

Shugaev MV, Gnilitskyi I, Bulgakova NM, Zhigilei LV (2017) Mechanism of single-pulse ablative generation of laser-induced periodic surface structures. Phys Rev B 96(20). https://doi.org/10.1103/physrevb.96.205429

Siegman A, Fauchet P (1986) Stimulated wood's anomalies on laser-illuminated surfaces. IEEE J Quantum Electron 22(8):1384–1403. https://doi.org/10.1109/jqe.1986.1073133

Sipe JE, Young JF, Preston JS, van Driel (1983) Laser-induced periodic surface structure. I Theory. Phys Rev 27(2):1141–1154. https://doi.org/10.1103/physrevb.27.1141

Sokolowski-Tinten K et al (2010) Short-pulse Laser Induced Transient Structure Formation and Ablation Studied with Time-resolved Coherent XUV-scattering. In: AIP conference proceedings, vol 1278, pp 373–379. International Symposium on High Power Laser Ablation 2010, Santa Fe. https://doi.org/10.1063/1.3507123

Stoian R, Mishchik K, Cheng G, Mauclair C, D'Amico C, Colombier J, Zamfirescu M (2013) Investigation and control of ultrafast laser-induced isotropic and anisotropic nanoscale-modulated index patterns in bulk fused silica. Opt Mater Express 3:1755–1768. https://doi.org/10.1364/OME.3.001755

Tdtam.com. Laser cutting of metal, https://tdtam.com, last accessed 14/12/2024

Tsibidis GD, Lingos P, Stratakis E (2022) Synergy of electromagnetic effects and thermophysical properties of metals in the formation of laser-induced periodic surface structures. Optics Lett Optic Ind 47(16):4251–4251. https://doi.org/10.1364/ol.466079

Ulsinc.com. Laser cutting, engraving & marking machines and equipment. https://www.ulsinc.com, last accessed 14/12/2024

Varlamova O, Reif J, Varlamov S, Bestehorn M (2010) The laser polarization as control parameter in the formation of laser-induced periodic surface structures: comparison of numerical and experimental results. Appl Surf Sci 257(12):5465–5469. https://doi.org/10.1016/j.apsusc.2010.11.157

Wu B, Zhou M, Li J, Ye X, Li G, Cai L (2009) Superhydrophobic surfaces fabricated by microstructuring of stainless steel using a femtosecond laser. Appl Surf Sci 256(1):61–66. https://doi.org/10.1016/j.apsusc.2009.07.061

Young JF, Sipe JE, van Driel (1984) Laser-induced periodic surface structure. III. Fluence regimes, the role of feedback, and details of the induced topography in germanium. 30(4):2001–2015. https://doi.org/10.1103/physrevb.30.2001

Zhang H, Colombier J-P, Witte S (2020) Laser-induced periodic surface structures: arbitrary angles of incidence and polarization states. Phys Rev B 101(24). https://doi.org/10.1103/physrevb.101.245430

Residual Stresses Induced by Manufacturing Processes: Methods for Their Determination

Lokman Yünlü and Emin Tamer

Abstract In the modern world, where industrialization is gaining momentum, manufacturing technologies constantly evolve to use engineering materials more efficiently. However, this process causes residual stresses to occur in materials as an inevitable result of manufacturing methods. Residual stresses are internal stresses that occur in the material without any external loading and are in equilibrium. These stresses can affect many material properties of a workpiece, from microstructure to surface roughness. Residual stresses that occur during manufacturing processes can have positive and negative effects on the performance of materials. For example, compressive residual stresses that occur under favorable conditions can increase the operational life of the material by increasing fatigue strength. In contrast, excessive tensile stresses can lead to crack formation, premature fatigue, and deformations. Analysis of residual stresses is crucial for understanding the material's behavior both during manufacturing and under end-use conditions. Therefore, understanding residual stress mechanisms, analyzing them with accurate measurement methods, and controlling them effectively are critical to producing reliable and long-lasting machine parts. Therefore, it will not be possible to take measures or produce solutions to the negative consequences that will occur without revealing the causes of residual stress in a meaningful way. For this purpose, residual stresses in the material can be quantitatively measured using various destructive and non-destructive methods. In addition, some computational methods, using simulation techniques such as finite element analysis (FEA), can be used to predict and optimize residual stresses, allowing researchers to make inferences about possible outcomes. This study comprehensively investigates the mechanisms of residual stresses and their effects on material properties, especially in common manufacturing processes such as additive manufacturing, machining, welding, heat treatment, and plastic forming.

L. Yünlü (✉)
Burdur Mehmet Akif Ersoy University, Burdur, Turkey
e-mail: lyunlu@mehmetakif.edu.tr

E. Tamer
TMF Solutions, Ankara, Turkey

V. Ivanov (ed.), *Emerging Technologies in Manufacturing and Materials Engineering*, Springer Tracts in Mechanical Engineering,
https://doi.org/10.1007/978-3-032-14914-5_11

It also provides a detailed overview of experimental and computational techniques for measuring and analyzing residual stresses. This book chapter aims to be an in-depth guide for engineers, researchers, and students seeking to better understand the complex nature of manufacturing processes and the critical effects of residual stresses on material performance.

Keywords Residual stress · Manufacturing processes · Measurement methods

1 Introduction

Residual stress is the trapped stress in a structure or composition. It persists even after all applied forces have been removed (Withers and Bhadeshia 2001). This stress, which influences the material's behavior, often occurs due to mechanical loads, thermal loads, or phase changes. The amount and state of stress may also continue to change during the transportation of the material (Totten et al. 2002; Fitzpatrick et al. 2005). However, it should be known that if the total stress in a material is to be calculated, it must be calculated by summing the trapped residual stress and the applied external stresses (Lozano et al. 2019). Determining the type of residual stress included in the calculation allows us to predict what behavior will occur in the material. Residual stresses can be classified into compressive (negative) and tensile (positive) stress types. Compressive residual stress is beneficial as it tends to increase fatigue strength and life, slow crack propagation, and increase resistance to environmental influences. Tensile residual stresses that start at the surface are undesirable because they reduce fatigue strength and life and increase crack propagation. The effect of tensile and compressive residual stresses on crack formation and propagation is shown graphically in Fig. 1.

Residual stresses are in a balanced state within the structure. The sum of the resultant forces and moments is zero. The residual stress distribution within a part is classified according to its magnitude (Nelson 2013). There are three types of residual stress magnitudes, as shown in Fig. 2. Type I residual stresses are generally between grains in the material structure. These stresses are also called macroscopic

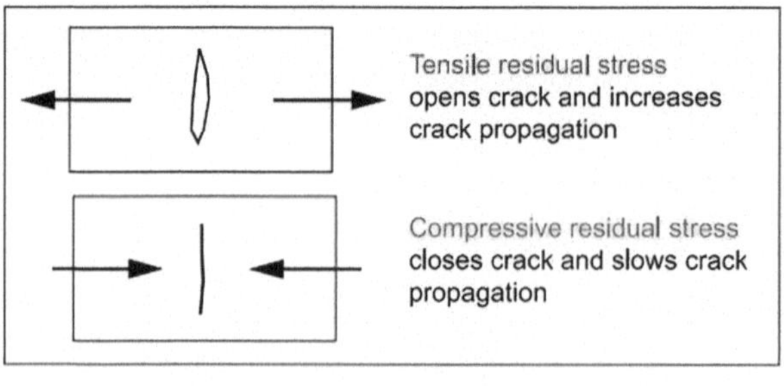

Fig. 1 Effect of tensile and compressive residual stress on the crack

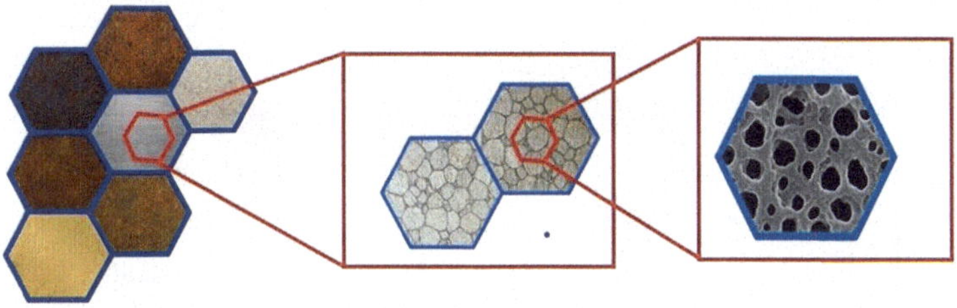

Fig. 2 Examples of the typical classification of residual stresses into three categories

residual stresses. Residual stresses of this type can be measured by almost all existing residual stress measurement methods. Type II residual stresses are called structural micro stresses and can spread over an area at most the grain size. These residual stresses occur between phases, physical properties, foreign materials, and the matrix. Type III residual stresses are minimal and usually at a distance of a few atoms within the grain (Bastola et al. 2023). In many applications, the presence of these residual stresses is neglected in residual stress measurements. When designing a new part from these three different residual stresses, Type I residual stresses must be considered (Daoud et al. 2025).

These three types of residual stresses occur after various manufacturing methods. We can briefly list these methods as welding, machining, casting, heat treatment, and forging.

2 Fundamentals and Challenges

2.1 Machining Process

Many manufacturing methods cause certain levels of residual stresses in parts. The magnitude and direction of residual stresses vary from process to process depending on loading conditions, material properties, and environmental conditions. It is difficult to define contact and external forces in machining compared to chipless forming processes (Brinksmeier et al. 1982). In addition, temperature effects cause residual stresses and trigger phase transformations (Ulutan and Ozel 2011). These effects act on the workpiece together or alone, and residual stresses occur in the part depending on the workpiece material and processing conditions (Jawahir et al. 2011). Different thermal and mechanical loads occur due to different cutting properties in different machining methods. For example, while the most critical factor causing residual stress formation in grinding processes is thermal loads, mechanical cutting forces are the most critical factors causing residual stress formation in turning and milling processes (Guo and Warren 2008). Since the loading conditions are different among machining methods, the distribution of residual stresses also

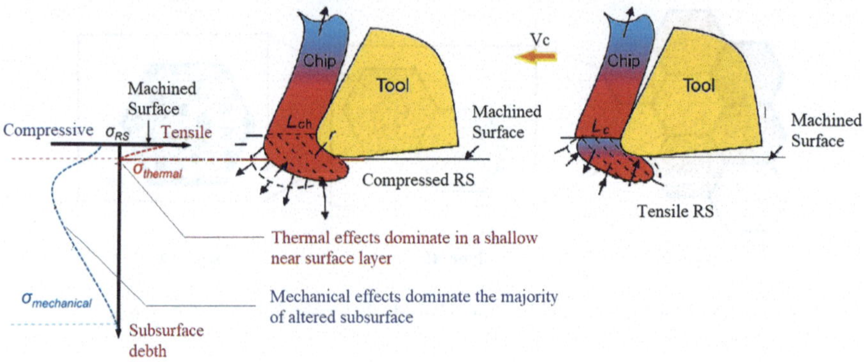

Fig. 3 Schematic representation of mechanical and thermal effects on residual stress generation; magnitude and depth of residual stress contributions of mechanical and thermal effects

differs. In milling, the tool periodically applies force to the workpiece during the cutting cycle. In this type of periodic contact, chip formation occurs in discontinuous and different ways, and the cutting forces change periodically. Variable cutting forces change the residual stress distribution in the workpiece (Ivanov et al. 2025). The milling tool geometry, feed rate, and main spindle speed affect the periodicity of the cutting forces. Each combination of cutting parameters creates its own residual stress distribution. As a result of the milling process, compressive residual stresses generally occur under the surface, and the residual stress decreases as the depth increases (Özel and Karpat 2005). The experimental studies observed that when up-milling is done, the residual stresses are in the compressive direction. When down-milling is done, the direction of the residual stresses is still compressive, but their magnitude is lower than in up-milling. Reducing the feed rate and cutting speed caused the residual stresses to change from compressive to tensile (Lauro 2018).

The roles of mechanical and thermal effects on residual stress formation in a general chip removal process are given below as a representative (Fig. 3).

2.2 *Plastic Forming Process*

The plastic forming method is a production method in which the metal is permanently changed and given the desired form. During this process, significant changes occur in the internal structure and mechanical properties of the material. Residual stresses occur within the material during and after the forging process and continue to exist even when no external load is applied (Dieter and Bacon 1976). During forging, the material is heated to high temperatures, then shaped by pressing or forging hammers and cooled in a controlled manner. Some of the main factors that cause the formation of residual stresses in this process are as follows:

Differences in Plastic Deformation; During plastic forming, different material regions undergo different levels of plastic deformation. Different amounts of

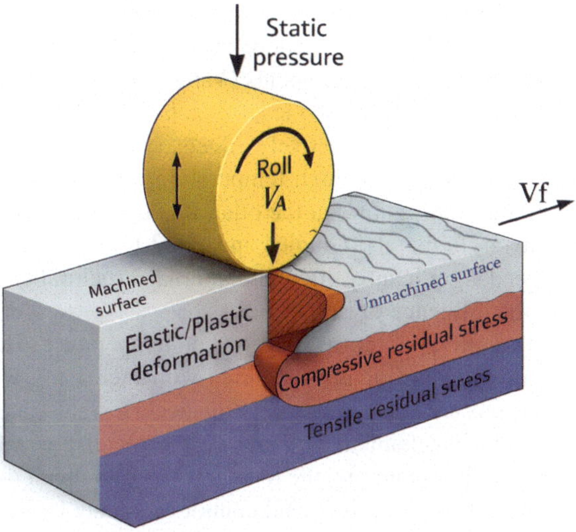

Fig. 4 Schematic illustration of the rolling process

deformation in the center and on the surface cause a difference in stress between the inner and outer regions (Ashby and Jones 2012).

Differences in Thermal Expansion and Cooling; At the temperature of plastic forming, the metal expands, but during cooling, the inner and outer layers contract at different speeds. While the inner parts cool slowly, the surface cools faster and creates tensile stresses. If the cooling is uneven, different magnitudes of residual stresses occur in the inner and outer regions.

Effects of Tool and Mold: During the forging process, the surfaces in contact with the tool are subjected to high pressure, while the inner parts are compressed less. This can create compressive stresses on the surface and tensile stresses in the inner parts.

Cooling Rates and Stress Distribution; Rapid cooling (immersion in water or oil) can create tensile stresses on the surface and compressive stresses in the inner parts. Controlled cooling or annealing can help reduce these stresses. The formation of residual stresses is inevitable in plastic forming methods, but these stresses can be controlled and optimized with appropriate manufacturing and finishing techniques (Tekkaya 2000). For example, the mechanism of residual stress formation using the plastic forming method is given below (Fig. 4).

2.3 Welding Process

Residual stresses in the welding process occur due to thermal expansion and contraction. Sudden temperature changes in the welded area cause the material to expand and contract locally, causing stress accumulation.

Heating metals to high temperatures causes internal structural changes in the base metal and may cause chemical reactions between the surrounding atmosphere and the hot or molten metal or alloy elements (Yang et al. 2025). On the other hand, chemical reactions also occur between the base metal, the filler metal, and the slag during this process. Temperature effectively affects chemical reactions; many reactions occur much more quickly at high temperatures. The diffusion process also depends on temperature. Diffusion, on the other hand, is the cause of many structural changes occurring in metals (Giudice and Sili 2022). Therefore, temperature and temperature speed significantly affect the internal structural changes occurring in the metal under the influence of heat.

Melting ability is another factor that depends on temperature. The ability of elements to dissolve in metals typically increases with increasing temperature, and when saturated solutions are cooled, precipitation occurs, significantly affecting the metal's properties. The mechanical properties of a metal always decrease with increasing temperature. For example, the tensile stress of a metal decreases to zero at the melting point; in this case, the metal cannot resist any tensile stress. Intense local heating and the subsequent rapid cooling process cause significant temperature differences during welding. The plastic deformations caused by this situation cause residual stresses in the metal (Colegrove et al. 2009) (Fig. 5).

Welded structures are widely used in many engineering applications, but are susceptible to fatigue cracks due to cyclic loading and stress concentrations at weld transitions. High-Frequency Mechanical Impact (HFMI), also known as High-Frequency Peening (HFP), is an advanced surface treatment method used to improve the fatigue performance of welded structures (Schajer 2013). This technique involves applying high-frequency mechanical impacts to weld transition zones and areas with high stress concentration, inducing beneficial compressive residual stresses and refining the material's microstructure. HFMI is widely applied in various industries, including structural engineering, automotive, aerospace, railway, shipbuilding, and the energy sector (Wang et al. 2022). This study examines the principles, mechanisms, applications, and advantages of HFMI, emphasizing its

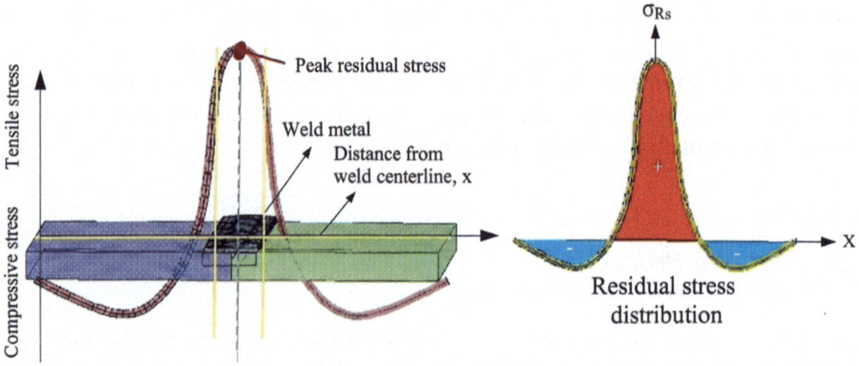

Fig. 5 A simplified function of the residual stress generated in a welding operation

potential to extend the service life of welded components and promote sustainability in engineering. High-Frequency Mechanical Impact (HFMI) is an innovative post-weld treatment technology that enhances the fatigue performance of welded structures by inducing compressive residual stresses, refining the microstructure, and reducing stress concentrations (Li et al. 2024a). Its application across multiple industries demonstrates its effectiveness in improving structural integrity and sustainability. Future research should focus on optimizing HFMI parameters for different materials and integrating automated systems to enhance industrial efficiency.

2.4 Additive Manufacturing Process

Additive manufacturing (AM) or three-dimensional (3D) printing technology is a manufacturing technology that deposits the relevant material in layers to obtain parts from data. AM enables rapid production of parts with complex geometries with a relatively short processing time compared to traditional manufacturing processes or subtractive manufacturing (Kalman et al. 2024). Parts produced with the 3D printing process (rapid prototyping) are highly suitable for making prototypes for scientific analysis and assembling final parts for various engineering applications, including automotive, aerospace, and medical. The main advantages of the AM process over subtractive manufacturing include rapid prototyping, relatively short cycle time, reduced material waste, production of complex-shaped geometries, high flexibility, increased functionality, and the production of lightweight parts that are impossible to achieve with traditional manufacturing processes.

AM technologies for metal 3D printing can be classified into two categories, considering their heat sources and raw materials. Powder bed fusion (PBF) and directed energy deposition (DED) are the two main categories of AM technologies. PBF involves AM techniques that use a bed of powder particles. A roller or spreader spreads a thin layer of powder across the base plate, and an electron beam or laser is used to melt the powder particles and scan the layer geometry. Selective laser sintering (SLS) is one of the prominent examples of a PBF 3D printing technology. Various materials can be used for the SLS manufacturing process, including metals, polymers, composites, and ceramics (Li et al. 2018). The formation of residual stress in metal AM technology can be explained as follows; During the printing process, the powder material is induced by thermal energy, which causes the solid to heat and cool differentially during solidification (Pant 2022). During the melting process, the expansion caused by the melting material causes tensile stresses. This situation also brings about changes in volume (Gibson et al. 2015). Similarly, during the solidification process, tensile and uncontrolled cooling can cause compressive stresses (Mercelis and Kruth 2006). These residual stresses can lead to part distortion, cracking, or reduced fatigue life if not properly managed (Vrancken 2012). Figure 6 illustrates the role and process of various additive manufacturing processes and supports in forming residual stress.

Although inevitable in the production process, residual stresses in additive manufacturing can be minimized with appropriate manufacturing strategies, heat

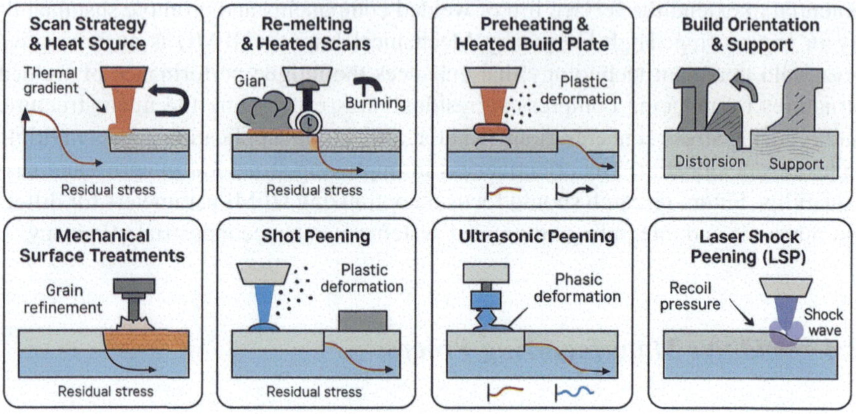

Fig. 6 Origin of residual stresses in additive manufacturing

treatments, and mechanical finishing processes. Residual stresses can be analyzed in advance using simulation and measurement techniques, and appropriate precautions can be taken. Therefore, careful selection of thermal management, scanning strategies, and heat treatment techniques in additive manufacturing processes is among the critical factors that increase part performance (DebRoy et al. 2018). Residual stresses significantly affect material properties such as fatigue, fracture, corrosion, and friction. Estimating the distribution and magnitude of residual stresses in a material is complex. It is essential to measure the stresses in mechanical parts accurately and precisely. The importance of residual stress depends on the type of application, the material, and the distribution and magnitude of residual stress. Various residual stress measurement and calculation methods depend on today's technology and science. The most common of these are explained below.

3 Ideas and Solutions

3.1 Introduction to Residual Stress Analysis

Why is residual stress analysis critical? Welding, grinding, peening, heat treatment, quenching, and many other industrial processes induce gradients within a material, leading to residual stress and affecting engineering product performance. The compressive residual stress, for instance, is known to counteract the effect of the external load on the component and reduce the fatigue life (Schajer 2013). For this reason, it is essential to understand the internal stresses in an engineering component. Traditional methods for gathering information about the stresses within a material use physical principles, including the linearly elastic stress, i.e., through strain measurement (Prime 2001; Gelatko et al. 2021). Stress analysis also employs the principle that the body forces and the surface traction balance the stress at a point in a continuous body. Experimental techniques make use of these principles to solve

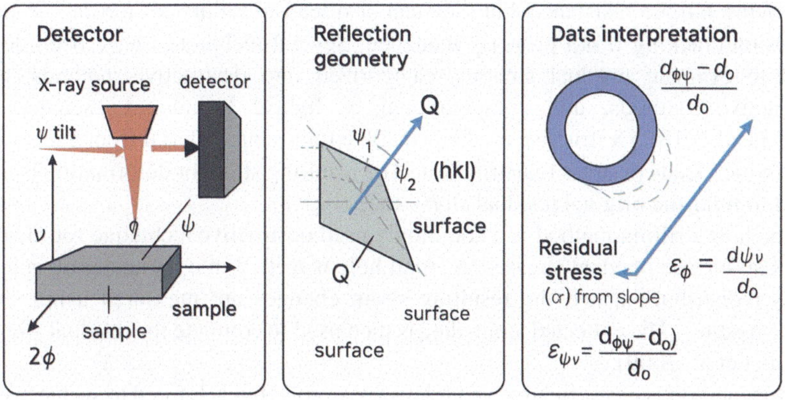

Fig. 7 Sketch illustrating the residual stress measurement using the $\sin^2\psi$ method in a reflection setup

stress (Sutton et al. 2009). In this chapter, methods used for Stokes streaming and resonance techniques, as well as diffraction analysis using X-rays, are discussed (Dulieu-Barton and Stanley 1998). Some recent advances in residual stress analysis, like the use of digital image correlation schemes and thermoelastic stress analysis, are also covered. The choice of the best method for a given problem is likely to be a trade-off between various factors, including the capability of the method, cost, the speed of measurement, and the acceptance of the technique by industrial companies, prototypes, etc. Below, in Fig. 7, you can briefly explain the $\sin^2\psi$ reflection method.

Setup (left panel): An X-ray beam strikes the specimen surface. The sample is tilted by an angle ψ and rotated in-plane by φ. For each ψ position, the diffraction peak position 2θ is recorded by the detector. Geometry (middle panel): For the chosen (hkl) planes, the scattering vector Q is normal to those planes. Changing ψ changes the angle between Q and the surface, making the measurement sensitive to the in-plane residual stress component along the φ direction. Data interpretation (right panel): From Bragg's law, the lattice spacing at each tilt is obtained and converted to strain, where d_0 is the stress-free spacing. Plotting ε{\phiψ} versus $\sin^2\psi$ yields (ideally) a straight line. The slope is converted to the in-plane residual stress σ-φ using the appropriate elastic constants (E, ν) or X-ray elastic constants. The sign of the slope indicates tension (+) or compression (−). The reflection geometry probes a near-surface layer (typically micrometers deep).

3.2 *Advanced Measurement Techniques for Comprehensive Residual Stress Analysis*

Residual stress measurement is crucial in understanding the internal stresses within a material after manufacturing processes such as welding, machining, heat treatment, and forming. These stresses can be beneficial in some cases, such as improving fatigue resistance, but they can also lead to failure, distortion, or stress corrosion cracking if not properly managed. Several techniques have been developed to evaluate residual stresses, categorized into destructive methods, non-destructive methods, and computational or hybrid techniques (Schajer and Whitehead 2018). Destructive methods involve removing material from the component, which leads to stress redistribution. The resulting strain or deformation is measured to infer the original residual stress state (Fig. 8).

The hole-drilling method is a commonly used destructive technique for measuring near-surface residual stresses. A small hole is drilled into the material, causing stress redistribution, and the resulting strain changes are measured using strain gauge rosettes. The collected strain data is then used to compute the residual stresses (Ficquet et al. 2009).

This method is cost-effective and relatively simple, but is limited to surface stress analysis and cannot measure bulk residual stresses. Another destructive method is the sectioning method. This method cuts the material, and the deformations are analyzed. A large piece of the material is cut into sections, and the shape changes are measured with the help of strain gauges, laser profilometry, or coordinate measuring devices. The original residual stress region is reconstructed based on the deformation data. This method is effective for large-scale stress measurement in all components, but it is a very destructive method. However, calculation corrections are required by changing the initial stress state. The ring-core method is a modified version of the hole-drilling technique that allows deeper residual stress analysis. In this method, a circular core is cut around the measurement region, which leads to stress relaxation recorded using strain gauges. This method provides higher accuracy for evaluating stresses measured in the upper layers of the material, but it is more tiring than hole-drilling (Schajer 2013). Special sectioning equipment is needed, and it is not suitable for thin-walled materials. The bending deflection

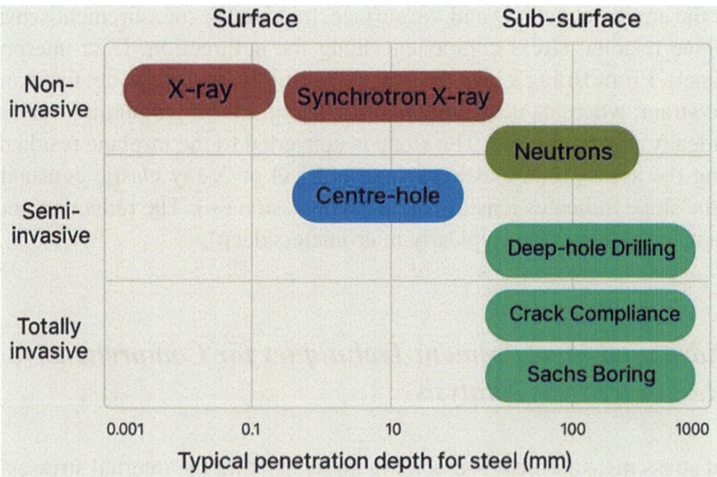

Fig. 8 Residual stress measurement methods in steel

method is another destructive method used for stress measurement in thin plates and beam-like structures. A component is sheared or cut, causing stress to be redistributed, and the resulting bending deformation is measured using dial indicators, optical sensors, or laser scanning. The residual stress field is estimated from the observed deflection patterns. This method is cost-effective and straightforward but is limited to specific geometries and assumes linear elasticity, which may not always be accurate (Fig. 9).

Non-destructive methods allow stress evaluation without damaging the component. These techniques are helpful for in-service monitoring and quality control. One widely used method is X-ray diffraction, which measures strain by analyzing shifts in diffraction angles of crystalline planes. This technique effectively measures surface stress but requires material calibration for accurate results. Neutron diffraction is another non-destructive technique that uses neutron beams to penetrate deep into the material, making it ideal for bulk stress measurement. However, it requires access to expensive neutron sources, which limits its widespread application.

Ultrasonic stress measurement is another non-destructive technique that evaluates wave velocity changes in materials, which correlate with residual stress states. This method is suitable for bulk stress analysis but can be affected by microstructural variations. Magnetic Barkhausen noise analysis is used for ferromagnetic materials such as steels. This method shows the variations in stress by taking into account the magnetic field variations under the influence of voltage. Analytical and hybrid algorithms use physical and digital solutions to predict residual stress. Finite element analysis is widely used to predict residual stresses in electric field applications. This technique requires programming calculations based on properties and manufacturing conditions, which are usually verified with discrete data for higher accuracy. Destructive measurement methods provide high accuracy but degrade the material, while non-destructive techniques allow in-service evaluation but may require complex calibration. A hybrid approach combining experimental and digital techniques often provides the most reliable and comprehensive residual stress analysis. Advances in this methodology will allow for the options of non-destructive

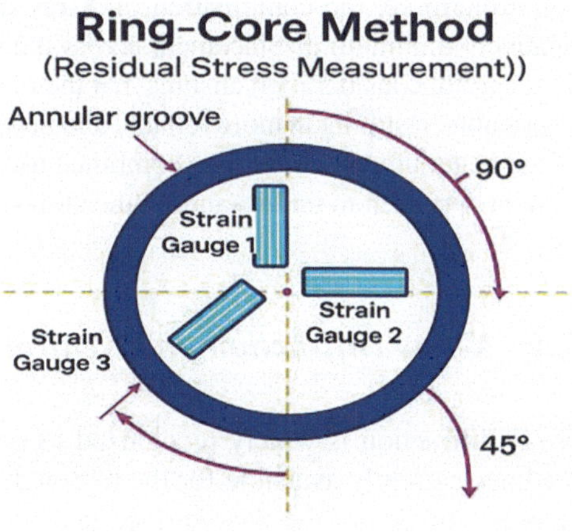

Fig. 9 Schematic view for the ring core method (Seers et al. 2021)

methods, the integration of artificial intelligence for automated analysis, and the expansion of hybrid systems. By optimizing these methods, industries can increase material diversity, minimize damage, and extend the operational life of engineered materials. Among these numerous methodologies, the hole-drilling technique is the oldest and most straightforward experimental approach consistently utilized for measuring residual stresses effectively within the materials science community. This well-established technique operates successfully as a purely nondestructive method, which is regarded as one of its most significant and essential advantages in residual stress assessment.

While this established technique can yield valuable insights into a material's detailed residual stress state under examination, it is essential to note that even more accurate and precise residual stress values can be attained by implementing complementary and additional measures (American Society for Testing and Materials 2020). These measures include applying specialized chemical solutions into the drilled hole, followed by a meticulous calculation of the stress distribution in the surrounding area. This calculation method carefully considers variations in stress along the cross-section of the material being measured, measuring it precisely with the help of displacement and strain equations (Prime 2001). In many scientific studies, the hole drilling method has been effectively and successfully applied to measure thin and thick weld material, demonstrating its remarkable versatility and adaptability in different scenarios (Noyan and Cohen 2013). This impressive feature has given accurate, reliable, and reproducible results in different experimental conditions and changes in calibration factors and analysis that may arise. The X-ray diffraction technique is a second widely recognized and highly accepted method for the experimental measurement of residual stress in various materials (Withers and Bhadeshia 2001). This method can be effectively implemented by carefully handling the X-ray film and using X-rays specially designed for precise measurements. Generally, this method gives more reliable and accurate results for residual stresses than many other techniques. However, the basic principle underlying the widely used hole drilling technique requires displacement measurements from multiple holes, thus significantly increasing the accuracy and precision of the results obtained (Schajer 2013).

Furthermore, the combination of X-ray diffraction and hole drilling requires a relatively minimum displacement across the surface area of the material under test. This careful consideration ensures that the readings taken are as precise and accurate as possible, resulting in more reliable and understandable data on the residual stresses of the material. As a result, the combined use of both methods provides a more convenient approach to understanding the stress distribution in various materials.

3.3 X-Ray Diffraction and Neutron Diffraction Method

X-ray diffraction is widely recognized as one of the most powerful and effective methods currently available for the thorough and detailed study and analysis of the residual stresses in the near-surface layers of various metals and numerous alloy systems (Dölle 1979). This sophisticated and intricate technique relies fundamentally on the complex change in the X-ray wavelength as it interacts intricately with the finely organized and meticulously structured crystal lattice arrangements of the materials undergoing rigorous investigation. According to Bragg's law, an essential foundational principle underlying this precise and vital method, the specific conditions necessary for constructive interference of X-ray reflection from two successive lattice layers of atoms cannot be satisfied if the distance d between these layers happens to become a mixed number. As a direct and critical result of this fundamental phenomenon, the intensity of the X-ray reflection will drop off abruptly, leading to significant implications for the analysis being performed and the ultimate accuracy of the results obtained through this procedure (Bragg 1913). Consequently, the reflection of X-rays by the various sets of atomic lattice planes can indeed be interpreted and treated as a complex wave phenomenon, necessitating careful and thorough consideration and rigorous analysis. Figure 10 shows a simplified schematic representation of Bragg's law and its application.

Blue circles represent the atoms, while blue and orange arrows represent the travel path of X-ray radiation. Coherent waves reflecting from the family of planes interfere with each other. When incident X-rays impinge forcefully upon these crystallographic planes at precisely defined and accurately measured angles θ, the reflecting planes will fulfill Bragg's law, which can be mathematically expressed as $n\lambda = 2\ d\ \sin\theta$. In this pivotal and crucial equation, λ represents the wavelength of the incident X-rays, θ is the angle of incidence that is meticulously controlled, and d denotes the distance between the successive atomic planes within the crystal

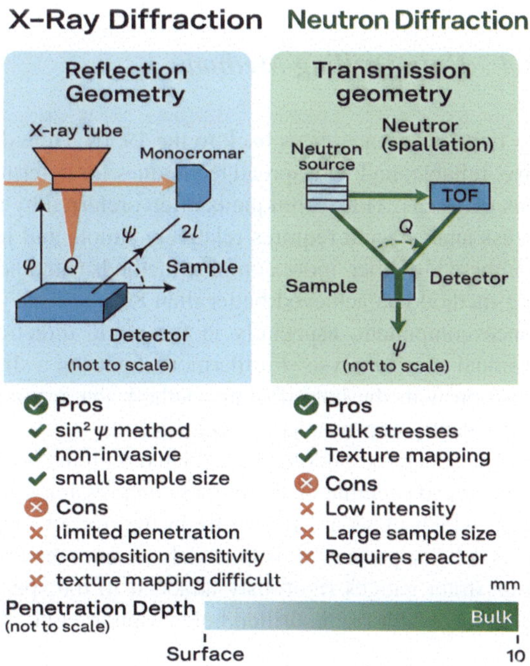

Fig. 10 Schematic illustration of a simplified model for Bragg's law

lattice structure. Therefore, X-ray diffraction emerges as a potent and undeniably invaluable tool for determining the stress state in both single and polycrystalline materials. This exceptional capability is achieved through the careful and methodical measurement of changes in the interatomic spacing. These critical measurements subsequently provide essential and indispensable information regarding the lattice strain components and the overall mechanical behavior of the materials studied, thereby enabling advanced research and fostering a deeper understanding of the material properties that are vital in various engineering applications (Cullity 1986).

The neutron diffraction method is highly efficient based on innovative calculation and analysis principles. It involves the rigorous and exact measurement of the diffraction angle, mainly in the narrow range between 120° and 150°. It also involves carefully analyzing the intensity of the rays produced during diffraction. Examining the measurements obtained can be calculated with a high degree of accuracy based on the precisely measured diffraction angle of the distance to the recorded lattice distance of diffractions in crystals where strain occurs. It is determined that the stress component bound to a particular direction within the crystal is directly proportional to the intensity of the diffraction peak corresponding to that particular plane standard. To facilitate this complex process, a significant amount of data can be efficiently collected by strategically using several detectors working together while simultaneously rotating the section under inspection. This precise calculation improves the accuracy of measurements and provides a better understanding of the behavior of materials under varying stresses and conditions.

3.4 Hole-Drilling Method

Its first appearance dates back to the 1930s. Hole drilling is one of the most sensitive, reliable, and widespread techniques for determining the residual stress in various materials. This technique is often preferred by analysts specializing in residual stress analysis and requires relatively simple and inexpensive equipment to set up compared to other, more complex methods (Stefanescu et al. 2006). The hole drilling method has achieved better than 80% accuracy in detecting the state in a distance component, especially in the circle direction, an essential factor in state residual stress analysis. Furthermore, by using a slit-crack extension meter in combination with the hole drilling method, the accuracy of this measurement can be significantly improved (Peng et al. 2021).

The actual methodology associated with the hole drilling technique requires the gradual and accurate machining of compression-compliant holes into the test specimen, which plays a critical role in the overall assessment of material properties. This delicate process aims to closely observe the in-plane displacements of three or four strain gauges rigorously attached to the specimen at equal angles radiating from the center of the drilled hole (Ammar et al. 2021). The specific holes created

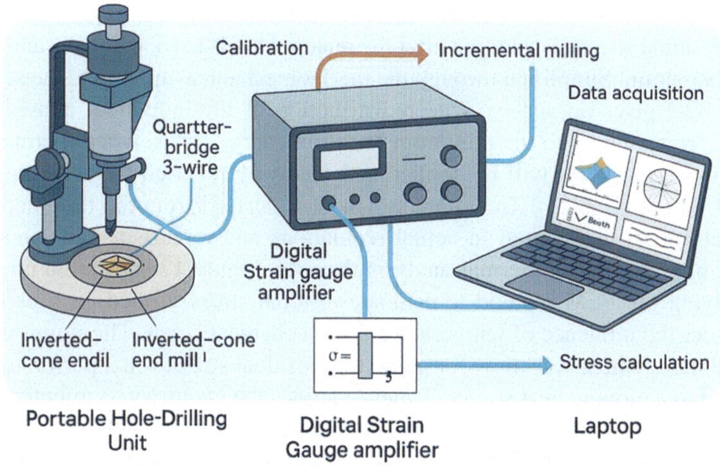

Fig. 11 Schematic illustration of the Hole-Drilling Method for residual stress measurement

during this elaborate procedure are characterized by circularly machined material losses, which initially effectively isolate the holes from each other, minimizing unwanted interactions between different drilling points. After this stage, the residual stress component in the plane is complexly related to the measured strain field. This complex strain field is not only a direct result of the drilling process. Still, it is also significantly influenced by the axial deformation of the material and the temperature fluctuations that can occur during the drilling process. This complex relationship is calculated through an advanced integral function that uses the measured stress-displacement relationships to obtain accurate and reliable results regarding the material's behavior under various stress conditions. Researchers and engineers can improve material performance and lifetime by predicting better interpretations of these interactions with today's artificial intelligence methods (Santus et al. 2024) (Fig. 11).

3.5 Finite Element Analysis in Residual Stress Prediction

Finite element analysis has proven to be a valuable tool for predicting part distortions and residual stresses resulting from heat treatment processes (Tabatabaeian et al. 2022). The finite element solutions for residual stresses are derived through material and body boundary conditions after the temperature and phase transformation history. Through the FEM, both elastic and elastoplastic models with isotropic or kinematic hardening have become popular in practice (De Baere et al. 2021). However, the FEM is just a tool and requires experience and judgment to be appropriately used in real engineering applications. There are two kinds of FEA used in residual stress predictions. The first kind is the seamless stress model, which provides a residual stress distribution after the material cools to room temperature without further strain. Simplified through the thermomechanical approach, the seamless stress model gives an approximate redistribution of the individual phases of the system corresponding to the minimum resulting energy. Thus, when thermal loads are involved, the loads will be applied and passively released from the bounding states (Ullah et al. 2024). The rate component is derived from the time-dependent stress relaxation factor used in actual regulations and represented in a nonlinear implicit line process by thermal analysis theories. Figure 12 shows the procedure for applying the FEM method to estimate residual stress formed on a perforated plate under the influence of temperature and mechanical stress. The figure summarizes a finite-element workflow for predicting residual stresses in a perforated plate subjected to a moving heat source. Panel A defines the geometry, symmetry planes, and a mesh refined along the heat path and around the hole; Panel B solves the transient heat-conduction problem with convection losses to generate the temperature history. Panel C couples this thermal field to an elasto-plastic mechanical model (with isotropic/kinematic hardening) to compute thermal strain, plasticity, and the resulting residual-stress map, which typically shows tensile zones near the heated/holed region balanced by surrounding compression. Panel D outlines the end-to-end process: gather temperature-dependent material data, set mesh and boundary conditions, run thermal and coupled mechanical analyses, and validate/calibrate against measurements (e.g., X-ray diffraction $\sin^2\psi$). Outputs include residual-stress contours and line profiles, with accuracy sensitive to mesh/time-step resolution and to heat-source and convective parameters.

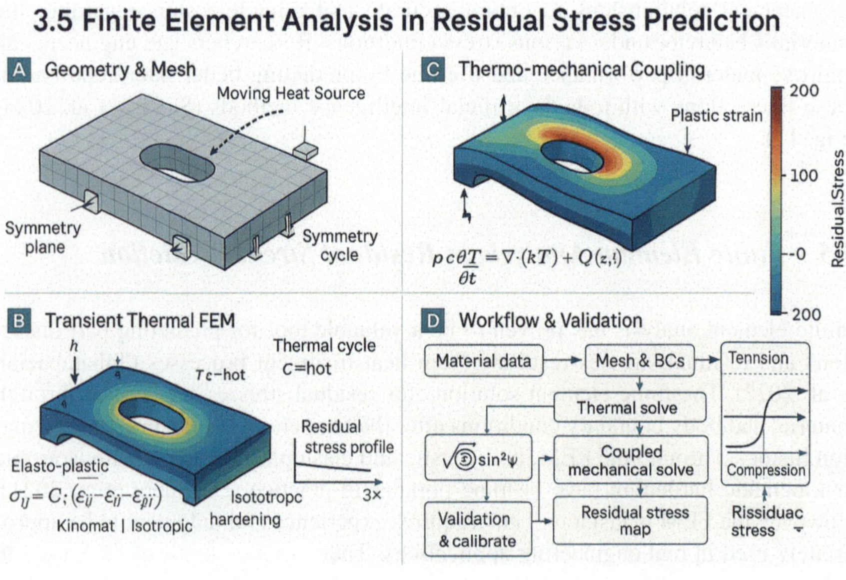

Fig. 12 FEM analysis process of residual stress on a plate under mechanical and thermal effects

However, thermal analysis neglects the magnitude and rate of distortions and residual stress development. The coupling of temperature and thermal translations of steel bars with loads, including torque, pressure, and thermal gradient, will influence the applied loadings and have phase transformation transients that affect the quality of the analysis results. As in submodeling and weld simulations, the thermal load distribution between elements in systems with slight load changes is determined by temperature and thermal expansions and contractions, which are thermal-scaled (Zobec and Klemenc 2021). The second kind is represented by the quasi-thermal analyses, in which the thermal history is approximated by a history with only a few increments, each covering a short period of heating and several days of cooling, and which models temperature distributed over the powder metallurgy part (Xia et al. 2025). The rapid heating is divided into several steps, with the shrinkage included, while the yield stress is modeled as a function of temperature and the maximum held back. For a much broader band, the solution could be searched for by plugging in the energy model's approach and studying the distribution of failure and residual stresses in the high-temperature and low-thermal strain domains (Shao et al. 2022). These applications focus on the solid phase's approach to the phase transformation cycle for solidification and primary solidification simulations, since such models are already available or well proven from simulations.

3.6 Applications of Residual Stress Analysis

The two traditional residual stress analysis applications are life prediction and dimensional instability. In both applications, the effects of depth on the state of stress are significant, so X-ray and neutron diffraction techniques have been popular. More recently, it has been realized that it is possible to use plastic strains to reduce the undesirable effects of volumetric strain in several materials production processes. Such processes are generally based on creating a limited compressive layer. These beneficial layer stresses may be called mechanochemical action and are essential in fuel claddings for pressurized water reactors and for fusion reactor first walls.

Several international symposia on residual stress have been held since the publication of several books on grinding and residual stress. A compendium of some of the contributions to these symposia was published in 2006. Some national standards have also been refined. Despite this recent progress, research topics continue proliferating, and two new topics summarizing these contributions were presented at a conference. The topics were residual stress effects on the growth rate of cracks and residual stress effects on crack closure. It is likely that, stimulated by these and other reports, research on residual stress will continue unabated for the foreseeable future.

Aerospace Industry

Aerospace structural components such as fuselage structures, wings, satellites, and launch vehicles are constructed using welding, adhesive bonding, milling, drilling, and riveting (Sun et al. 2025). These processes produce martensite phase transformation surrounding the heat-affected zones of the material components, which introduces residual stress and strain. Residual stress and strain significantly reduce the lifespan of the structure. In particular, the launch, flight, and reentry phases produce high thermal loads in the structure, substantially reducing the component's lifespan (Ramineni et al. 2024). To assess the residual stress and strain, X-rays are used to estimate their effect, while contour and hole drilling interfere with the component's surface and damage the near-violent regions (Li et al. 2024b). The semi-circular notch rounding process is a significant consideration in the aerospace industry because it utilizes a notch to reduce the edge gradient effect by modifying the transition radius. However, because the material is deformed, the width loss is compensated for by the setup force of the material (Ye et al. 2023). Since the force introduced buckles the material, stress is produced. Since it may be greater than the material's yield strength, it is essential to assess the functionality of the welded region due to possible failure during testing.

Automotive Industry

In the automotive area, residual stresses of various components and systems are interesting from different viewpoints. Significant results concerning the effects of residual stresses on the service performance of various automotive parts, components, and systems have already been obtained by researchers (Chiqueti et al. 2023). The automotive industry deals with almost all the manufacturing processes, and known applications are performed for die-quenched engine blocks, wheel disks, forged chassis parts, gas, laser, and hybrid-laser-cut sheet components, etc. (Jo et al. 2024). The specific aluminum engine block investigated is a cast article which, after the necessary machining steps, is stress-relieved, quenched by driving liquid transient heat transfer, and stress-relieved again.

The objectives for applying a die-quenching treatment, as mentioned by the industrial collaborator of this research, are the reduction of the cycle time, a better product form or geometric stability of the cast part, avoiding delamination or cracking problems, and the overall lack of the necessary final stress-relief operations apart from those performed immediately after the casting (Kim et al. 2023). At this stage, it should be mentioned that the first stress-relieving cycle occurs at 170 °C. In comparison, the temperature and pressure necessary to perform the second cycle are approximately 400 °C and 4 MPa, respectively (Alipooramirabad et al. 2024). This implies an initial cost analysis that entails both money and energy savings. In relation to this, the die quenching process seems very promising, as shown by early implementations to different automotive components like engine blocks, cylinder heads, composite sheet components, etc.

Biomedical Applications

With the advent of new minimally invasive surgical techniques in medicine, many improvements in these techniques have become possible, such as cutting and sewing in a hernia operation (Vanaei et al. 2024). These methods generally use high-tech stapling or bioabsorbable synthetic sutures that have replaced the traditional needle and thread. However, due to the great difficulty in controlling the suture's strain, the internal tissues of the patients may be subjected to excessive stresses that can lead to necrosis. To improve this problematic issue, metal or bioabsorbable external rings have been used to hold the internal tissue layers together and maintain them at a desired tension (Fardin et al. 2022). Biomechanical tests done with these wires or staples showed the importance of these imposed residual stresses. To truly simulate the internal uniaxial residual tension existing in the biological tissue, it became necessary for the metallic staples to perform a cutting operation against the tissue, imposing some permanent residual tension in the biological fibers. These new studies have shown the values of imposed permanent residual stresses necessary to avoid unnecessary tissue damage. However, these implanted metal staples may tend to become a permanent implanted skeleton, although they should be gradually absorbed by the body (Hrubanová et al. 2025). Amorphous wire staples have been proposed to associate flexible staples in the early moments, but rigid enough to prevent the hernia, warts, or blood vessel hole from reopening during the tissue regeneration. The main disadvantage of these wires (high residual permanent stress) is compensated for using staples made from fully stiff, brittle materials (Giudici and Spronck 2022).

3.7 Significance of Residual Stress Analysis in Material Engineering

Using suitable methods to analyze the residual stresses is crucial in managing and controlling the detrimental impacts of residual stress in a structure. Residual stress analysis is also needed to understand other material processes, such as phase transformation, metallurgical changes, thermal treatment, welding, shot peening, etc. This chapter shares a brief overview of the fundamental concepts of residual stress. Then, the most commonly used residual stress analysis methods are explained in detail. The significance of residual stress analysis is presented with innovative applications, including the residual stress analysis of a patient-specific curved tibial implant and numerical modeling, stress measurement, and magnitude validation of a 4.5 MW wind turbine shaft before and after a fatigue test. Remember that residual stresses are also of engineering importance in that they can cause catastrophic failures if not recognized and considered.

The detection, measurement, and explanation of residual stresses are critical in designing and analyzing engineering components and structures. The term residual means that the stresses are self-balanced; these are an intrinsic feature caused by processes that have occurred during the birth, growth, and fabrication of the materials, and these are independent of any previously applied loads. Consequently, residual stresses are associated with permanent deformations and other irreversible material changes. Residual stresses are an essential subject of many engineering disciplines, such as aerospace, civil, mechanical, nuclear, materials, biological, and medical engineering, and cause significant problems and catastrophic failures in engineering structures, like fatigue cracking, buckling, creep, stress corrosion cracking, loss of material strength, microfracture, and dimensional instability.

3.8 Challenges and Limitations in Residual Stress Analysis and General Evaluation

Reliable measurement of residual stresses, utilizing one or a combination of the various existing techniques, is a key element in materials engineering. In cold forming operations, residual stresses are generated due to the unequal transformation of cross-section portions of the material, leading to dimensional deviations that manufacturers widely recognize as one of the most frequent and severe problems of cold-formed parts. Local residual stresses, combined with macroscopic load due to the tip load, may also lead to manufacturing difficulties, crack formation, or further problems in subsequent operations. Therefore, evaluating and minimizing local residual stresses in cold-formed parts has significantly enhanced cold-forming technology. In this light, manufacturers and researchers widely research and apply residual stress analysis in cold-formed parts. The various existing methods display certain limitations, and under certain circumstances, they provide erroneous results. Consequently, the selection of the appropriate method and the in-depth examination of the existing problems are essential for the minimization of errors and the provision of reliable measurement.

4 Conclusions

4.1 Future Trends and Developments in Residual Stress Analysis

Residual stress analysis continues to be a very active and fundamental area of research. In practice, many present-day needs are satisfied through the use of publicly available computer programs. It is clear, however, that even as the basic methods evolve, the new techniques are still primarily semi-deterministic; that is, degrees of sensitivity based on sequences of crystal planes, treatments for the masking of X-ray sources, or the effects of curvature of the specimen. Polycrystal-based Fourier analysis and even the t-resolution concepts all make clear that, except for the most fundamental global strain measurements, the applicability of the higher-order methods relies heavily on the expectation of the kind of residual stress behavior typically seen in engineering practice. These approaches are increasingly crucial for multi-axial residual stress measurements using laboratory-based X-ray/Synchrotron diffraction techniques and non-diffractive optical methods.

There is good reason to be optimistic that future work will involve models based on pioneering work performed at synchrotron X-ray facilities, including residual elastic stress distribution analysis based on micro-beam diffraction and finite element modeling. However, It is unclear whether these novel methods are suitable for the kinds of high-sensitivity measurements widespread in non-destructive evaluations on actual components, whose residual stress distribution is far more complex than is typically found in the academic literature. In addition, relatively new problems such as identifying and mitigating micro-stresses in fuel cells and multi-layer thin films or even lattice destruction in materials undergoing irradiation damage will continue to yield complex challenges for nondestructive techniques. Since the increased imaging contrast associated with elastic strains provides an essential constituent of high-resolution imaging for the current state of the art and developments in imaging, much more significant than smaller enhanced strain signals are anticipated in imaging areas such as high-energy X-ray computed tomography using extended monochromatic beams. Applications for both residual stress and non-residual stress three-dimensional analysis will continue to be essential motivators in the field of residual stress analysis using diffraction. The theoretical chromatography and coherence problems in these applications will become increasingly important. Other significant areas of opportunity are in micro/mechanical plasma display panels, X-ray waveguides, and micro-structured radiation detectors for high-energy synchrotron and laser radiation. Future challenging questions will center on high-stress materials exhibiting reduced acoustic propagation; i.e., future crack spallation problems will demand diffraction and crack propagation by synchrotron micro-computed tomography.

In addition, it is apparent that interdisciplinary techniques will continue to develop and that residual stress behavior will continue to motivate and be motivated by work in fracture mechanics, mechanical fatigue, heat transfer, and the thermo-mechanical responses of different classes of materials. To develop new methods or applications, new coordinate systems and problems in the size effects of plasticity are likely to receive valuable insights incidentally. Recent work validating residual stress and coherent structural models and evidently contributing to the improved modeling of strain profiles observed at synchrotron facilities recently involved remote fatigue crack growth, heat treatment, and plastic bonding in precipitation hardening stainless steel steels and microscopic fretting fatigue of stainless steel. An entirely different pathway of innovation in residual stress analysis uses micro-structural models built up from the long history of materials engineering, but have rich residual stress-related elastic strain behaviors.

After the residual stress measurement and determination methods mentioned above, it is possible to reduce or eliminate the effect of residual stresses in the part by applying various methods.

These can be briefly listed as follows;

- Heat Treatments (Annealing, Stress Relief Annealing); Internal stresses can be reduced with controlled heat treatments. Mainly stress relief annealing minimizes residual stresses with long-term heating at low temperature.
- Vibration Stress Relief; Irregular stresses within the material can be reduced by applying mechanical vibrations.
- Surface Treatments (Sandblasting, Shot Blasting); The fatigue life of the material can be increased by creating compressive stresses on the surface.

4.2 Future Perspectives

- While current strategies provide effective means to mitigate residual stress, future research should focus on:
- Real-Time Residual Stress Monitoring: Advanced in situ monitoring techniques, such as infrared thermography and acoustic emission analysis, can help detect and manage stress development during printing.
- Advanced Simulation Techniques: Machine learning and AI-driven modeling can optimize AM parameters for residual stress reduction.
- Hybrid Manufacturing Approaches: Combining additive and subtractive processes like hybrid AM-CNC machining can improve stress management and part quality.
- New Material Developments: Engineering novel metal alloys with enhanced thermal properties can further reduce residual stress formation in AM components.

Acknowledgments The research was partially supported by the International Association for Technological Development and Innovations.

References

Alipooramirabad H, Kianfar S, Paradowska A, Ghomashchi R (2024) Residual stress measurement in engine block—an overview. Int J Adv Manuf Technol 131(1):1–27. https://doi.org/10.1007/s00170-024-13071-3

American Society for Testing and Materials (2020) Standard test method for determining residual stresses by the hole-drilling strain-gage method. ASTM International

Ammar MM, Shirinzadeh B, Lai KZ, Wei W (2021) On the sensing and calibration of residual stresses measurements in the incremental hole-drilling method. Sensors 21(22):7447. https://doi.org/10.3390/s21227447

Ashby MF, Jones DR (2012) Engineering materials 1: an introduction to properties, applications and design. Elsevier

Bastola N, Jahan MP, Rangasamy N, Rakurty CS (2023) A review of the residual stress generation in metal additive manufacturing: analysis of cause, measurement, effects, and prevention. Micromachines 14(7):1480. https://doi.org/10.3390/mi14071480

Bragg WH (1913) The reflection of X-rays by crystals.(II.). Proc Royal Soc London Series A Contain Papers Math Phys Character 89(610):246–248

Brinksmeier E, Cammett JT, König W, Leskovar P, Peters J, Tönshoff HK (1982) Residual stresses—measurement and causes in machining processes. CIRP Ann 31(2):491–510
Chiqueti CM, Manini RG, Gomes BM, dos Santos M (2023) Relation of compression residual stress on pre-tensioned surface in parabolic leaf spring submitted to the stress shot peening process in carbon steel. SAE Technical Paper 2023-36-0354. https://doi.org/10.4271/2023-36-0354
Colegrove P, Ikeagu C, Thistlethwaite A, Williams S, Nagy T, Suder W, Pirling T (2009) Welding process impact on residual stress and distortion. Sci Technol Weld Join 14(8):717–725. https://doi.org/10.1179/136217109X406938
Cullity BD (1986) Elements of X-ray diffraction. Hsin Yueh
Daoud ME, Taha I, Abdelgawad A (2025) Mapping residual stresses in AlSi10Mg alloy fabricated by powder bed fusion-laser beam method. Sci Rep 15:24784. https://doi.org/10.1038/s41598-025-09316-0
De Baere D, Van Cauwenbergh P, Bayat M, Mohanty S, Thorborg J, Thijs L, Hattel JH (2021) Thermo-mechanical modelling of stress relief heat treatments after laser-based powder bed fusion. Addit Manuf 38:101818. https://doi.org/10.1016/j.addma.2020.101818
DebRoy T, Wei HL, Zuback JS, Mukherjee T, Elmer JW, Milewski JO, Zhang W (2018) Additive manufacturing of metallic components–process, structure and properties. Prog Mater Sci 92:112–224. https://doi.org/10.1016/j.pmatsci.2017.10.001
Dieter GE, Bacon D (1976) Mechanical metallurgy, vol 3, New York, pp 43–53
Dölle H (1979) The influence of multiaxial stress states, stress gradients and elastic anisotropy on the evaluation of (Residual) stresses by X-rays. J Appl Crystallogr 12(6):489–501
Dulieu-Barton JM, Stanley P (1998) Development and applications of thermoelastic stress analysis. J Strain Analy Eng Design 33(2):93–104
Fardin VP, Paula VG, Bonfante EA, Coelho PG, Bonfante G (2022) Residual stress estimated by nanoindentation in pontics and abutments of veneered zirconia fixed dental prostheses. J Appl Oral Sci 30:e20210475. https://doi.org/10.1590/1678-7757-2021-0475
Ficquet X, Chidwick L, Gilles P, Joly P (2009) Residual stress measurement on a narrow gap dissimilar metal weld pipe. In: International conference on offshore mechanics and arctic engineering, vol 43468, pp 193–207. https://doi.org/10.1115/OMAE2009-79918
Fitzpatrick ME, Fry AT, Holdway P, Kandil FA, Shackleton J, Suominen L (2005) Determination of residual stresses by X-ray diffraction, Measurement good practice guide no. 52. National Physical Laboratory, UK
Gelatko M, Hatala M, Vandžura R, Botko F (2021) Longitudinal critically refracted (LCR) ultrasonic wave for residual stress measurement. In: IOP conference series: materials science and engineering, vol 1199. IOP Publishing, p 012060
Gibson I, Rosen D, Stucker B (2015) Direct digital manufacturing. In: Additive manufacturing technologies: 3D printing, rapid prototyping, and direct digital manufacturing. Springer, New York, pp 375–397. https://doi.org/10.1007/978-1-4939-2113-3_16
Giudice F, Sili A (2022) A theoretical approach to the residual stress assessment based on thermal field evaluation in laser beam welding. Int J Adv Manuf Technol 123:2793–2808. https://doi.org/10.1007/s00170-022-10247-7
Giudici A, Spronck B (2022) The role of layer-specific residual stresses in arterial mechanics: analysis via a novel modelling framework. Artery Res 28:41–54. https://doi.org/10.1007/s44200-022-00013-1
Guo YB, Warren AW (2008) The impact of surface integrity by hard turning vs. grinding on fatigue damage mechanisms in rolling contact. Surf Coat Technol 203(3–4):291–299. https://doi.org/10.1016/j.surfcoat.2008.09.005
Hrubanová A, Lisický O, Sochor O, Bednařík Z, Joukal M, Burša J (2025) Layer-specific residual strains in human carotid arteries revealed under layer separation. PLoS One 20(4):e0308434. https://doi.org/10.1371/journal.pone.0308434
Ivanov V, Evtuhov A, Dehtiarov I, Trojanowska J (2025) Fundamentals of manufacturing engineering using digital visualization. Springer Nature, p 110

Jawahir IS, Brinksmeier E, M'saoubi R, Aspinwall DK, Outeiro JC, Meyer D, Jayal AD (2011) Surface integrity in material removal processes: recent advances. CIRP Ann 60(2):603–626. https://doi.org/10.1016/j.cirp.2011.05.002

Jo W, Woo I, Mikami Y, An G (2024) Residual stress characteristics in spot weld joints of high-strength steel: influence of welding parameters. Appl Sci 14(24):2076–3417. https://doi.org/10.3390/app142411971

Kalman O, Nazim A, Pavlenko I, Ivanov V (2024) Advantages of 3d printing in the education process. In: EAI international conference on automation and control in theory and practice. Springer Nature Switzerland, Cham, pp 3–17

Kim Y, Kolesov S, Kim H, Choi K, Lee M-G (2023) Residual stress development and thermo-elasto-plastic distortion in brake discs. Tribol Int 190:109056. https://doi.org/10.1016/j.triboint.2023.109056

Lauro CH (2018) Study of micro-cutting in the finishing of a difficult-to-cut biomaterial. Universidade de Aveiro Portugal

Li C, Liu ZY, Fang XY, Guo YB (2018) Residual stress in metal additive manufacturing. Proc Cirp 71:348–353

Li X, Liu J, Wu H, Miao K, Wu H, Li R, Liu C, Fang W, Fan G (2024a) Research progress of residual stress measurement methods. Heliyon 10:e28348. https://doi.org/10.1016/j.heliyon.2024.e28348

Li R, Ju G, Zhao X, Zhang Y, Li Y, Hu G, Lin D (2024b) Simulation of residual stress and distortion evolution in dual-robot collaborative wire-arc additive manufactured Al-Cu alloys. Virtual Phys Prototyp 19(1):e2409390. https://doi.org/10.1080/17452759.2024.2409390

Lozano DE, Totten GE, Bedolla-Gil Y, Guerrero-Mata M, Carpio M, Martinez-Cazares GM (2019) X-ray determination of compressive residual stresses in spring steel generated by high-speed water quenching. Materials 12:1154. https://doi.org/10.3390/ma12071154

Mercelis P, Kruth JP (2006) Residual stresses in selective laser sintering and selective laser melting. Rapid Prototyp J 12(5):254–265. https://doi.org/10.1108/13552540610707013

Nelson D (2013) Experimental methods for determining residual stresses and strains in various biological structures. Exp Mech 54(4):695–708. https://doi.org/10.1007/s11340-013-9806-6

Noyan IC, Cohen JB (2013) Residual stress: measurement by diffraction and interpretation. Springer

Özel T, Karpat Y (2005) Predictive modeling of surface roughness and tool wear in hard turning using regression and neural networks. Int J Mach Tools Manuf 45(4–5):467–479

Pant P (2022) Residual stress in additive manufacturing: control using orientation and scan strategies. PhD dissertation, Linköping University Electronic Press. https://doi.org/10.3384/9789179292935

Peng Y, Zhao J, Chen LS, Dong J (2021) Residual stress measurement combining blind-hole drilling and digital image correlation approach. J Constr Steel Res 176:106346. https://doi.org/10.1016/j.jcsr.2020.106346

Prime MB (2001) Cross-sectional mapping of residual stresses by measuring the surface contour after a cut. J Eng Mater Technol 123(2):162–168. https://doi.org/10.1115/1.1345526

Ramineni L, Almotari A, Ali M, Algamal A, Qattawi A (2024) Residual stress mapping in heat-assisted additive manufacturing of IN 718: An X-ray diffraction study. J Mater Eng Perform 33(8):4124–4135. https://doi.org/10.1007/s11665-024-09269-x

Santus C, Neri P, Romoli L, Cococcioni M (2024) Residual stress determination with the hole-drilling method on FDM 3D-printed precurved specimen through digital image correlation. Appl Sci 14(10):3992. https://doi.org/10.3390/app14103992

Schajer GS (2013) Practical residual stress measurement methods. Wiley, Chichester. https://doi.org/10.1002/9781118402832

Schajer GS, Whitehead PS (2018) Relaxation type residual stress measurement methods. In: Hole-drilling method for measuring residual stresses, Synthesis/SEM lectures on experimental mechanics. Springer, Cham. https://doi.org/10.1007/978-3-031-79713-2_2

Seers B, Tomlinson R, Fairclough P (2021) Residual stress in fiber reinforced thermosetting composites: a review of measurement techniques. Polym Compos 42(4):1631–1647. https://doi.org/10.1002/pc.25934

Shao Z, Zhang C, Li Y, Shen H, Zhang D, Yu X, Zhang Y (2022) A review of non-destructive evaluation (NDE) techniques for residual stress profiling of metallic components in aircraft engines. Aerospace 9(10):534. https://doi.org/10.3390/aerospace9100534

Stefanescu D, Truman CE, Smith DJ, Whitehead PS (2006) Improvements in residual stress measurement by the incremental centre hole drilling technique. Exp Mech 46(4):417–427. https://doi.org/10.1007/s11340-006-7686-8

Sun W, Ma Y, Li P, Zhang W (2025) Residual stress distribution and its effect on fatigue crack path of laser powder bed fusion Ti6Al4V alloy. Aerospace 12(2):103. https://doi.org/10.3390/aerospace12020103

Sutton MA, Orteu JJ, Schreier H (2009) Image correlation for shape, motion and deformation measurements: basic concepts, theory and applications. Springer Science & Business Media

Tabatabaeian A, Ghasemi AR, Shokrieh MM, Marzbanrad B, Baraheni M, Fotouhi M (2022) Residual stress in engineering materials: a review. Adv Eng Mater 24(3):2100786. https://doi.org/10.1002/adem.202100786

Tekkaya AE (2000) State-of-the-art of simulation of sheet metal forming. J Mater Process Technol 103(1):14–22

Totten GE, Howes MAH, Inoue T (2002) Handbook of residual stress and deformation of steel. ASM International. ISBN: 978-0-87170-729-0

Ullah I, Akinlabi ET, Songmene V (2024) Thermo-metallo-mechanical based phase transformation modeling for high-speed milling of Ti–6Al–4V through stress-strain and temperature effects. J Mater Res Technol 30:894–909. https://doi.org/10.1016/j.jmrt.2024.03.096

Ulutan D, Ozel T (2011) Machining induced surface integrity in titanium and nickel alloys: a review. Int J Mach Tools Manuf 51(3):250–280. https://doi.org/10.1016/j.ijmachtools.2010.11.003

Vanaei S, Hashemi M, Solouk A, Asghari Ilani M, Amili O, Hefzy MS, Elahinia M (2024) Manufacturing, processing, and characterization of self-expanding metallic stents: a comprehensive review. Bioengineering 11(10):983

Vrancken B (2012) Heat treatment of Ti6Al4V produced by selective laser melting: microstructure and mechanical properties. J Alloys Compd 541:177–185. https://doi.org/10.1016/j.jallcom.2012.07.022

Wang Q, Maryum P, Yuxin Liu Y (2022) An overview of techniques for measuring residual stress in metal matrix composites. J Mater Sci Chem Eng 10:59–77. https://doi.org/10.4236/msce.2022.1010005

Withers PJ, Bhadeshia HKDH (2001) Residual stress. Part 1—Measurement techniques. Mater Sci Technol 17(4):355–365. https://doi.org/10.1179/026708301101509980

Xia T, Cui P, Song T, Liu X, Liu Y, Zhu J (2025) An investigation of heat treatment residual stress of type I, II, III for 8Cr4Mo4V steel bearing ring using FEA-CPFEM-GPA method. Meta 15(5):548. https://doi.org/10.3390/met15050548

Yang T, Wei X, Zhou J, Jiang H, Liu X, Man Z (2025) The formation mechanism of residual stress in friction stir welding based on Thermo-mechanical coupled simulation. Symmetry 17(6):917. https://doi.org/10.3390/sym17060917

Ye Y, Zhang Y, Huang T, Zou S, Dong Y, Ding H, Vasudevan VK, Ye C (2023) A critical review of laser shock peening of aircraft engine components. Adv Eng Mater 25(16):2201451. https://doi.org/10.1002/adem.202201451

Zobec P, Klemenc J (2021) Application of a nonlinear kinematic-isotropic material model for the prediction of residual stress relaxation under a cyclic load. Int J Fatigue 150:106290. https://doi.org/10.1016/j.ijfatigue.2021.106290

Effect of Hard Alloy Surface Modification Process by Titanium, Carbon, and Nitrogen

Nadiia Kharchenko and Vitalii Ivanov

Abstract This chapter examines the influence of modern innovative surface modification on the composition, structure, and properties of the HG30 hard alloy. The proposed technology consists of two consecutive stages: diffusion nitriding in an ammonia environment, followed by diffusion metallization with titanium in a chlorine environment. Theoretical calculations were conducted, and practical experiments were carried out using metallographic, durometric, X-ray structural, and micro-X-ray spectral analysis methods. Thermodynamic analysis of the physicochemical conditions for the complex saturation of the HG30 hard alloy with carbon, nitrogen, and titanium enabled a theoretical approach to developing the saturation method. As a result of diffusion nitriding at a temperature of 540 °C for 36 hours in an ammonia environment, some nitrogen solubility in the cobalt binder was observed. Subsequent diffusion titanizing at 1050 °C for 2 hours formed a two-layer protective coating. The phase composition of the coating after the complex two-stage saturation was identified as TiC and TiN. The coating layers were continuous and nearly non-porous. The thickness of the carbide layer was twice that of the nitride layer, measuring 4 μm. The nitride layer adhered tightly to the hard alloy base. The combination of the nitride-carbide coating positively affected the gradual reduction of the surface layer microhardness. The gradual change in microhardness of the protective coating layers after nitrogentitanizing was observed, with values of 34.0 GPa for the outer layer and 27.2 GPa for the base layer.

Keywords Process innovation · Industrial growth · Diffusion coating · Titanium carbide · Titanium nitride · Microhardness

---

N. Kharchenko (✉)
Sumy State University, Sumy, Ukraine
e-mail: n.harchenko@pmtkm.sumdu.edu.ua

V. Ivanov
Sumy State University, Sumy, Ukraine

Technical University of Košice, Prešov, Slovak Republic

V. Ivanov (ed.), *Emerging Technologies in Manufacturing and Materials Engineering*, Springer Tracts in Mechanical Engineering,
https://doi.org/10.1007/978-3-032-14914-5_12

1 Introduction

The development and implementation of protective coatings on tool materials is a pressing issue due to the growing need to enhance wear resistance and extend the service life of products under modern production conditions. Research in protective coatings is closely aligned with several sustainable development goals defined by the United Nations, particularly Goal 9—Industry, Innovation, and Infrastructure (The Sustainable Development Goals, n.d.) and related to the sustainable asset life cycle management (Jasiulewicz-Kaczmarek et al. 2023).

All spheres of our lives are, in one way or another, closely connected with various directions of industrial development. Today, we can speak not only about the interrelation but also about the dependence of our living standards on the level of industrial development. The development of industry concerns not only the life of an individual but also that of society and the state as a whole (Ivanov et al. 2025). For an individual in modern life, industrial progress is reflected in the daily use of household items and conveniences. At the same time, progress on the scale of a state or humanity as a whole is no less critical. Above all, such development has positive economic consequences: the creation of new jobs, the emergence of a broad new social stratum with higher levels of education, skills, and competencies, and improvements in the safety and environmental sustainability of production.

Special attention should be paid to producing tool materials and technologies that enhance their performance characteristics. The development and implementation of new methods and technologies for material processing, particularly protective coating technologies, contribute to the creation of innovative materials that improve both efficiency and durability of products. In this way, urgent technical challenges are addressed, and a significant contribution is made toward achieving global objectives.

Protective coatings ensure resistance to frictional contact, exposure to aggressive external environments, and other damaging factors. As a result, they can enhance a wide range of product properties, including wear, corrosion, and heat resistance.

Surface treatment processes help optimize the structure of the surface layers of base materials, leading to improved mechanical properties of products. Applying modern technologies and analytical methods provides a better understanding of the relationship between coating structure and its performance characteristics, opening up new opportunities for creating advanced, innovative materials.

2 Fundamentals and Challenges

The study of the processes of applying protective coatings made of carbides and carbonitrides of transition metals necessitates the use of a complex of modern, highly sensitive research methods. A comprehensive analysis will allow a thorough analysis of the phenomena that occur due to the application of coatings.

Metallographic, durometric, X-ray structural, and spectral analysis methods are used.

Among the existing materials for tool manufacturing (steels, hard tool alloys, ceramics, etc.) (Dvornik and Zaitsev 2018; Rizzo et al. 2020; Mills 1996), the optimal combination of availability, high productivity, environmental friendliness, and cost-effectiveness is occupied by hard tool alloys. Several methods can improve the properties and characteristics of hard tool alloys. In particular, some works address the issue of the potential of microstructural design of the base and surface of hard tool alloys (Dvornik and Zaitsev 2018; Roebuck et al. 2001). Thus, hard tool alloys with ultrafine-grained and extra fine-grained grains have a more balanced combination of hardness and toughness than alloys with standard and large grain sizes (Dvornik and Zaitsev 2018). The doping of the cobalt bond with elements (Re and Ru) can also be attributed to the microdesign of the base, which will ensure an increase in the heat resistance of the entire alloy (Dvornik and Zaitsev 2018).

The development of universal grades of hard alloys with protective coatings will significantly improve the operational properties of the tool. The most common methods of applying protective layers and coatings are chemical vapor deposition (CVD) (Rizzo et al. 2020; Sun et al. 2021), physical vapor deposition (PVD) (Rizzo et al. 2020; Mills 1996; Glushkova 2023), electrospark alloying (ESA) (Haponova et al. 2024), various high-energy saturation methods, and diffusion chemical-thermal treatment (pack cementation) (Mévrel et al. 1986). Each of the methods has its advantages and disadvantages (Rizzo et al. 2020). One of the ways to eliminate the disadvantages is the use of combined saturation methods. Thus, in work (Godse and Santhanam 1996), it is proposed that the disadvantages of the CVD method be compensated for by adding the PVD method. Multilayer coatings of titanium and aluminium compounds obtained using this technology will prevent a decrease in the strength of complex alloy tools. The unavailability of a transition zone is the main disadvantage of single-layer protective coatings obtained by PVD methods compared to pack cementation methods. The subsequent diffusion saturation of such coatings leads to the formation of a transition zone (Loskutova et al. 2022). The formation of a transition zone provides better adhesion to the base. The combination of diffusion titanizing with the preliminary application of a titanium nitride layer by the PVD method is recommended as a wear-resistant protective coating on tool materials. In the case of the formation of a multilayer diffusion coating, in addition to increasing adhesion to the base, diffusion saturation leads to a gradual change in properties. A multilayer protective coating can provide a gradient of properties in the volume of the geometric body of the tool. Multilayer coatings obtained using diamond as a base layer with protective CVD layers of TiC and Al_2O_3 provide high adhesion under voltage and can be recommended as wear-resistant (Koepf et al. 2002). The combination of flame spraying and arc-heating spraying of coatings and their electrical contact treatment (Lopata et al. 2020) also contributes to an increase in the degree of adhesion of the coating to the base.

The formation of multilayer coatings using various methods and technologies of saturation leads to an increase in the complex properties. Thus, in works (Degula et al. 2015; Dehula et al. 2024), it is proposed that the saturation process be carried out with two carbide-forming elements (titanium and chromium) in one technological cycle. The resulting carbide multilayer coating has high adhesion to the base. Multilayer coatings can be carbide and contain nitride layers (Pavlenko et al. 2021).

Given the existing practical experience, carbides and nitrides of transition metals of groups IV-VI of the periodic system are highly effective in protective coatings for hard tool alloys. In particular, coatings based on titanium carbide and nitride (TiC, TiN), as well as aluminium oxide (Al_2O_3) (Šnajdar et al. 2024; Ge et al. 2024), attract special attention. Such advantages characterize these materials as high microhardness, low coefficient of friction with steel, and significant barrier properties of multilayer compositions based on TiC and TiN.

A promising and little-studied direction in pack cementation is the creation of coatings on the surface of hard tool alloys through two- or three-component saturation. It is predicted that such coatings will have better properties than single-layer ones, combining and enhancing their positive characteristics.

In the case of using two- or three-component saturation with elements that have a high ability to carbide formation (for example, using Zr and Ti; Ti and V; Ti, V, and Cr, etc.), carbide layers doped with additional elements are formed (Dabees et al. 2022; Lande 2024).

The main advantage of such coatings is a significant increase in the service life, which depends on the coating type, the base material's composition, and operating conditions.

3 Ideas and Solutions

The work implements an innovative technology for the surface modification of tool materials. The comprehensive technology consists of two consecutive stages: nitriding and subsequent contact diffusion titanizing.

VK8 hard alloy (HG30 standard DIN, %, wt: 92 (WC) + 8 (Co)) was chosen as the base material. This alloy is inexpensive and has found wide application in industry.

Ammonia was used as the starting reagent for nitriding; for diffusion metallization (Table 1), titanium powder with a particle size of 100–300 μm, charcoal, and carbon tetrachloride (CCl_4) were used. In this case, carbon tetrachloride serves two functions. First, upon dissociation, CCl_4 forms titanium chlorides of varying valence, which aids in the mass transfer of titanium to the treated surface of the

Table 1 Rational composition of starting reagents and technological parameters of the titanizing process

Composition of the saturating mixture	Quantity CCl_4, mL/m^2	Pressure, mmHg	Temperature, °C	Duration, h
50 g/m^2 Ti, 15 g/m^2 charcoal	4–5	10^{-1}	1050	2

samples. Second, CCl_4, like charcoal, acts as an additional carbon carrier, which helps prevent the formation of a decarburized zone beneath the coating.

Nitriding was carried out in production conditions using known technology in a shaft furnace in an ammonia environment with a degree of dissociation of 25–35% at a temperature of 540 °C for 36 hours. Diffusion titration was carried out in a closed reaction space under reduced gas phase pressure of the gas phase in a particular installation based on a shaft furnace «SShOL 1.1.6/12». The isothermal holding time when applying a protective coating to the surface of HG30 hard alloys was 2 hours, and the temperature was −1050 °C.

A thermodynamic approach was used to analyze the physicochemical conditions of diffusion saturation processes. This approach theoretically allows the equilibrium composition of an isolated system to be determined (Loskutova et al. 2024). During the diffusion saturation processes, the equilibrium in the closed system is somewhat different from the real one. At the same time, the analysis of literature sources (Siegle 1984; Sereda et al. 2020; Naji et al. 2015; Zhang et al. 2024; Kattner and Campbell 2009; Sun 2024) showed the feasibility and correctness of the proposed approach in various branches of metallurgy. For a specific equilibrium system, the predicted phase composition of the coating will be determined by the hypothetical composition of the condensed state of the system. Based on thermodynamic modelling, it is possible to determine the main parameters of the pack cementation process of the surface of parts to obtain the required chemical composition of the coating.

The calculation method determines the self-regenerating composition of the gas mixture in the installation, the chemical composition of the powders of the diffusing elements, their weight, and the number of halogens. The calculations specify the temperature and total pressure and use reference data on the reactants' thermodynamic characteristics and the coating elements' diffusion characteristics (Sun 2024).

The Astra.4 software package (Outotec Oy, Pori, Finland), which includes a thermodynamic database, was employed to calculate chemical reactions under specific process conditions and determine the systems' equilibrium composition. The program enables the determination of the equilibrium state of a closed system corresponding to maximum entropy. The equilibrium was estimated in the temperature range of 250–1150 °C for systems containing titanium, chlorine, carbon, cobalt, tungsten, and nitrogen—as starting components and oxygen and nitrogen—as air components in different proportions. The pressure in the system was considered constant throughout the time. The possibility of more than 500 substances in the reaction space in gaseous and condensed media was considered when determining the equilibrium composition. The database of the «Astra. 4» program is based on reference literature on the thermodynamics of individual substances and therefore includes a significant number of inorganic compounds. The program allows for a series of calculations, considering all possible binary phase diagrams and enabling the addition of new compounds (not included in the standard configuration) to the database.

Thus, the results of the thermodynamic analysis of closed systems will allow us to correctly approach the selection of technological parameters of the process and the probable phase composition of the surface of nitrided alloys after titanizing in a chlorine environment.

The phase composition of the coatings investigated in this study was analyzed using a Rigaku Ultima IV diffractometer with Cu Kα radiation ($\lambda = 0.1541841$ nm). Measurements were carried out in the Bragg–Brentano geometry, enabling the recording of diffraction lines within the 2θ range of 20–90° with a step size of $\Delta 2\theta = 0.04°$. The operating conditions were set to 40 kV tube voltage and 40 mA current, with an exposure time of 2 s per point. Phase identification and lattice parameter determination were performed using the PowderCell 2.4 software package based on the Rietveld refinement method (full-profile analysis). The obtained diffraction patterns were compared with reference patterns from the ICDD PDF-2 database.

The images of the structures and the corresponding micro-X-ray spectra were taken on a scanning electron microscope REMMA-101A (JSC «Selmi», Sumy, Ukraine). Qualitative and quantitative spectral analysis of samples with carbonitride coatings was performed along the thickness of the diffusion layers on a scanning electron microscope «JSM-6490LV» (Tokyo Boeki Ltd., Tokyo, Japan). Metallographic analysis is carried out by the microscope «Neophot-21» (Carl Zeiss Jena GmbH, Oberkochen, Germany). Microhardness and coating thickness are determined according to the standard ASTM E92-82(2003) (ASTM E92-82(2003) 2023) by Vickers hardness tester LHVS-1000Z (Chongqing Leeb Instrument Co., Ltd., Chongqing, China). An indenter is a quadrangular diamond pyramid with an angle at the top of 136°. The characteristics of microhardness are determined as follows, kgf/mm^2 (Loskutova et al. 2009):

$$H = 1{,}854\frac{P}{D^2} \qquad (1)$$

where P = load, and D—imprint diagonal, μm.

Measurements of the thickness and microhardness of the diffusion layers were performed in at least 15–20 view areas. The measurement error is 0.1 GPa.

The composition of the theoretically determined gas and condensed phases of some systems involving Cl, N, C, Co, W, Ti, and O is given in Table 2. Evaluation of the condensed state will provide an opportunity to find out the expected phase composition of the coating.

Table 2 Equilibrium phase composition of systems during diffusion nitrogentitanizing of HG30 hard alloy (T = 250–1150 °C, P = 10^2 Pa)

№	Components of the system/ quantity, mole	Composition of the reaction space	
		gas phase	condensed phase
1	Cl-N-C-Co-W-Ti = 2-0,5-2-0,1-0,2-2	Cl, N_2, CoCl, $CoCl_2$, $TiCl_2$, $TiCl_3$, $TiCl_4$.	Co, WC, $TiCl_3$, TiN, TiC.
2	O-Cl-N-C-Co-W-Ti = 1-2-0,5-2-0,1-0,2-2	Cl, N_2, CO, CO_2, CoCl, $CoCl_2$, $TiCl_2$, $TiCl_3$, $TiCl_4$, $TiOCl_2$.	Co, WC, Ti_4O_7, $TiCl_3$, TiN, TiC.

Analysis of the obtained data shows that the composition of the gas and condensed phases is determined by the closed system's initial components and the process's temperature (Figs. 1 and 2).

For the system Cl-N-C-Co-W-Ti = 2-0.5-2-0.1-0.2-2 (Table 2, Fig. 1). in the composition of the gas phase, the formation of titanium and cobalt chlorides of different stoichiometry is characteristic. In addition, the release of molecular nitrogen and atomic chlorine is observed. At the same time, the maximum partial pressure is observed in titanium chlorides $TiCl_4$ and $TiCl_3$. At temperatures above 550 °C, the formation of cobalt chlorides is observed, which may indicate the occurrence of

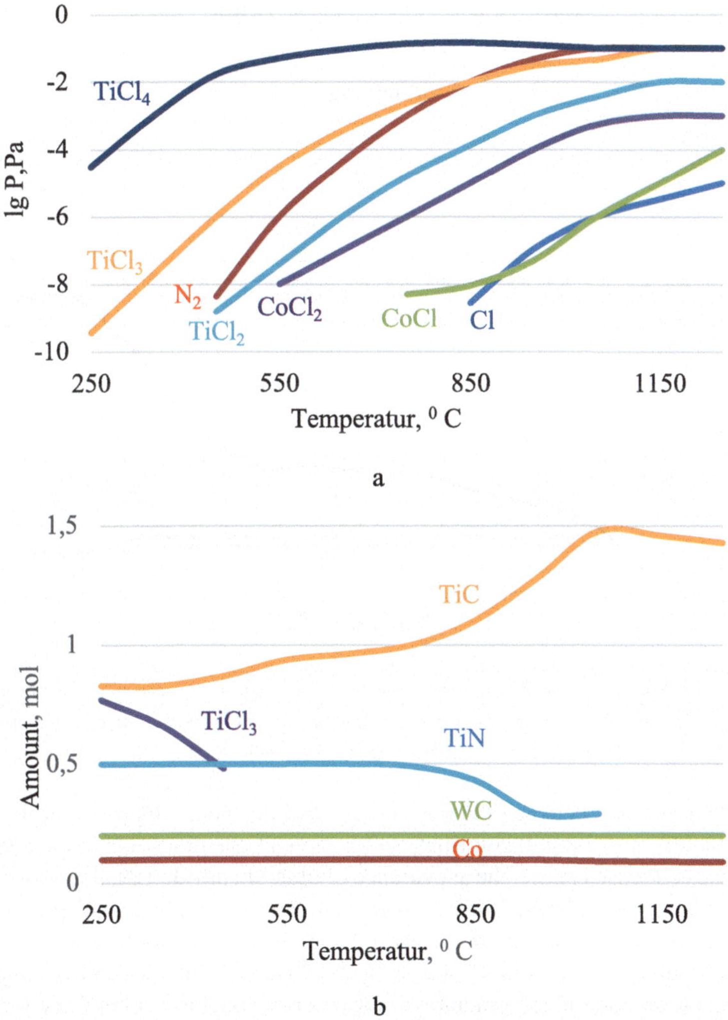

Fig. 1 Equilibrium phase composition (gas compounds (**a**) and condensed phases (**b**)) of the system Cl-N-C-Co-W-Ti = 2-0,5-2-0,1-0,2-2 (system 1, Table 2)

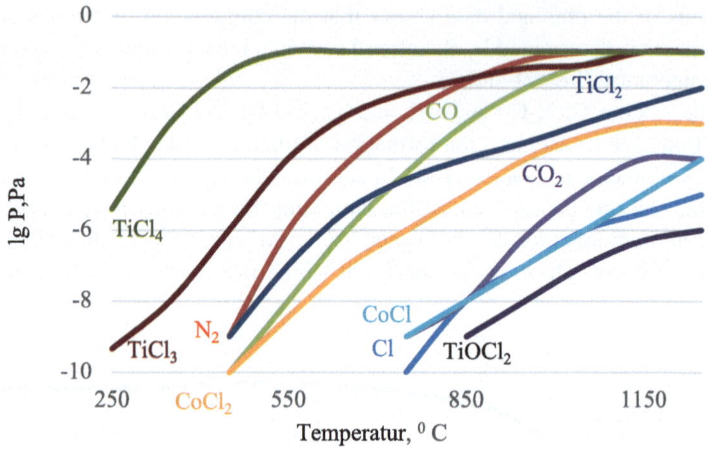

a

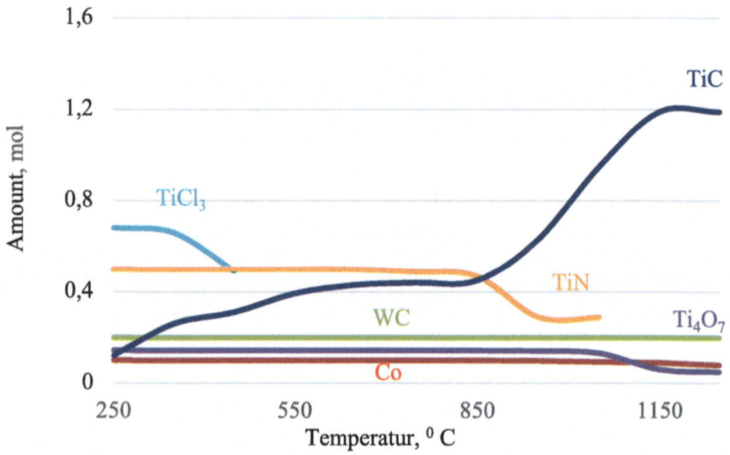

b

Fig. 2 Equilibrium phase composition (gas compounds (**a**) and condensed phases (**b**)) of the system O-Cl-N-C-Co-W-Ti = 1-2-0,5-2-0,1-0,2-2 (system 2, Table 2)

exchange processes between the base matrix and the saturating medium. It is predicted that this process will contribute to forming a transition zone. In the composition of the condensed phase, the appearance of titanium nitride TiN, titanium carbide TiC, and titanium chloride $TiCl_3$ is recorded. The latter phase will probably be deposited on the surface of the retort. Changing the saturation temperature can change the quantitative ratio of phases in the coating. Thus, with increasing temperature, the amount of TiC gradually increases from 0.83 to 1.48 mol at a temperature of 1050 °C. The formation of TiN in the condensed phase is observed up to a temperature of 1050 °C, while its amount does not exceed 0.5 mol. Starting from a temperature of 750 °C, the amount of TiN decreases slightly. Therefore, the process temperature should not exceed 1000–1050 °C for the simultaneous formation of titanium carbide and titanium nitride.

When oxygen is added to the system in the gas phase, carbon and chlorine oxides of different valences are formed (Table 2. Figure 2.). The addition of oxygen to the system composition has almost no effect on the partial pressures of titanium and cobalt chlorides but leads to the formation of titanium oxychloride ($TiClO_2$) and carbon oxides (CO, CO_2) in the gaseous state. The condensed phase was also found to contain titanium oxide Ti_4O_7. In addition, a decrease in titanium carbide TiC is noted for system 2 in the condensed state compared to system 1 without oxygen. The maximum value of the titanium carbide content in the system with oxygen does not exceed 1.19 mol; the amount of titanium nitride does not change.

Thus, the addition of oxygen to the composition is undesirable. Oxygen in the system leads to a decrease in titanium carbide and to the additional formation of titanium oxide Ti_4O_7. Titanium oxide can be located as individual inclusions in the coating and lead to embrittlement. Oxygen removal should be provided during saturation, which can be achieved by preliminary degassing of the working space.

Diffusion nitriding in an ammonia environment is one of the most common methods of saturating the surface of alloys with nitrogen. As a result of surface nitriding, the mechanical properties of tool materials increase. Today, the issue of not only studying the properties and characteristics of alloys (Siegle 1984; Smirnov et al. 2020; Yuda et al. 2024; Cruz et al. 2022; Aktaş Çelik et al. 2023; González-Pociño et al. 2024) but also the issue of modelling and controlling the nitriding process is relevant. Many works are devoted to obtaining products with a specific structure of nitrided layers and specified properties (Michalski et al. 2005; Cavaliere et al. 2009; Spies et al. 2001).

The results of the X-ray structural analysis of the HG30 tool hard alloy after nitriding (Fig. 3) coincide with the literature data (Janisch et al. 2013) and indicate that nitride phases are not formed in the surface layer. At the same time, the phase composition of the HG30 hard alloy surface after nitriding does not differ from the composition of the surface of untreated samples and consists of tungsten carbide WC and Co. Analysis of the obtained data indicates that nitrogen does not dissolve in tungsten carbides under these saturation conditions. At the same time, it is known that cobalt, dissolving in a small amount of nitrogen, forms nitride Co_3N and Co_2N (Massalski et al. 1990). A specific solubility of nitrogen in cobalt can be indirectly confirmed by comparing the sizes of its crystal lattices before and after nitriding. Thus, according to data (Nishizawa and Ishida 1983), the crystal lattice parameter of cobalt that was not subjected to treatment is a = 0.251 nm, c = 0.409 nm. In turn, the corresponding parameters of cobalt HG30 hard alloy after nitriding in an ammonia environment at a temperature of 540 °C for 36 hours are a = 0.26040 nm and c = 0.67561 nm. The lattice parameters of cobalt increased slightly after nitriding, which can be explained by the low solubility of nitrogen in it.

X-ray structural and metallographic surface analysis methods of the hard alloy HG30 after nitriding did not reveal a diffusion layer. In turn, micro-X-ray spectral

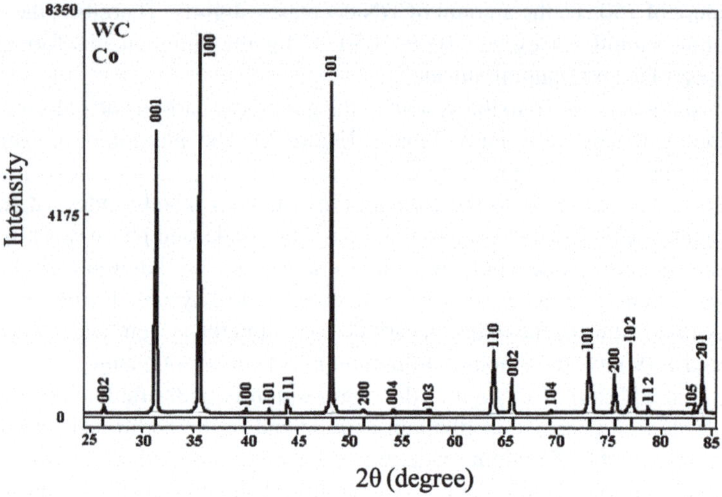

Fig. 3 X-ray diffraction pattern of HG30 obtained by nitriding (T = 540 °C, τ = 36 h)

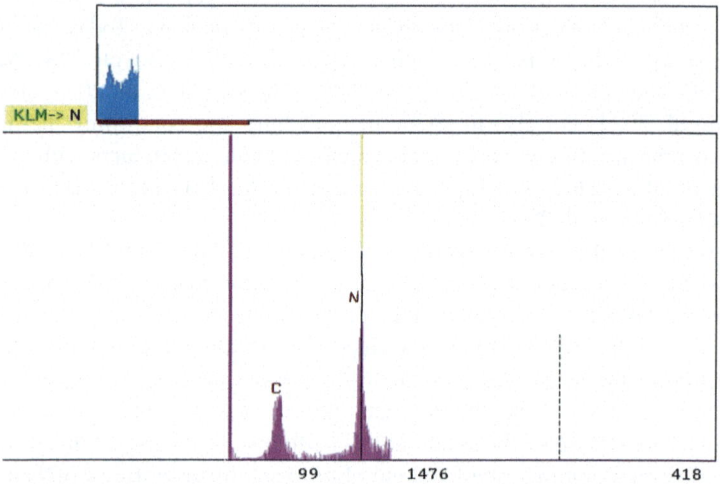

Fig. 4 X-ray spectra of the surface of HG30 after nitriding (T = 540 °C, τ = 36 h)

analysis of nitrided samples of hard tool alloys revealed a small amount of nitrogen on the surface (Fig. 4).

In this work, nitriding of the HG30 hard alloy was implemented as a preparatory stage for the subsequent production of nitride compounds and layers based on them. It is assumed that nitrogen dissolved in the cobalt bond of the hard alloy as a result of nitriding will be released upon further heating in a chlorine environment and form a nitride of the saturating component. Transition metal nitrides have a unique

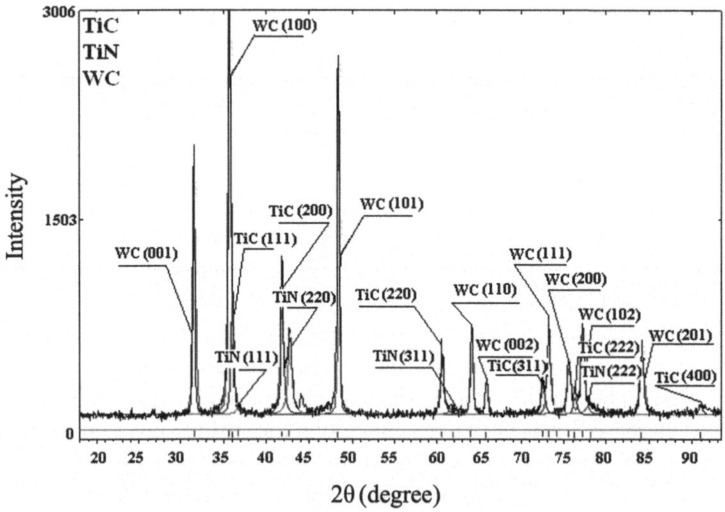

Fig. 5 X-ray diffraction pattern of HG30 obtained by nitrogentitanizing: nitriding (T = 540 °C, τ = 36 h); titanizing (T = 1050 °C, τ = 2 h)

Table 3 Results of X-ray diffraction analysis of HG30 after pack cementation

Processing type, temperature, °C saturation time, hours	Phase composition of the surface	Lattice period, nm	
		a	c
Nitriding, 540; 36	WC	0.2906	0.2836
	Co	0.2604	0.6756
Nitrogentitanizing	TiC	0.4325	–
Nitriding 540; 36	TiN	0.4226	–
Titanizing 1050; 2	WC	0.2904	0.2837

set of properties. The use of Ti nitride is promising in the production of tools, machine parts, and equipment.

X-ray structural analysis (Fig. 5 and Table 3) of the HG30 hard alloy determined the formation of titanium nitride TIN and titanium carbide TIC.

According to known ideas, titanium carbide's lattice period can vary according to the carbon content within wide limits. Thus, for the compound $TiC_{0,3}$ the lattice period can be 0.4295 nm, and for the compound $TiC_{0,85}$ is 0.4332 nm (Pierson 1996). It should be noted that the lattice period of the titanium carbide phase formed in the coatings depends on the production method and the base material's chemical composition (Pierson 1996). It can be assumed that the TiC formed after nitrogentitanizing is described by the formula $TiC_{0,7}$. The difference in the crystal lattice periods is due to the conditions of their formation. As a result, the crystal lattice period of TiC in the coating after nitrogentitanizing is found to be smaller than that in the titanizing coating (Khyzhnyak et al. 2018). The titanium nitride TiN layer formed first during the nitrogentitanizing coating of the HG30 inhibits the diffusion of carbon to the surface.

X-ray spectral analysis determined carbon, titanium, tungsten, iron, and cobalt on the surface of the HG30 hard alloy after nitrogentitanizing (Fig. 6 and Table 4).

A significant amount of Co on the surface of HG30 after nitrogentitanizing is due to the formation of the CoTi phase. In this case, the Co of the base diffuses to the surface, and thus, the probability of the formation of the Co_3W_3C or Co_6W_6C phases decreases. Furthermore, it is known that increasing the isothermal holding time during titanizing for more than 2 hours leads to the formation of a thin decarburized zone based on the Co_6W_6C carbide (Khyzhnyak et al. 2018).

Metallographic analysis has established that, as a result of nitrogentitanizing, a two-layer coating is formed on the surface of the hard alloy HG30: titanium carbide TiC is located on the outside, under which a layer of titanium nitride TiN is directly adjacent to the base. It has been determined that the carbide layer on the surface of the hard alloy HG30 appears as a white stripe that is not etched in a 3–4% solution of nitric acid in alcohol. This layer is homogeneous in structure (Fig. 7).

The nitride layer, even without etching, exhibits a characteristic yellowish-golden color. This layer was identified as titanium nitride (TiN) using a combination of X-ray structural, spectral, and metallographic analysis methods. Titanium nitride with a near-stoichiometric composition exhibits a characteristic yellow-golden color. In work (Vaz et al. 2005), it was stated that the color of titanium nitride changes with increasing nitrogen content from grey (typical of a high titanium content) to light yellow (typical of titanium nitride Ti_2N), golden (TiN), brown-bronze,

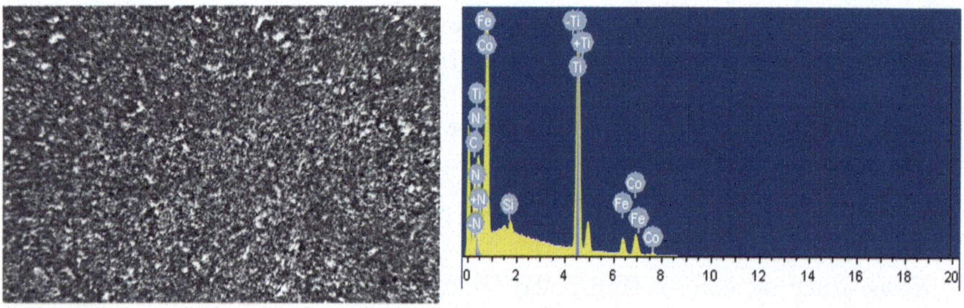

a b

Fig. 6 Microstructure, ×100 (**a**) and X-ray spectra (**b**) of the surface of HG30 after nitrogentitanizing: nitriding (T = 540 °C, τ = 36 h); titanizing (T = 1050 °C, τ = 2 h)

Table 4 Distribution of chemical elements of HG30 after pack cementation

	Content of elements, %wt.				
Processing type, temperature, °C; saturation time, hours	C	N	Ti	Fe*	Co
Nitrogentitanizing	1.2	20.3	70.0	0.5	8.0
Nitriding 540; 36					
Titanizing 1050; 2					

*The source of Fe is the retort material (12Cr18Ni10Ti)

Fig. 7 Microstructure of HG30 after nitrogentitanizing: nitriding (T = 540 °C, τ = 36 h); titanizing (T = 1050 °C, τ = 2 h), ×200

and even red (in the presence of oxygen). Therefore, the color can be used as an indirect indicator for classifying experimentally obtained titanium nitride (TiN) based on its stoichiometric phase composition (Leclair 1998). The approximate stoichiometric composition of titanium nitride, TiN, can be established by referring to the results of the X-ray structural analysis of the HG30 alloy after nitrogentitanizing. Cubic titanium nitride has a wide range of homogeneity from $TiN_{0.38}$ to $TiN_{1.2}$ (Lyakishev 1997; Zhurakovskii and Dzeganovskii 1965). The lattice parameter of titanium nitride obtained after nitrogentitanizing at a temperature of 1050 °C for 2 h on the HG30 surface is 0.4226 nm, which is slightly smaller than the lattice parameter of the corresponding nitride of stoichiometric composition (0.4260 nm) (Gorel'ik et al. 1994). Based on this, it can be concluded that the structure of titanium nitride on HG30 has a pronounced lack of nitrogen.

It is known that the high microhardness of carbides of transition metals of groups IV-VI is due, first of all, to the high bond strength between metal atoms and, secondly, to the wedging of slip planes in the lattice of penetrating atoms by carbon atoms (Ivan'ko 1974). Many factors influence the microhardness value. Coatings obtained by PVD, CVD, or diffusion methods on the same base have different microhardness values. Thus, according to (Loskutov et al. 1991), the microhardness of TiC can vary within 15.6–49.0 GPa. The latter value is significantly higher than the microhardness of titanium carbide (TiC) obtained through powder metallurgy methods, which is 28.5 GPa (Sarkar 2016). The authors of (Loskutov et al. 1991) explain the wide range of microhardness values of carbide layers by the presence of substitutional and penetrating impurities in the carbide phases, the peculiarities of the structural state within the homogeneity region, and the result of surface deformation during the manufacture of the samples.

The thickness of the protective layers on the surface of hard alloys after complex processing (Table 5) is somewhat less than on steels (Minkevich et al. 1979). This can be explained by the lower carbon content in the hard alloy's base and nitrogen after nitriding. As is known, the carbon of the base takes an active part in the formation of coatings. Under such conditions, during the metallization of hard alloys, the

Table 5 Results of X-ray diffraction and durometric analyses of HG30 hard alloy after nitrogentitanizing: nitriding (T = 540 °C, τ = 36 h); titanizing (T = 1050 °C, τ = 2 h)

Phase	Thickness, μm	Microhardness, GPa
TiC	4.0	34.0
TiN	2.0	27.2
WC	–	–

main contribution to the process of forming the carbide phase is made by carbon from charcoal and carbon tetrachloride.

It is known (Loskutov et al. 1991) that, in any case, the optimal total thickness of the protective coating on hard alloys does not exceed 15.0 μm. Only under these conditions can the inherent high brittleness of ultra-hard materials and coating delamination be avoided. Thus, the protective coating obtained from nitrogentitanizing of the HG30 is optimal: the thickness of the outer carbide layer is 4.0 μm, while the thickness of the inner coating layer is twice as small and amounts to 2.0 μm (Table 5).

The reduced carbon content in titanium carbide (TiC) formed on the HG30 hard alloy surface affects the coating's microhardness. The microhardness of titanium carbide (TiC) on the HG30 hard alloy surface was found to be 34.0 GPa (Table 5). In turn, the microhardness of the titanium nitride (TiN) layer on the surface of the hard alloy HG30 is 27.2 GPa (Table 5). The obtained values are slightly lower than those of the corresponding carbide and nitride on the surface of steels (Minkevich et al. 1979; Hizhnyak et al. 2009).

4 Conclusions

The paper presents a modern, innovative technology for modifying the surface of HG30 hard alloy. The technology combines two diffusion processes: nitriding in an ammonia environment and subsequent diffusion saturation with titanium in a chlorine environment.

Investigation of the physicochemical conditions for the nitrogentitanizing process in equilibrium phase compositions of closed systems involving titanium, carbon, nitrogen, chlorine, and oxygen in the temperature range of 250–1150 °C at a pressure of 1×10^2 Pa revealed that the primary condensed phases are TiN and TiC, with their quantities varying depending on the temperature. As the temperature increases, the probability of the existence of the TiN nitride phase decreases, while the TiC carbide phase increases.

Nitriding of the HG30 alloy (T = 540 °C, τ = 36 hours in an ammonia atmosphere with a dissociation level of 25–35%) does not form a nitrided coating on the surface. However, some nitrogen dissolution in the cobalt binder is observed.

During nitrogentitanizing of HG30 (T = 1050 °C, $\tau = 2$ hours), a two-layer coating is formed. The TiC carbide layer forms on the outer side of the diffusion coating, while the TiN nitride layer is directly adjacent to the base.

The total thickness of the resulting coatings is 6 μm. The outer TiC-based layer is 4 μm thick, appearing as a white stripe with a homogeneous structure. The inner TiN-based layer, with a characteristic yellowish-golden color, has a thickness of 2 μm. The microhardness of titanium carbide (TiC) was found to be 34.0 GPa, while the microhardness of the TiN nitride layer is 27.2 GPa.

Acknowledgments This research was partially supported by the International Association for Technological Development and Innovations. The research was funded by the EU NextGenerationEU through the Recovery and Resilience Plan for Slovakia under project no. 09I03-03-V01-00094. This work was supported by the projects VEGA 1/0453/24, KEGA 012TUKE-4/2024 granted by the Ministry of Education, Research, Development and Youth of the Slovak Republic. This work was supported by the Slovak Research and Development Agency under the Contract no. VV-MVP-24-0190.

References

Aktaş Çelik G, Atapek Ş, Polat Ş, Obrosov A, Weiß S (2023) Nitriding effect on the tribological performance of CrN-, AlTiN-, and CrN/AlTiN-coated DIN 1.2367 hot work tool steel. Materials 16(7):2804. https://doi.org/10.3390/ma16072804

ASTM E92-82(2003) (2023) Standard test method for Vickers hardness of metallic materials. ASTM International, West Conshohocken, PA

Cavaliere P, Zavarise G, Perillo M (2009) Modeling of the carburizing and nitriding processes. Comput Mater Sci 46(1):26–35. https://doi.org/10.1016/j.commatsci.2009.01.024

Cruz J, Corregidor V, Nunes B, Alves L, Colaço R, Alves E (2022) Surface characterization and performance evaluation of nitrogen implanted coinage dies. Processes 10(3):479. https://doi.org/10.3390/pr10030479

Dabees S, Mirzaei S, Kaspar P, Holcman V, Sobola D (2022) Characterization and evaluation of engineered coating techniques for different cutting tools–review. Materials 15(16):5633. https://doi.org/10.3390/ma15165633

Degula A, Govorun T, Kharchenko N, Khyzhnyak V, Karpets M, Myslyvchenko O, Smetanin R (2015) Investigation of the phase and chemical compositions of complex carbide coatings. Metallophys New Mater 37(11):1461–1476. https://doi.org/10.15407/mfint.37.11.1461

Dehula A, Kharchenko N, Hovorun T, Sytnikov V, Khaniukov K, Varakin B (2024) Structure and properties of multilayer coatings obtained by chromotitanizing. J Nano- Electron Phys 16(4):04031-1–04031-4. https://doi.org/10.21272/jnep.16(4).04031

Dvornik M, Zaitsev A (2018) Variation in strength, hardness, and fracture toughness in transition from medium-grained to ultrafine hard alloy. Russ J Non-ferrous Met 59:563–569. https://doi.org/10.3103/S1067821218050024

Ge F, Xia Z, Yuan H, Guo S, Hu Z, Guan J, Cai J, Guan Q, Lyu P (2024) Microstructure and properties of TiN/TiCN/Al_2O_3/TiN coating enhanced by high-current pulsed electron beam. Coatings 14(4):378. https://doi.org/10.3390/coatings14040378

Glushkova D (2023) Research on new plasma deposition regimes. Bull Kharkiv Nat Automob Highway Univ 1(103):58–58. https://doi.org/10.30977/BUL.2219-5548.2023.103.0.58

Godse R, Santhanam A (1996) Composite CVD + PVD coatings. Mater Sci Eng A 209(1–2):384–388. https://doi.org/10.1016/0921-5093(95)10138-1

González-Pociño A, Alvarez-Antolin F, Peral-Martinez L (2024) Effects of nitriding and thermal processing on wear and corrosion resistance of vanadis 8 steel. Coatings 14(8):1066. https://doi.org/10.3390/coatings14081066

Gorel'ik S, Skakov Y, Rastorguev L (1994) X-ray and electron-optical analysis. MISA

Haponova O, Tarelnyk V, Marchenko S, Tarelnyk N, Konoplianchenko I (2024) The development of nanostructuring method metal surfaces by electrospark alloying. In: Nanocomposite and nanocrystalline materials and coatings: microstructure, properties and applications. Springer Nature, Singapore, pp 181–199. https://doi.org/10.1007/978-981-97-2667-7_7

Hizhnyak V, Kurilo N, Kostenko D, Rudenko P (2009) Formation and wear resistance of titanium carbonitride coatings on steels. Bull SumDU Tech Sci (4):188–194. http://essuir.sumdu.edu.ua/handle/123456789/720

Ivan'ko A (1974) The microhardness of transition-metal carbides. In: Samsonov GV (ed) Refractory carbides. Studies in soviet science. Springer, New York. https://doi.org/10.1007/978-1-4684-8598-1_44

Ivanov V, Evtuhov A, Dehtiarov I, Trojanowska J (2025) Fundamentals of manufacturing engineering using digital visualization, Springer Tracts in Mechanical Engineering. Springer, Cham. https://doi.org/10.1007/978-3-031-74360-3

Janisch D, Lengauer W, Eder A, Dreyer K, Rödiger K, Daub H, Van Den Berg H (2013) Nitridation sintering of WC–Ti (C, N)–(Ta, Nb) C–Co hardmetals. Int J Refract Met Hard Mater 36:22–30. https://doi.org/10.1016/j.ijrmhm.2011.12.013

Jasiulewicz-Kaczmarek M, Antosz K, Zhang C, Ivanov V (2023) Industry 4.0 technologies for sustainable asset life cycle management. Sustainability 15(7):5833. https://doi.org/10.3390/su15075833

Kattner U, Campbell C (2009) Invited review: modelling of thermodynamics and diffusion in multicomponent systems. Mater Sci Technol 25(4):443–459. https://doi.org/10.1179/174328408X372001

Khyzhnyak V, Loskutova T, Kalashnikov T et al (2018) Producing multilayer coatings from the gas phase with the participation of TiC and TiN compounds on the hard alloy VK8. J Superhard Mater 40:170–178. https://doi.org/10.3103/S1063457618030036

Koepf A, Haubner R, Lux B (2002) Multilayer coatings containing diamond and other hard materials on hardmetal substrates. Int J Refract Met Hard Mater 20(2):107–113. https://doi.org/10.1016/S0263-4368(02)00008-2

Lande D (2024) Tool coating development: A historical overview. Int J Adv Res Sci Commun Technol 4(6):155–161. https://doi.org/10.48175/IJARSCT-17624

Leclair P (1998) Titanium nitride thin films by the electron shower process. Massachusetts Institute of Technology

Lopata O, Lopata V, Solovykh Y, Posviatenko N (2020) Increasing the wear resistance of steel gas-thermal coatings by electrocontact treatment. Probl Friction Wear 3(104):4–14. https://doi.org/10.18372/0370-2197.3(104).18973

Loskutov V, Khizhnyak V, Kunits'kyi Y, Kindrachuk M (1991) Diffusion carbide coatings. Tekhnika, K

Loskutova T, Sigova V, Khizhnyak V, Loskutov V (2009) Complex carbide coatings on metals and alloys. Sumy State University, Sumy

Loskutova T, Hatala M, Pogrebova I, Nikitina N, Bobina M, Radchenko S, Kharchenko N, Kotlyar S, Pavlenko I, Ivanov V (2022) Composition, structure, and properties of Ti, Al, Cr, N, C multilayer coatings on AISI W1-7 alloyed tool steel. Coatings 12:616. https://doi.org/10.3390/coatings12050616

Loskutova T, Scheffler M, Ivanov V, Pohrebova I, Kononenko Y, Bobina M, Kharchenko N, Bartoszuk M, Pavlenko I (2024) Physicochemical conditions of boron–siliconizing of molybdenum-based alloys in chlorine and fluorine medium. Metals 14:302. https://doi.org/10.3390/met14030302

Lyakishev N (1997) Phase diagrams of binary metal system. Metallurgiya, Moscow

Massalski T, Okamoto H, Subramanian P, Kacprzak L (1990) Binary alloy phase diagrams. ASM International. https://doi.org/10.31399/asm.hb.v03.9781627081634

Mévrel R, Duret C, Pichoir R (1986) Pack cementation processes. Mater Sci Technol 2(3):201–206. https://doi.org/10.1179/mst.1986.2.3.201

Michalski J, Tacikowski J, Wach P et al (2005) Formation of single-phase layer of γ′-nitride in controlled gas nitriding. Metal Sci Heat Treat 47:516–519. https://doi.org/10.1007/s11041-006-0023-0

Mills B (1996) Recent developments in cutting tool materials. J Mater Process Technol 56:16–23. https://doi.org/10.1016/0924-0136(95)01816-6

Minkevich A, Zakharov B, Toné É et al (1979) Titanium carbide coatings on steels and hard alloys. Met Sci Heat Treat 21:453–458. https://doi.org/10.1007/BF00780483

Naji A, Galetz M, Schütze M (2015) Improvements in the thermodynamic and kinetic considerations on the coating design for diffusion coatings formed via pack cementation. Schütze Mater Corros 66:863–868. https://doi.org/10.1002/maco.201407810

Nishizawa T, Ishida K (1983) The Co (Cobalt) system. Bull Alloy Phase Diagr 4:387–390. https://doi.org/10.1007/BF02868089

Pavlenko I, Zajac J, Kharchenko N, Duplák J, Ivanov V, Kostyk K (2021) Estimation of wear resistance for multilayer coatings obtained by nitrogenchroming. Metals 11:1153. https://doi.org/10.3390/met11081153

Pierson HO (1996) Handbook of refractory carbides and nitrides. properties, characteristics, processing and applications, 1st edn ISBN: 9781455778164

Rizzo A, Goel S, Luisa Grilli M, Iglesias R, Jaworska L, Lapkovskis V, Novak P, Postolnyi BO, Valerini D (2020) The critical raw materials in cutting tools for machining applications: a review. Materials 13(6):1377. https://doi.org/10.3390/ma13061377

Roebuck B., Gee MG, Morrell R (2001) Hardmetals—microstructural design, testing and property maps, 245–266

Sarkar R (2016) Refractory technology: fundamentals and applications, 1st edn. CRC Press. https://doi.org/10.1201/9781315368054

Sereda B, Khina B, Kruglyak I (2020) Thermodynamic analysis of the gas environment in the formation of diffusion coatings using composite saturating charges. Metallurgy Collect Sci Work 2(43):43–49. https://doi.org/10.26661/2071-3789-2020-1-06

Siegle L (1984) Thermodynamics and kinetics of pack cementation processes. In: Kossowsky R, Singhal SC (eds) Surface engineering, NATO ASI series, vol 85. Springer, Dordrecht. https://doi.org/10.1007/978-94-009-6216-3_25

Smirnov L, Zayakin O, Zhuchkov V et al (2020) Structure and properties of nitrided ferroalloys. Russ Metall 2020:39–44. https://doi.org/10.1134/S0036029520010139

Šnajdar M, Ćorić D, Sakoman M (2024) Comparative study of multilayer hard coatings deposited on WC-Co hardmetals. Coatings 14(6):674. https://doi.org/10.3390/coatings14060674

Spies H-J, Reinhold B, Wilsdorf K (2001) Gas nitriding–process control and nitriding non-ferrous alloys. Surf Eng 17(1):41–54. https://doi.org/10.1179/026708401101517593

Sun X (2024) Thermodynamic investigation of high-temperature coatings: microstructure, oxidation and interdiffusion. Doctoral dissertation, Linköping University Electronic Press. https://doi.org/10.3384/9789180757737

Sun L, Yuan G, Gao L et al (2021) Chemical vapour deposition. Nat Rev Method Prim 1:5. https://doi.org/10.1038/s43586-020-00005-y

The Sustainable Development Goals (n.d.) United Nations. https://www.un.org/en/

Vaz F, Ferreira J, Ribeiro E, Rebouta L, Lanceros-Méndez S, Mendes JA, Alves E, Goudeau P, Rivière JP, Ribeiro F, Moutinho I, Pischow K, de Rijk J (2005) Influence of nitrogen content on the structural, mechanical and electrical properties of TiN thin films. Surf Coat Technol 191(2–3):317–323. https://doi.org/10.1016/j.surfcoat.2004.01.033

Yuda A, Arifin A, Yani I, Oemar B (2024) Titanium nitriding: a systematic literature review. Key Eng Mater 978:25–34. https://doi.org/10.4028/p-vZ6A7I

Zhang Z, Zhang J, Zhao X, Cheng X, Liu X, Zhang Q (2024) Thermodynamic simulation calculations of phase transformations in low-aluminum Zn-Al-Mg coatings. Materials 17(11):2719. https://doi.org/10.3390/ma17112719

Zhurakovskii E, Dzeganovskii V (1965) X-ray spectral investigation of the homogeneity region of titanium nitride. Powder Metall Met Ceram 3:400–405. https://doi.org/10.1007/BF00774241

Effect of Consecutive Boron–Siliconizing on the Structure, Composition, and Protective Properties of TZM Molybdenum Alloy

Tetiana Loskutova and Michael Scheffler

Abstract Molybdenum refractory alloys exhibit high creep resistance, but their high-temperature stability is limited due to the formation of MoO_3 on the surface, which undergoes sublimation. Thermal stability enhancement can be achieved by applying protective coatings resistant to elevated temperatures. This study investigated the effect of sequential boron–siliconizing on the structure, phase composition, and chemical composition of the TZM molybdenum alloy. Coatings were applied using a two-step contact powder method. In the first step, boronizing was performed at 950 °C for 8 hours in a saturating mixture of 25 wt.% amorphous boron, 70 wt.%.% Al_2O_3, and 5 wt.% NH_4Cl. In the second step, siliconizing was done at 1150 °C for 48 hours in a mixture of 35 wt.% silicon powder, 62 wt.%.% Al_2O_3, and 3 wt.% NH_4Cl. It was found that the two-step boron–siliconizing produces a diffusion coating with a thickness of 230–252 μm, composed of $MoSi_2$, Mo_5Si_3C, Mo_5Si_3, and MoB_2 phases. The surface layer based on $MoSi_2$ contains point inclusions of boron, titanium, and zirconium, forming MoB and $(Ti,Zr)B_2$-type phases. The silicon content in the $MoSi_2$ layer is approximately 33 wt.%. The microhardness of the surface layers is 14.6–16.8 GPa. The high-temperature stability of the coatings was evaluated after oxidation at 800 °C for 48 hours. The results demonstrate the potential of these coatings to enhance the high-temperature performance of the TZM molybdenum alloy.

Keywords Coating · Pack cementation · Molybdenum silicide · Molybdenum boride · Heat resistance

T. Loskutova (✉)
National Technical University of Ukraine «Igor Sikorsky Kyiv Polytechnic Institute», Kyiv, Ukraine

Otto-von-Guericke University Magdeburg, Magdeburg, Germany
e-mail: tetiana.loskutova@ovgu.de

M. Scheffler
Otto-von-Guericke University Magdeburg, Magdeburg, Germany

V. Ivanov (ed.), *Emerging Technologies in Manufacturing and Materials Engineering*, Springer Tracts in Mechanical Engineering,
https://doi.org/10.1007/978-3-032-14914-5_13

1 Introduction

Innovative structural materials work under the conditions of simultaneous action of contact loads, elevated temperatures, and action of aggressive environments. For this reason, structural materials must have a high set of properties: combined wear resistance, chemical stability at elevated temperatures, and high corrosion resistance. Molybdenum and alloys based on it, along with a high melting point, high mechanical properties at elevated temperatures, and a low thermal expansion coefficient, are characterized by low heat resistance in oxidizing environments (Zhang et al. 2023; Perepezko 2018; Gatzen et al. 2020; Litovchenko et al. 2012). These requirements are due to the formation of molybdenum oxide, MoO_3, leading to mechanical disintegration during alloy exposure to hot air; the oxide sublimes at temperatures around 730 °C. This process leads to premature destruction and degradation of samples made of molybdenum and alloys based on it. By changing the phase and chemical composition, the surface structure of the material through pack cementation, it is possible to significantly improve its operational properties, such as hardness, heat resistance, corrosion resistance, etc. (Perepezko 2018; Litovchenko et al. 2012). It is possible to combine the surface's high operational properties and the favourable properties of the base by obtaining coatings. Chromium, silicon, aluminium, and boron are the main elements contributing to the increase in heat resistance of alloys (Gatzen et al. 2020; Litovchenko et al. 2012; Efremenko et al. 2023; Zhou et al. 2001; Lange et al. 2015a; Pohrebova et al. 2023).

2 Fundamentals and Challenges

Silicon-based protective coatings have become widely used to increase the heat resistance of molybdenum and alloys based on it. The simplest method of obtaining them is carrying out the process using pack cementation with activators (Gatzen et al. 2020; Litovchenko et al. 2012; Pohrebova et al. 2023). As a result of this treatment, a coating is formed, which usually consists of molybdenum silicides $MoSi_2$ and Mo_5Si_3. The thickness of these coatings depends on the mode of silicification and is essential for proper operation. On the one hand, thick coatings will prevent oxidation of the base material for a longer period.

On the other hand, an increase in their thickness leads to a rise in the energy of elastic deformation, and therefore, the probability of the appearance of cracks increases (Gatzen et al. 2020). When silicifying molybdenum, a mixed mechanism of coating formation is implemented. This mechanism consists of a counter-diffusion of silicon and the base material. Given that the diffusion rate of silicon is an order of magnitude higher than the rate of molybdenum diffusion, only silicon diffusion can be taken into account in practice (Litovchenko et al. 2012).

The high heat resistance of silicide coatings in oxidizing environments is explained by their thermodynamic tendency toward passivation, which is realized by forming a dense and adherent silicon dioxide (SiO_2) layer. This layer effectively limits oxygen diffusion to the metallic substrate. At the same time, the coefficient of inhibition of corrosion by molybdenum silicides will be determined by the rate of formation of this film. In turn, the diffusion rate will be determined by the temperature and concentration of silicon within the homogeneity region of the $MoSi_2$ phase (Litovchenko et al. 2012). In addition, silicate coatings are characterized by the formation of through cracks. Different coefficients of thermal expansion of molybdenum and silicon cause the formation of cracks. In turn, the destruction of silicide coatings on molybdenum and alloys based on it occurs due to the crystallization of an amorphous film of SiO_2 in places where microcracks appear on the surface and the consecutive formation of molybdenum oxide, MoO_3, in the cracks. The stability and protective properties of the SiO_2-based oxide film under high-temperature oxidation can be enhanced by introducing alloying elements such as Cr (Pohrebova et al. 2023; Fox – Rabinovich et al. 2004), B (Zhai et al. 2022), Al (Qiao et al. 2021; Majumdar and Sharma 2011), Hf (Kiryukhantsev-Korneev et al. 2022), Zr (Kiryukhantsev-Korneev et al. 2022), and Y (Qiao et al. 2022) into the coating composition. These elements promote passivation and reduce the rate of oxygen diffusion through the oxide layer. Complex coatings can be obtained with simultaneous saturation with two elements due to consecutive two-stage saturation (Gatzen et al. 2020; Litovchenko et al. 2012; Nishimoto et al. 2003; Akinc et al. 1999; Fox-Rabinovich et al. 2008; Perepezko et al. 2008). Works (Perepezko 2018; Gatzen et al. 2020; Perepezko and Sakidja 2010; Lange et al. 2015b; Tang et al. 2008; Wang et al. 2013) show the results of studies of the structure, phase, and chemical composition of coatings obtained by simultaneous saturation with silicon and boron on the surface of molybdenum and alloys based on it. The coating was applied for 48–50 hours at a temperature of 900–1000° C in a fluorine-containing environment. It was established that the obtained coatings consist of two zones: the first is based on molybdenum silicides $MoSi_2$, Mo_5Si_3, and the second is based on molybdenum borides MoB, Mo_2B. Separated inclusions based on molybdenum boride MoB were observed in the first coating zone. The thickness of the first zone is 100.0–110.0 μm. The second zone is considerably thinner, measuring no more than 10.0–12.0 μm. The structure of the coatings obtained under the conditions of consecutive saturation with boron and silicon is comparable to that obtained during simultaneous saturation (Perepezko 2018; Tang et al. 2008; Wang et al. 2013; Wang et al. 2017). It has been demonstrated that MoB and Mo_2B-based layers act as diffusion barriers and prevent the diffusion of silicon into the base. Oxidation leads to the formation of a layer based on Mo_5SiB_2, which is formed between the Mo_5Si_3 and MoB phases. The Mo_5SiB_2 layer performs additional barrier functions and slows the transition from $MoSi_2$ to the Mo_5Si_3 phase, which has low oxidation resistance (Wang et al. 2013). In addition, dispersed inclusions of MoB in the layer based on molybdenum disilicide, $MoSi_2$, further contribute to forming a dense protective film based on SiO_2. Research (Wang et al. 2013) showed increased service life of samples with coatings obtained by complex saturation with silicon and boron compared to single-component silicification. This is explained by forming barrier layers based on

Mo_5SiB_2 and MoB. The diffusion coefficient of these compounds is much lower than that of Mo_5Si_3.

Thus, the application of protective coatings to molybdenum and its alloys is accompanied by changes in their microstructure, phase composition, and chemical composition, which directly affect the service properties of the material. Among the effective approaches to improving the oxidation resistance of molybdenum and its alloys, silicon-containing coatings additionally alloyed with elements that enhance the thermal stability of the SiO_2 oxide film are of particular interest. An additional contribution to oxidation resistance is provided by the formation of barrier layers, which may develop during the coating deposition process and subsequent oxidation. This work investigated the effect of combined sequential boron–siliconizing in chlorine-containing environments on the structure, phase, and chemical compositions and the oxidation resistance of TZM molybdenum alloy.

3 Ideas and Solutions

The molybdenum alloy TZM (in wt% %: 99.34 (Mo) + 0.5 (Ti) + 0.08 (Zr) + 0.08 (C)) was chosen as the object to be investigated. This alloy has found wide industrial use. A bar with a diameter of 15 mm was cut from washers with a thickness of 5 mm. The samples were degreased before coating deposition.

The contact powder method provided the coating materials in alumina containers in two stages. At the first stage, boronizing was carried out in a saturating mixture of the following composition: 25% wt. of amorphous boron, 5% wt. of NH_4Cl, and 70% wt. of Al_2O_3. The process was carried out at a temperature of 950 °C for 8 hours. In the second stage, siliconizing was carried out in a saturating mixture of the following composition: 35% wt. of silicon powder and 3% wt. NH_4Cl, 62% wt. Al_2O_3. During the process, powders of saturating elements were used, namely silicon crystal powder with a crystalline size of 325 mesh and amorphous boron powder. The chemical purity of the initial powders was 99% or above. An argon atmosphere was used to prevent the oxidation of samples during this pack cementation processing.

Microstructural analysis was carried out using a Carl Zeiss SIGMA 300 scanning electron microscope in phase contrast mode after preparing cross-sections of the samples, which allowed examination of the coating cross-sections. The chemical composition of the coatings was determined by X-ray spectroscopic microanalysis using a Carl Zeiss SIGMA 300 scanning electron microscope equipped with an Oxford Instruments X-MaxN energy-dispersive spectrometer and INCA 5.05 software. The analysis was performed under the following conditions: accelerating voltage of 5 kV and absorbed current of 0.8×10^{-9} A. The expected error of measurements with quantitative XMR did not exceed 9%. The content of the elements was determined by stepwise scanning of the sample under an electronic probe in the direction from the surface of the coating to the depth of the base matrix with a step of 1–2 μm. The phase composition of the coatings was determined by X-ray diffraction analysis using a Rigaku Ultima IV diffractometer with monochromatic Cu Kα radiation (λ = 1,541,841 nm). The operating parameters were: voltage 40 kV, current 40×10^{-3} A, and exposure time ranging from 30 min to 3 h. X-ray diffraction pattern analysis and lattice parameter calculations were performed using the Rietveld method in the PowderCell 2.4 software. Ось більш академічно відшліфований варіант перекладу для наукової статті: Microhardness measurements were performed using a PMT–3 device under a load of 0.25 N. Heat resistance was evaluated based on the mass change of the samples after oxidation at 800 °C for 1 to 48 hours under natural air conditions.

It was established that consecutive boron–siliconizing of the TZM molybdenum alloy leads to forming diffusion layers with a total thickness of 230 μm to 252 μm. The results of phase analyses, chemical composition, structure, and some properties of the coatings investigated in the work are shown in Table 1, Figs. 1, 2, 3 and 4. X-ray structural analysis showed that a coating is formed on the surface, consisting of phases based on $MoSi_2$, MoB, Mo_5Si_3C, (Ti, Zr)B_2 (Fig. 1 and Table 1). Molybdenum disilicide $MoSi_2$ is the main phase of the surface layer of the protective coating. The content of $MoSi_2O$ in the coating is 87.76 wt. %.

Metallographically, it was determined that the resulting coating consists of three layers with a clear dividing line between individual zones (Fig. 2a). The presence of cracks with a periodicity of 100–240 μm was found throughout the cross-section of the coatings. A significant thickness of the obtained coatings (up to 252 μm) will increase internal stresses in the coating and lead to the formation of cracks (Perepezko 2018; Efremenko et al. 2023). On the outside of the diffusion coating is a layer with a thickness of 175 μm to 180 μm, consisting of separate zones that differ in structure. The surface zone of this layer (Fig. 2a, Fig. 2b) consists of small equiaxed grains, the size of which does not exceed 1.5–2.0 μm; the thickness of this

Table 1 Phase composition and characteristics of diffusion coatings obtained on TZM molybdenum alloy after boron–siliconizing

Processing type, temperature, °C; saturation time, hours	Coverage zone	Phase	Phase content, % wt.	Lattice period, Å	Thickness, μm	Micro-hardness, GPa
Boron-izing, 950;8 Silicon-izing; 1050; 48	I	$MoSi_2$	87.76	a: 3.2024–3.2024 c: 7.8436–7.8436	175–180	16.8–14.6
		MoB	6.35	a: 3.1151–3.1152 c: 16.9259–16.9261		
		Mo_5Si_3C	1.64	a: 7.2619- c: 5.0102-		
		(ZrTi)B_2	4.25	a: 3.0880–3.0909 c: 3.3650–3.3710		
	II	Mo_5Si_3			10–12	19.1–20.5
	III	MoB			45–60	27.0–28.8
	Base	Mo				4.5

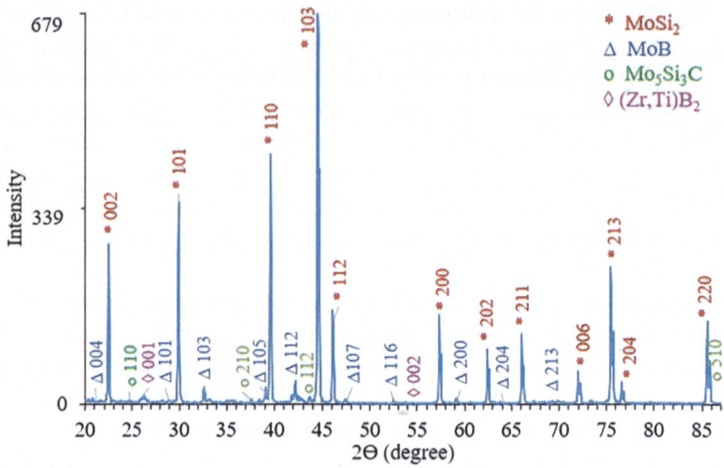

Fig. 1 X-ray diffractogram of the surface of a TZM alloy after boron–siliconizing; CuK_α radiation, wavelength is 0.1541841 nm

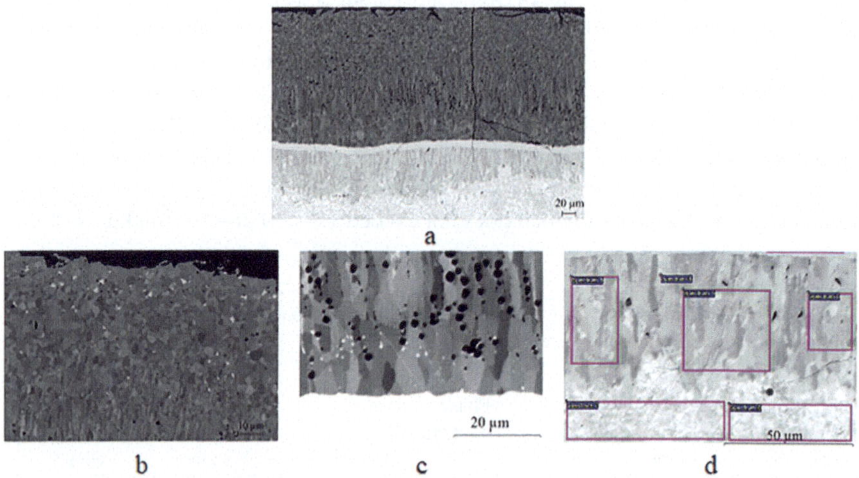

Fig. 2 Microstructure of TZM alloy after consecutive boron–siliconizing. (**a**) general appearance of the coating, (**b**) surface zone of the coating (zone of tiny (crystallite) equiaxed and columnar crystals), (**c**) surface zone of the coating (columnar crystals), (**d**) zone of jagged crystals

zone is about 45% of the total thickness of the surface layer. Some dark, almost black inclusions were observed in the zone in the intergranular boundaries of this microstructure. Still, their number is insignificant in the radial direction to the base (Fig. 2 a and c). The number of dark inclusions increases simultaneously. The thickness of the columnar zone is approximately 35% of the total thickness of this layer. Below it is a zone consisting of larger (up to 10 μm to 15 μm) equiaxed grains

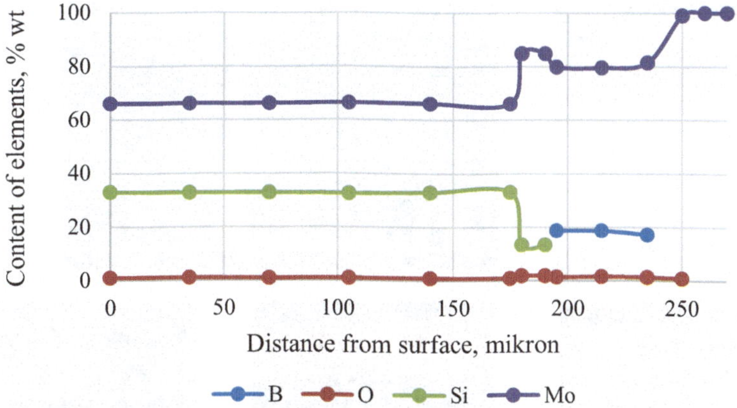

Fig. 3 Distribution of elements on the intersection of cross sections of coatings obtained during consecutive boron–siliconizing of TZM alloy

(Fig. 2a). No dark inclusions were observed in this zone. The second layer is a solid, almost white zone of 10 μm to 12 μm (Fig. 2a). A layer with a thickness of 45 μm to 60 μm and a pronounced toothed structure is directly adjacent to the base (Fig. 2 a and d).

The distribution of saturating and base elements along the coating thickness on the TZM molybdenum alloy was analyzed (Figs. 3 and 4). It was found that ~66 wt. % of molybdenum and ~ 33 wt. % of silicon are dissolved in the first coating layer, which, according to the Mo-Si phase diagram (Okamoto 2011), corresponds to the phase based on molybdenum disilicide, $MoSi_2$. In addition, separate point inclusions of oxygen, zirconium, and boron were found in this layer. The content of these elements is not significant and does not exceed ~1 wt. %. The obtained data are in accordance with the data from the X-ray structural analysis. There are no differences in the chemical composition across the cross-section of the first coating layer. The difference in microstructure is likely due to different deposition conditions during the deposition time. First, larger equiaxed grains are formed at the boundaries of the junction of the grains of the second coating layer. After that, a zone of columnar crystals is formed due to the temperature gradient at the boundary of the coating–gas medium. Lastly, during cooling, a zone of small equiaxed grains is formed. Dark point inclusions in the coating, which are most pronounced in the zone of columnar crystals, correspond to a phase with an increased oxygen content, the amount of which does not exceed 26 wt. %; the source of oxygen in coatings can be residual air in the saturating mixture during pack cementation.

The presence of individual inclusions with increased oxygen content is undesirable. These inclusions can lead to embrittlement of coatings (Lyakishev 1996). The second coating layer corresponds to the phase containing 84.8 wt. % of molybdenum, and 13.3 wt. % of silicon. We interpreted this compound as Mo_5Si_3. In addition, up to 1.8 wt. % oxygen is observed in this coating layer. The formation of this layer should have a favourable character due to the reduction of diffusion

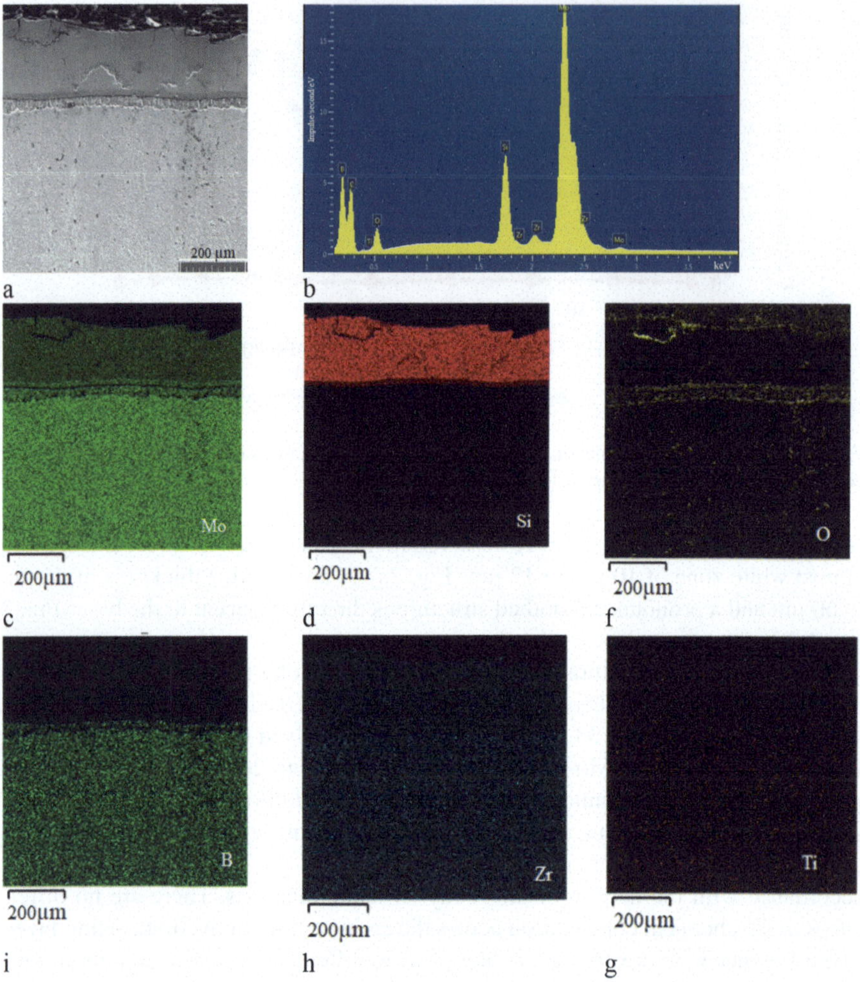

Fig. 4 Microstructure in back-scattering mode (**a**), spectral analysis (**b**), distribution of elements along the plane of the molybdenum alloy TZM after consecutive boron–siliconizing (**c**–**g**)

coefficients (Litovchenko et al. 2012). The third coating layer corresponds to the phase that contains 79–80 wt. % of Mo and 19 wt. % of B. This phase corresponds to molybdenum diboride MoB_2 as described in the Mo-B phase diagram (Chernega et al. 2015). The obtained data are in good agreement with the results of microstructural analysis, according to which, for this phase, a jagged structure is observed, characteristic of boride coatings (Chernega et al. 2016; Samsonov and Epic 1974).

The conducted studies revealed that the compounds on the coating surface comprise both saturating elements and elements of the TZM alloy, namely Si, B, Mo, Ti, and Zr. In addition, oxygen was detected in the coating. The analysis of coatings in the characteristic X-ray radiation (Fig. 4) made it possible to recognize that the maximum amount of silicon is concentrated in the surface layer based on $MoSi_2$. In addition, separate, evenly spaced inclusions with the participation of boron, zirconium, and titanium are observed in this layer. The principal amount of oxygen is concentrated in the layer based on Mo_5Si_3 and on the boundary of separation of MoB_2 - the base. In the outer layer based on $MoSi_2$, oxygen was found in the form of separate inclusions or in places where cracks appear. The molybdenum diboride layer performs a barrier function, inhibiting silicon diffusion into the base (Fig. 4). The analysis results of the distribution of the elements obtained in the characteristic X-ray radiation are well in accordance with the data above.

The microhardness of the first coating layer is 15–17 GPa, the second is 19–20.5 GPa, and the third is 27–29 GPa. The higher microhardness of layers based on the molybdenum silicides $MoSi_2$ and Mo_5Si_3 compared to stoichiometric ones (Loskutova et al. 2024) can be explained by two factors: (a) by the fact that in the process of diffusion saturation, the coating is alloyed with elements of the base (titanium and zirconium), and (b) point-dispersed boride inclusions are separated (Chabak et al. 2022) along the cross-section of the coating.

Thus, the obtained data on the structure, phase, and chemical composition of the coatings agree with those given in (Perepezko 2018; Tang et al. 2008; Wang et al. 2013, 2017). However, there is an increase in the total thickness of the obtained coatings to 230–252 μm, which is probably due to higher saturation temperatures and the use of the chlorine activator NH_4Cl, which ensures the formation of a more significant number of saturating halides and accelerates the mass transfer process (Lyakishev 1996).

Investigations of the TZM alloy without treatment and after boron–siliconizing for heat resistance showed the protective effect of the coatings before oxidation at a temperature of 800 °C. The results of heat resistance studies are shown in Figs. 5 and 6. For the TZM alloy without a coating at the chosen experiment temperature, a linear dependence of the change in the weight of the samples over time was found

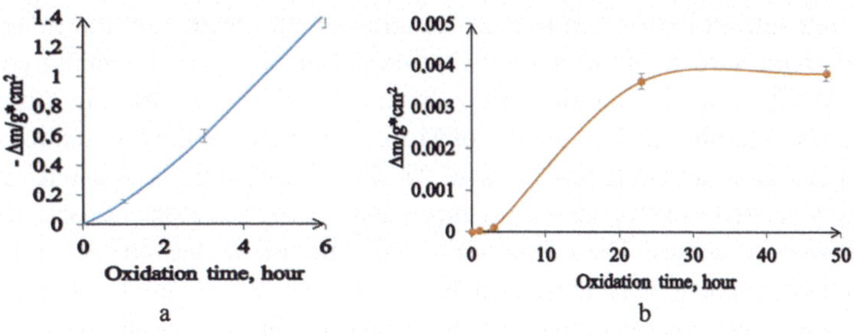

Fig. 5 Oxidation kinetics of TZM alloy without treatment (**a**) and after boron–siliconizing (**b**). The oxidation temperature was 800 °C

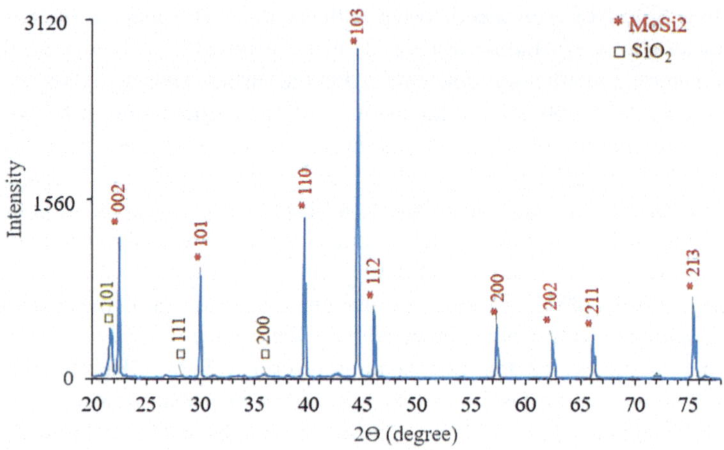

Fig. 6 X-ray diffractogram of the surface of TZM alloy with coatings obtained by consecutive boron–siliconizing after settling at a temperature of 800 °C for 48 hours (CuKα radiation, wavelength—0.1541841 nm)

(Fig. 5a). This dependence indicates the retardation of the chemical stage of molybdenum oxidation. This is due to the formation of an oxide film, $MoO_{3,}$ on its surface, which is characterized by high volatility, and complete evaporation of the original alloy is observed already after 6 hours of testing.

Another oxidation mechanism is characteristic of the TZM alloy with complex coatings after boron–siliconizing (Fig. 5b). During the first 3 hours of testing, almost no change in the mass of the samples was observed, which is probably related to the occurrence of two processes. In places where pores in the coating are located, the interaction of oxygen with molybdenum results in the formation of a molybdenum oxide, $MoO_{3,}$ SiO_2-based film. However, in the first hours of oxidation, the film does not yet form a continuous protective layer capable of completely isolating the surface. An increase in the oxidation time leads to the formation of a continuous SiO_2-based film (Fig. 5), which indicates the diffusion mechanism of the process and self-inhibition due to the formation of a protective film of SiO_2. X-ray structure on the surface of the coating after oxidation phases based on molybdenum silicide, $MoSi_2$, and silicon oxide, SiO_2, are fixed within 48 hours (Fig. 6); a phase based on MoO_3 is absent. The amount of $MoSi_2$, compared to the non-oxidized coating, decreases and amounts to 75.7 wt%. %. The decrease in the amount of $MoSi_2$ is probably related to its partial degradation and acts as a silicon source during further diffusion processes. The formation of silicon oxide on the surface of the coatings ensures high heat resistance. Visually, the surface of the samples with complex boron–siliconizing coatings after 48 hours of oxidation remains smooth; no visible chips and peeling effects were found.

4 Conclusions

The effect of sequential boron and silicon saturation in chloride-containing environments on the microstructure, phase composition, and elemental composition was analyzed. The influence of boron–siliconizing on the high-temperature oxidation resistance of the TZM molybdenum alloy was also investigated. Based on previous theoretical studies by the authors (Lyakishev 1996), the silicification processes were carried out using the chloride-based activator NH4Cl at 1150 °C, promoting thicker coating formation.

It was found that boron–siliconizing of the TZM alloy leads to the formation of a diffusion coating on its surface, with a thickness of 230.0–252.0 μm. The phase composition of the coating after boron–siliconizing is: $MoSi_2$, Mo_5Si_3C, Mo_5Si_3, and MoB_2. In the surface layer of the coating based on $MoSi_2$, point inclusions based on boron, titanium, and zirconium were observed, which led to the formation of phases based on MoB and $(Ti, Zr)B_2$. The silicon content in the surface layer of $MoSi_2$ (32–33 wt.% %) is sufficient for forming a continuous heat-resistant SiO_2 film during oxidation. This film ensures the heat resistance of the alloy at a temperature of 800 °C.

The layer based on MoB_2 performs barrier functions and blocks the penetration of silicon into the base. Dispersed inclusions of MoB in the layer based on molybdenum disilicide, $MoSi_2$, should additionally contribute to forming a dense protective film based on SiO_2.

The higher microhardness of surface layers based on the molybdenum silicides MoSi2 and Mo5Si3 (16.8–14.6 GPa and 19.1–20.5 GPa, respectively) is due to the presence of dispersed point inclusions based on boron, titanium, and zirconium, which are uniformly located along the cross-section of the coatings. This should contribute to increasing the stability of the alloy when working under erosive and abrasive conditions.

The obtained coatings can be recommended for practical use on products made of the molybdenum alloy TZM, which works at 800 °C for up to 48 hours. High heat resistance is due to forming a uniform protective coating based on SiO_2 during oxidation.

The method of molybdenum alloy surface protection proposed in the paper is cost-efficient. Materials for implementing this method are not in short supply. Technological methods are easily implemented in the conditions of innovative processing. This makes it possible to recommend them for simultaneous exposure to mechanical loads and elevated temperatures on metal parts. Ongoing research will be performed to combine pack cementation processing with further innovative coating processes to increase the oxidation resistance of Mo-based alloys.

Acknowledgments The authors would like to thank the German Research Foundation (DFG; SCHE 628-22-1, Grant ID 550760674) and the International Association for Technological Development and Innovations for financial support. The authors would also like to thank Volodymyr Taran for technical suggestions and his tireless assistance in carrying out experimental work, and Manja Krueger and Georg Hasemann for numerous suggestions and discussions of the results (both are the TZM specialists at the IMFW). VT, MK, and GH are with the IWTM.

References

Akinc M, Meyer M, Kramer M, Thom A, Huebsch J, Cook B (1999) Boron-doped molybdenum silicides for structural applications. Mater Sci Eng A 261(1–2):16–23. https://doi.org/10.1016/s0921-5093(98)01045-4

Chabak Y, Shimizu K, Efremenko V, Golinskyi M, Kusumoto K, Zurnadzhy V, Efremenko A (2022) Microstructure and phase elemental distribution in high-boron multi-component cast irons. Int J Miner Metall Mater 29(1):78–87. https://doi.org/10.1007/s12613-020-2135-8

Chernega S, Polyakov I, Medova I, Krasovskyi M (2015) Structure and properties of complex boron coatings on carbon steel. Metallofiz Noveishie Tekhnol 37(6):751–761. https://doi.org/10.15407/mfint.37.06.0751

Chernega S, Poliakov I, Krasovskiy M (2016) Increasing the wear resistance of the T15K6 hard alloy by boriding and complex saturation with boron and copper. J Superhard Mater 38(3):190–196. https://doi.org/10.3103/S1063457616030060

Efremenko V, Chabak Y, Shimizu K, Golinskyi M, Lekatou A, Petryshynets I, Efremenko B, Halfa H, Kusumoto K, Zurnadzhy V (2023) The novel hybrid concept on designing advanced multi-component cast irons: effect of boron and titanium (Thermodynamic modelling, microstructure and mechanical property evaluation). Mater Charact 197:112691. https://doi.org/10.1016/j.matchar.2023.112691

Fox – Rabinovich G, Weathrly G, Wilkinson D (2004) The role of chromint in protective scale formation during oxidation of ternary TiAlCr allays inair. Intermetallics 12:165–180. https://doi.org/10.1016/j.intermet.2003.09.014

Fox-Rabinovich G, Endrino J, Bearet B, Aguirre M, Veldhuis S, Quinto D, Bauer C, Kovalev A, Cray A (2008) Effect of temperature of annealing below 900°C on structure, properties and toll life of an AlTiN coating under various cutting conditions. Surf Coat Technol 202:2985–2992. https://doi.org/10.1016/j.surfcoat.2007.10.034

Gatzen C, Smokovych I, Scheffler M, Krüger M (2020) Oxidation-resistant environmental barrier coatings for Mo-based alloys. Adv Eng Mater:2001016. https://doi.org/10.1002/adem.202001016

Kiryukhantsev-Korneev P, Sytchenko A, Sviridova T, Sidorenko D, Andreev N, Klechkovskaya V, Polˇcak J, Levashov E (2022) Effects of doping with Zr and Hf on the structure and properties of Mo-Si-B coatings obtained by magnetron sputtering of composite targets. Surf Coat Technol 442:128141. https://doi.org/10.1016/j.surfcoat.2022.128141

Lange A, Heilmaier M, Sossamann T, Perepezko J (2015a) Oxidation behavior of pack-cemented Si-B oxidation protection coatings for Mo-Si-B alloys at 1300°C. Surf Coat Technol 266:57–63. https://doi.org/10.1016/j.surfcoat.2015.02.015

Lange A, Heilmaier M, Sossamann T, Perepezko J (2015b) Oxidation behavior of pack-cemented Si–B oxidation protection coatings for Mo–Si–B alloys at 1300°C. Surf Coat Technol 266:57–63. https://doi.org/10.1016/j.surfcoat.2015.02.015

Litovchenko S, Beresnev V, Drobyshevskaya A, Turbin P (2012) Silicide coatings on molybdenum: production, structure, properties. Phys Surf Eng 10(2):110–137. https://www.researchgate.net/publication/313888183

Loskutova T, Scheffler M, Ivanov V, Pogrebova I, Kononenko Y, Bobina M, Kharchenko N, Bartoszuk M, Pavlenko I (2024) Physicochemical conditions of boron–siliconizing of molybdenum-based alloys in chlorine and fluorine medium. Metals 14(3):302. https://doi.org/10.3390/met14030302

Lyakishev N (1996) Phase diagrams of binary metallic systems. Mashynostroenye

Majumdar S, Sharma I (2011) Oxidation behavior of $MoSi_2$ and $Mo(Si, Al)_2$ coated Mo-0.5Ti-0.1Zr-0.02C alloy. Intermetallics 19:541–545. https://doi.org/10.1016/j.intermet.2010.12.002

Nishimoto T, Izumi T, Hayashi S, Narita T (2003) Two-step Cr and Al diffusion coating on TiAl at high temperatures. Intermetallics 11(3):225–235. https://doi.org/10.1016/S0966-9795(02)00233-9

Okamoto H (2011) Mo-Si (Molybdenum-Silicon). J Phase Equilib Diffus 32:176. https://doi.org/10.1007/s11669-010-9843-0

Perepezko J (2018) High temperature environmental resistant Mo-Si-B based coatings. Int J Refract Met Hard Mater 71:246–254. https://doi.org/10.1016/j.ijrmhm.2017.11.033

Perepezko J, Sakidja R (2010) Oxidation-resistant coatings for ultra-high-temperature refractory Mo-based alloys. JOM 62(10):13–19. https://doi.org/10.1007/s11837-010-0148-x

Perepezko JH, Rioult F, Sakidja R (2008) Oxidation performance of high temperature Mo-Si-B alloys and coatings. Mater Sci Forum 595-598:1065–1074. https://doi.org/10.4028/www.scientific.net/MSF.595-598.1065

Pohrebova I, Yantsevych K, Loskutova T, Kharchenko N, Hovorun T (2023) Structure and protective properties of complex chromosilicide diffusion coatings on steel 20. J Nano- Electron Phys 15(5):05031. https://doi.org/10.21272/jnep.15(5).05031

Qiao Y, Chen T, Guo X (2021) Diffusion barrier effect of Al_2O_3 layer at the interface between Mo-Si-B coating and Nb-Si based alloy. Corrosion Commun 4:45–56. https://doi.org/10.1016/j.corcom.2021.10.003

Qiao Y, Zhang X, Guo X (2022) Y_2O_3 film as a diffusion barrier between $MoSi_2$ coating and Nb-Si based alloy. Corros Sci 196:110044. https://doi.org/10.1016/j.corsci.2021.110044

Samsonov H, Epic A (1974) Refractory coatings. Metallurgy

Tang Z, Thom A, Kramer M, Akinc M (2008) Characterization and oxidation behavior of silicide coating on multiphase Mo–Si–B alloy. Intermetallics 16:1125. https://doi.org/10.1016/j.intermet.2008.06.014

Wang Y, Wang D, Yan J, Sun A (2013) Preparation and characterization of molybdenum disilicide coating on molybdenum substrate by air plasma spraying. Appl Surf Sci 284:881–888. https://doi.org/10.1016/j.apsusc.2013.08.029

Wang Y, Yan J, Wang D (2017) High temperature oxidation and microstructure of $MoSi_2$/MoB composite coating for Mo substrate. Int J Refract Met Hard Mater 68:60–64. https://doi.org/10.1016/j.ijrmhm.2017.06.008

Zhai R, Song P, Huang T, Li C, Hua C, Huang W, Li Q, Zheng B, Lu J (2022) Microstructure and oxidation behaviour of $MoSi_2$ coating combined MoB diffusion barrier layer on Mo substrate at 1300°C. Ceram Int 47:10137–10146. https://doi.org/10.1016/j.ceramint.2020.12.162

Zhang Y, Zhou X, Cheng H, Geng Z, Li W (2023) Fabrication and oxidation resistance of a novel $MoSi_2$-ZrB_2-based coating on Mo-based alloy. Materials 16(16):5634. https://doi.org/10.3390/ma16165634

Zhou C, Yang Y, Gong S, Xu H (2001) Effect of Ti–Al–Cr coatings on the high temperature oxidation behavior of TiAl alloys. Mater Sci Eng 307:182–187. https://doi.org/10.1016/S0921-5093(00)01956-0

Computational Approaches in Manufacturing

Criteria for Ensuring the Stability of Stationary Technological Systems

Alexandr Orgiyan and Anna Balaniuk

Abstract In this chapter, the basic definitions are introduced, and a review of typical problems of machine tool dynamics is provided for the technological system lathe-device-tool-workpiece (LDTW), the parameters of which are constant in time. The methods of solving known problems are presented in the form necessary for their subsequent transformation. The following issues are considered in the work: the structure of the dynamic system of the LDTW, physical and mathematical models of the dynamic system, dynamic effects and processes in the LDTW system, the dynamic characteristic of the cutting process, and the stability of the dynamic system with one and two degrees of freedom according to the Nyquist and Rauss-Hurwitz criteria. The formation of an equivalent single-loop structure of the LDTW and a number of typical sources of force and kinematic effects on the elastic-dissipative-inertial system of the machine are considered. The force factors affecting the LDTW system are divided into internal and external. External influences and the processes they initiate in a closed dynamic system are divided according to their time form into deterministic and random. Thus, the LDTW system is assumed to be stationary, deterministic, and linearizable. The parameters of internal effects depend on the properties of the components of a specific dynamic system LDTW: UDIS (elastic-dissipative-inertial system) and WP (working process). The parameters of internal effects depend on the properties of the components of a specific dynamic system. The elastic-dissipative-inertial system includes body parts, parts, and units of the main motion and translational motion drives, a tool, a workpiece, together with the corresponding devices, and movable and fixed connections of the elements. The transition to systems of ordinary differential equations is achieved by discretization of the physical model of the system, i.e., by transition to a model with lumped parameters. Static and kinematic characteristics of technological systems are given.

Keywords Stationary technological system · Criteria of stability · Cutting process · Physical and mathematical models · Dynamic system

A. Orgiyan · A. Balaniuk (✉)
Odesa Polytechnic National University, Odesa, Ukraine

V. Ivanov (ed.), *Emerging Technologies in Manufacturing and Materials Engineering*, Springer Tracts in Mechanical Engineering,
https://doi.org/10.1007/978-3-032-14914-5_14

1 Introduction

Simple analysis of the process systems shows that only auto-oscillations in non-linear dynamic machine systems and vibrations caused by machine effects can occur under the constant parameters of the LDTW dynamic system (Wu et al. 2003; Ftorek et al. 2018). All other excitation sources of oscillations in machines are related to the variability of dynamic system parameters (Alammari et al. 2015). Methods of calculation and experimental investigation of stability and oscillations in such systems have been developed to justify recommendations for improvement of the structures of metal cutting machines, design of special tools, and choice of processing modes (Alifov 2020; Ramesh et al. 2013).

The choice of a physical model of a real object consists of simplifying its properties, which facilitates the solution of the task. Still, with the necessary accuracy, the characteristics of the object are preserved (Basovich and Arogeti 2021). Simplifications can consist of typical material resistance assumptions about the simplicity of body shape, the inconsistency and uniformity of material, as well as system discretization, linearization of characteristics, neglecting of parameter variability, in the transition from probabilistic problem to deterministic, in neglecting part of the forms of fluctuations or impacts, etc. (Belov et al. 2017). This work assumes that the LDTW system is stationary, deterministic, and linearizable. This subsection refers to a stationary, deterministic, and linearized LDTW system. Local interaction of EDIS with the cutting process is allowed (Oborskyi et al. 2023).

A stationary dynamical system of a machine is a mechanical system whose behavior is described by differential equations of motion, and which has constant (time-invariant) properties. A stationary dynamic system characterizes the machine's behavior under the influence of external forces, loads, and disturbances, provided that the parameters of the machine (mass, hardness, and damping) remain constant over time (Ozlu and Budak 2007; Parmar et al. 2023; Grigorian 2022).

Mathematical criteria such as the Rausa-Gurgwitz and Nyquist criteria are widely used to analyze the resilience of dynamic systems in mechanical engineering (Butikov 2018). These criteria allow for assessing the system's resistance to external influences and fluctuations, and to analyse the vibration resistance of stationary and non-stationary dynamic systems.

Oscillations are one of the main factors affecting the accuracy and quality of processing. To ensure high accuracy and quality, the fluctuations should be reduced (Ivanov et al. 2019). This will increase equipment productivity, reduce accidents, and improve working conditions.

Important characteristics of the dynamic system are its own frequencies of oscillations and forms of oscillations, which depend on the physical properties of the structure (mass, rigidity, and damping) (Reith et al. 2016).

2 Fundamentals and Challenges

In addition to the problems of limiting the level of oscillations according to the performance, accuracy, and reliability criteria of the technological system, the dynamics of fine separation also considers the problems of proper use of vibrations during cutting. The study of cutting kinematics with given vibrations, the investigation of the particularities of the process of particle formation, wear, and quality of processing served to justify the conditions for effective application of vibration cutting (Vnukov et al. 2023; Smolenicki et al. 2014). The mutual movement of the elastic elements in one of the oscillating movements causes changes in the parameters of the other oscillating contour, and parametric fluctuations can be triggered.

To describe the dynamics of systems, differential equations of second order are often used, describing changes in the movement of parts of the system under the action of external forces. In the simplest case, the equation of motion looks like this:

$$m\frac{d^2x}{dt^2}+b\frac{dx}{dt}+cx=F(t) \tag{1}$$

where

m—mass of the system;
b—damping factor (friction);
c—rigid system;
x—displacement;
$F(t)$—external force acting on the system;
t—time

So, the sign of stationarity can be defined as follows: under the action of disturbance (for example, when cutting material), the system begins to fluctuate, but its parameters—mass, stiffness, and damping—remain unchanged. Oscillations can be attenuated or supported by external forces (forced oscillations).

Main causes of oscillations of stationary technological systems: one of the main reasons; unbalance (imbalance) of rotating elements (eg, spindle, cutting tool, cartridge and so on); luffing and gaps in the assemblies (for example, in guides and bearings); uneven cutting forces (when changing material characteristics, cutting speed and geometry of the tool); resonance phenomena; technological factors (improper cutting modes, tool wear); misalignment; external effects (from surrounding machines, from foundation, etc.) (Barois et al. 2014; Butikov 2018).

Note that the study of dynamic interactions between elements of technological systems is a reserve to increase processing accuracy and durability of equipment.

The research aims to develop methods for calculating dimensionless stability criteria of stationary technological systems according to Nyquist and Routh-Hurwitz.

The objectives are as follows: to determine the stability of a one-dimensional technological system with a single degree of freedom using the Raun-Gurgwitz criterion; to determine the influence of parameters of a stationary system of part-tool on stability (in the example of a two-dimensional technological system); to perform a graphical interpretation of the inequalities of the Raun-Gurwitz criterion for a two-dimensional system.

3 Ideas and Solutions

The structure of the LDTW dynamic system is shown in Fig. 1, is multi-contour, closed, and has the following properties (Oborskyi et al. 2021):

- the states of the elastic-dissociative-inertial system (EDIS) and various working processes (WP) that make up the dynamic system are described by vectors which play the role of input signals for one element and output for another;

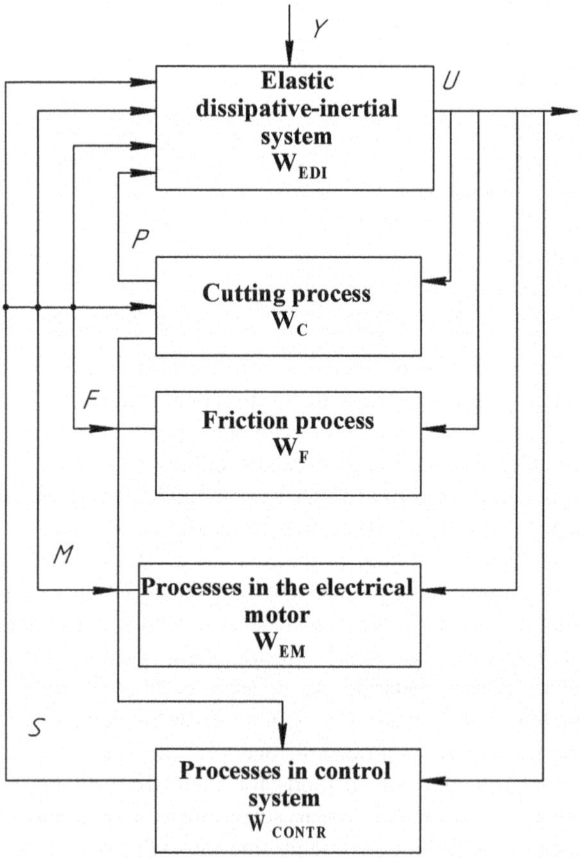

Fig. 1 Diagram of the dynamic system LDTW

- EDIS and WP are subject to controlling and stimulating effects, as well as interference;
- In addition to the force effects of WP on EDIS, there are also more slowly changing thermal effects;
- The interaction of work processes is only possible through EDIS.

The vectors acting on EDIS: cutting force P, friction force F, motor torque M and control force S are formed in WP links under the action of movements. The combined action of these force factors on the EDIS link generates a generalized displacement vector u at the exit of this link. Transition matrices that define the dependence of output signals χ_{out} on input χ_{inp}, describe the dynamic properties of individual LDTW system links: EDIS—W_{EDI}, cutting process—W_c, friction process—W_F, electric motor process—W_{EM}, control process—$W_{control}$.

The transition matrices are related to the transformations of the Laplace output and input signals of the linear system

$$L[\chi_{out}] = W(S)L[\chi_{inp}]8;8\ \chi_{out}(S) = W(S)\chi_{inp}(S) \tag{2}$$

In a scalar view

$$\begin{aligned}\chi_{out.j}(S) &= W_{j1}(S)\chi_{inp\,1}(S) + W_{j2}(S)\chi_{inp\,2}(S) + \ldots + \\ &+ W_{jn}(S)\chi_{inp\quad n}(S),\ j = 1,2,\ldots,m.\end{aligned} \tag{3}$$

We will use $\chi_{out.\,jk}$ to represent the value $\chi_{out.\,j}$ if only one input coordinate $\chi_{inp\,.k}$ is different from zero. Then (3) gives an expression of a transition matrix element

$$W_{jk}(S) = \chi_{out.jk}(S) / \chi_{inp\,\text{E}.k}(S) \tag{4}$$

which is the basis of the experimental definition of this matrix.

By giving the complex argument S a purely imaginary value $S = i\omega$, where ω—the circular frequency of the oscillations, we get an expression of the amplitude-phase frequency characteristic (APFC) of the system element

$$W(i\omega) = W(S)\big|_{s=i\omega} \tag{5}$$

The external action on EDIS Y is dependent on it and force factors

$$u = W\quad(\text{Y} + \text{P} + F + \text{M} + S) \tag{6}$$

which in turn depend on displacement

$$\begin{aligned}\text{P} &= -W_C u, \quad F = -W_F u, \\ M &= -W_{EM} u, \ S = -W_{control} u.\end{aligned} \tag{7}$$

By setting (7) to (6) and allowing equality with respect to u, we get

$$u = \left[I + W_{EDI} \left(W_C + W_F + W_{EM} + W_{control} \right) \right]^{-1} \mathrm{Y}, \tag{8}$$

where the expression of the closed-system transition matrix on Y input comes from

$$W_{\mathrm{Y}} = \left[I + W_{ED} \left(W_C + W_F + W_{EM} + W_{control} \right) \right]^{-1} W_{EDI}. \tag{9}$$

Since this paper explores the dynamic processes in cutting due to their influence on the accuracy of processing, it is advisable to switch from the multi contour system shown in the Fig. 1, to single-circuit system: EDIS link combines with "friction process", "electric motor process" and "control process" in one link—"equivalent elastic-dissipative-inertial system" (EEDIS), interacting (Fig. 2b) with "cutting process" (CP).

To construct the stability regions of a closed dynamic system, this work defines the features of the cutting process during fine boring, which are reflected in the dynamic characteristics.

The stability of the dynamic system AIDS determines the operability of the technological system—the absence of undermining and self-oscillations during cutting. Stability is manifested in the attenuation of transient processes, while in unstable states deviations from equilibrium increase. Mathematical theory provides criteria for stability.

Rational field of application of the criterion of Raun-Gurgwitz—construction of regions of stability (D-breakdowns) in space of parameters of multidimensional dynamical systems. As an example, let's look again at the simplest one-dimensional model (Fig. 3).

We further assume that the cutting process is actually stable, i.e., the formation of chips occurs without disrupting their continuity and without forming a build-up. The simplest assumption about the dependence of the change in cutting force *P* on the change in the thickness of the cut *a* that caused it

$$\mathrm{P} = \mathrm{K}_p a \tag{10}$$

is called the static cutting characteristic, and K_p, *N*/*m* is called the cutting rigidity.

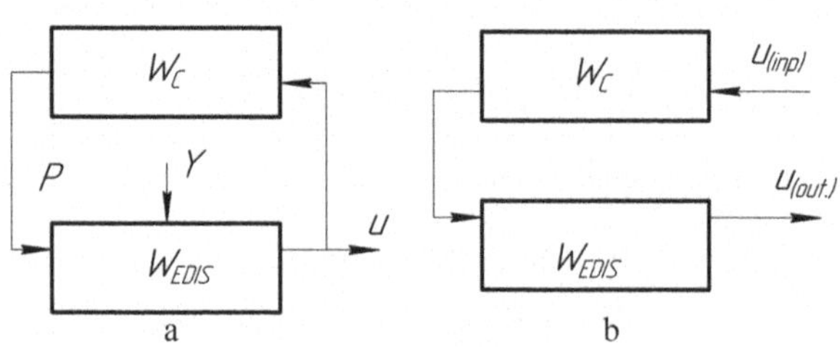

Fig. 2 Formation of equivalent single-circuit structure of LDTW system: (**a**) equivalent closed single-circuit system; (**b**) calculated open single-circuit system

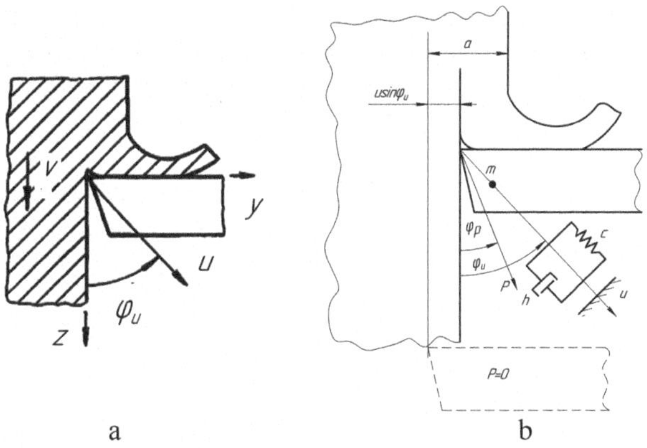

Fig. 3 Calculation diagram of a one-dimensional model of a dynamic system. (**a**) coordinates in the cutting zone, (**b**) calculation scheme

This relationship, applicable in a number of machine tool dynamics problems, does not describe at least two important properties of the cutting process that should be reflected in the dynamic characteristic:

1. Inertia of the process—phase lag of *P* from *a;*
2. Dependence of *P* on the rate of change of the cutting thickness.

Combining these two characteristics of the cutting process, we write a differential equation for a simple one-dimensional model (Fig. 3a), where *u* is the coordinate in the direction of the oscillations developing in the system, and *P* is the main component of the cutting force.

$$T_p\dot{\mathrm{P}} + \mathrm{P} = -k_p \sin\varphi_u u + k_s \cos\varphi_u \dot{u} \tag{11}$$

Here the chip formation time constant T_p is a measure of the inertia of the cutting process. The angles φ_u and φ_P project the directions of oscillations (*u*) and cutting forces (*P*) onto the direction of the main movement, which occurs at a speed of *v* (Fig. 3b).

The force $k_s \cos\varphi_u \dot{u}$ is defined as the sum of the change in the friction force on the back surface of the cutter—$k_{Tp} dz/dt$ and the change in the cutting force $k_p s \cos\varphi_u \dot{u} / v$, caused by the change in the cross-section of the cut layer when the feed *s* changes by one revolution. Therefore, the relationship is valid

$$k_s = k_p s / v - k_{Tp} \tag{12}$$

from which it follows that the cutting speed coefficient can take values of any sign depending on the relationship between the physical and kinematic parameters of the cutting process.

Thus, the dynamic characteristic of cutting describes the totality of phenomena of plastic deformation of chips, friction on tool surfaces, as well as the kinematic dependences of the thickness and cross-section of the cut layer on changes in the speed of the main movement and feed. Let us write the system of equations of motion

$$\begin{aligned} m\ddot{u} + h\dot{u} + cu &= \cos\left(\varphi_u - \varphi_p\right)P, \\ \mathrm{T}_p\dot{P} + \mathrm{P} &= k_p\left(a - u\sin\varphi_u\right) + k_s\dot{u}\cos\varphi_u. \end{aligned} \tag{13}$$

To obtain the stability condition in a criteria way it is rational to go to dimensionless parameters:

time

$$\tau = t\left(c / m\right)^{1/2}; \tag{14}$$

dissipation coefficient

$$\eta = h / \left(cm\right)^{1/2}; \tag{15}$$

constant clotting time

$$\theta = \mathrm{T}_p\left(c / m\right)^{1/2}; \tag{16}$$

the EDIS connection ratio with the cut process on the move

$$\gamma = k_k k_c k_p k_{2p} / c; \tag{17}$$

EDIS Speed Cut Ratio

$$\beta = k_k k_c k_s k_{2s} / \left(cm\right)^{1/2}. \tag{18}$$

where $k_{2p} = \cos\varphi_u$ and $k_{2s} = \sin\varphi_u$ are the projecting coefficients.

After moving to dimensionless parameters, find equations for dimensionless constant components of displacement and force

$$u_{const} = \mathrm{P}_{const}; \quad \mathrm{P}_{const} = 1 - \gamma u_{const}, \tag{19}$$

for variable components

$$\begin{aligned} d^2u_{var} / d\tau^2 + \kappa du_{var} / d\tau + u_{var} &= \mathrm{P}_{var}, \\ \theta d\mathrm{P}_{var} / d\tau + \mathrm{P}_{var} &= -\gamma u_{var} + \beta du_{var} / d\tau, \end{aligned} \tag{20}$$

where $\gamma = k_p c^{-1}\sin\varphi_u\cos\left(\varphi_u - \varphi_p\right); \beta = k_s\left(cm\right)^{-1/2}\sin\varphi_u\cos\left(\varphi_u - \varphi_p\right)$.

Write down for equation (17) the coefficient of characteristic equation: $a_0 = \theta$; $a_1 = 1 + \theta\eta$; $a_2 = \theta + \eta - \beta$; $a_3 = 1 - \gamma$.

In addition to the condition $\Delta_2 = a_1a_2 - a_0a_3 > 0$ that leads to known inequality $\gamma + \beta(\theta^{-1} + \eta) < \gamma_0$ at $\gamma_0 = \eta(\theta + \theta^{-1} + \eta)$, the Raun-Gurgwitz criterion also requires that a_i are positive conditions $a_0 > 0$, $a_1 > 0$ are trivial.

Condition $a_3 > 0$ or $\gamma > -1$ prevents aperiodic instability—with $\gamma \to -1$ unlimited increase of the permanent part of the solution $P_{const} = (1 + \gamma)^{-1}$.

Figure 4 shows that $a_2 > 0$ at $a_3 > 0$ and $\Delta_2 > 0$.

Therefore, for the stability of the system (13) enough

$$\gamma > -1 \text{ and } \gamma + \beta\left(\theta^{-1} + \eta\right) < \gamma_0. \tag{21}$$

Often, the properties of a dynamic system are determined by two lower forms of EDIS oscillations. This problem can arise when formally moving to normal coordinates or when physically forming the EDIS model, for example, when studying the interaction through the process of cutting bending oscillations with the torsional or interaction of the subsystems of the part and the tool, shown on Fig. 5.

We write the equations of motion for such a system, assuming the increment of cutting force from speed.

$$\begin{aligned} &m_1 \ddot{u}_1 + h_1\dot{u}_1 + c_1u_1 = \cos\left(\varphi_{u1} - \varphi_p\right)\ ; \\ &m_2 \ddot{u}_2 + h_2\dot{u}_2 + c_2u_2 = \cos\left(\varphi_{u1} - \varphi_p\right)\ ; \\ &T_p dP / dt + P = -k_p\left(u_1 \sin\varphi_{u1} + u_2 \sin\varphi_{u2}\right). \end{aligned} \tag{22}$$

Denote

$$\omega_j = \left(c_j m_j^{-1}\right)^{1/2};\ \Omega = \omega_1\omega_2^{-1};\ \theta = \omega_1 T_p;$$

$\eta_j = h_j\left(c_j m_j\right)^{-1/2}$; $\gamma_j = k_p c_j^{-1} \sin\varphi_{uj} \cos\left(\varphi_{uj} - \varphi_p\right)$ and go to (22) dimensionless variables $\tau = t\omega_1$; $u_{oj} = u_1 k_p \sin\varphi_{uj}$.

For the resulting system of equations

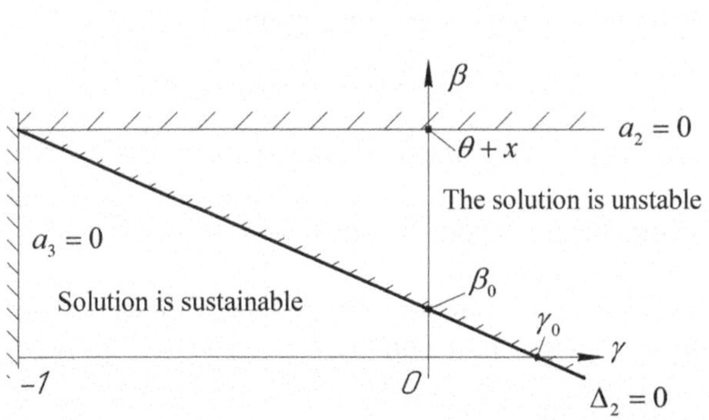

Fig. 4 Geometric meaning of the Rauza-Gurgwitz criterion for one-dimensional systems

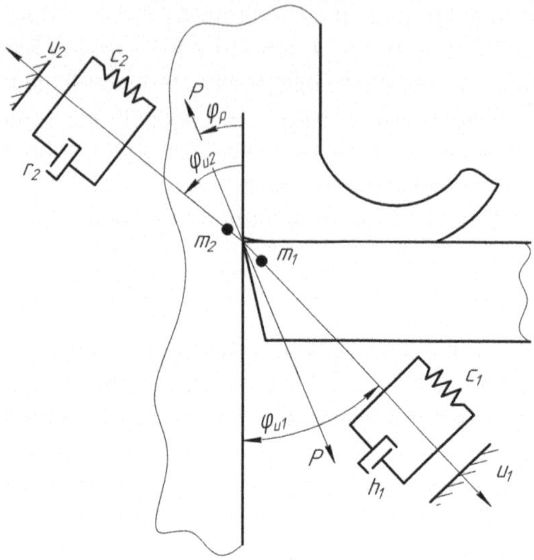

Fig. 5 Subsystems of parts and tools, interacting through the cutting process

$$
\begin{aligned}
&d^2u_{01}/d\tau^2+\eta_1du_{01}/d\tau+u_{01}=\gamma_1\mathrm{P},\\
&\Omega^2d^2u_{02}/d\tau^2+\Omega\eta_2du_{02}/d\tau+u_{02}=\gamma_2\mathrm{P},\\
&\theta d\mathrm{P}/d\tau+\mathrm{P}=-u_{01}-u_{02},
\end{aligned}
\tag{23}
$$

find the coefficients of the characteristic equation:

$$
\begin{aligned}
&a_0=\Omega^2\theta, a_1=\Omega^2+\Omega\theta\left(\Omega\eta_1\eta_2\right),\\
&a_2=\Omega\left(\Omega\eta_1+\eta_2\right)+\theta\left(1+\Omega^2+\Omega\eta_1\eta_2\right),\\
&a_3=\Omega^2\left(1+\gamma_1\right)+1+\gamma_2+\Omega\eta_1\eta_2+\theta\left(\eta_1+\Omega\eta_2\right),\\
&a_4=\kappa_1\left(1+\gamma_2\right)+\Omega\eta_2\left(1+\gamma_1\right)+\theta, a_5=1+\gamma_1+\gamma_2.
\end{aligned}
\tag{24}
$$

From expressions (24) it can be seen that at any meaningful values of the system parameters (23) remain positive a_0, a_1 and a_2. Therefore, the Raun-Gurgwitz conditions are reduced to the following inequality system:

$$a_1>0; a_2>0; a_3>0; \Delta_2=a_1a_2-a_0a_3>0; \tag{25}$$

$$\Delta_4=\left(a_4a_3-a_2a_5\right)\left(a_1a_2-a_0a_3\right)-\left(a_1a_4-a_0a_5\right)^2>0. \tag{26}$$

In practice it is convenient to apply the graphical interpretation of conditions (26), (25) on the γ_1 γ_2 plane (Fig. 6).

Each of the inequalities (25) defines a half-plane, with $a_j (j = 3, 4, 5)$ being positive at points above the lines, cutting off the axes γ_1 and γ_2 of the segments γ_{1j} and γ_{2j}

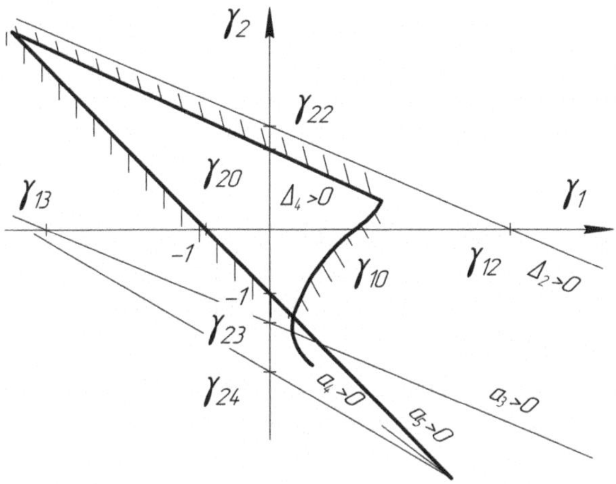

Fig. 6 Graphical interpretation of inequalities of the Raus-Gurgwitz criterion for a two-dimensional system

$$\begin{aligned}
\gamma_{13} &= -\left[1+\Omega^2+\theta\left(\eta_1+\Omega\eta_2\right)\right]/\Omega^2;\\
\gamma_{23} &= -1-\Omega^2-\theta\left(\eta_1+\Omega\eta_2\right);\quad \gamma_{15}=-1;\\
\gamma_{14} &= -\left(\kappa_1+\Omega\eta_2+\theta\right)/\Omega\eta_2;\\
\gamma_{24} &= -\left(\kappa_1+\Omega\eta_2+\theta\right)/\eta_1;\quad \gamma_{25}=-1.
\end{aligned} \tag{27}$$

Value Δ_2 positive at points below the line cutting on the axis of the segment

$$\begin{gathered}
\gamma_{12}=\left(\eta_1+\Omega^{-1}\eta_2\right)\left(\eta_1+\Omega^{-1}\eta_2+\theta^{-1}\right)\\
+\theta\left[\eta_1\eta_2\left(\Omega^{-1}\eta_1+\Omega^{-2}\eta_2\right)+\eta_1+\Omega^{-3}\eta_2\right]\\
\gamma_{22}=\Omega^2\gamma_{12}.
\end{gathered} \tag{28}$$

The study of condition (26) is more complicated, because the expression Δ_4 is a third-degree polynomial relative to parameters γ_1 and γ_2.

The line $\Delta_4=0$ can be obtained from the condition of the periodic solution of the system of equations (23) in the form of a parametrically determined curve

$$\begin{aligned}
\gamma_1(v) &= \left[\left(v^2-1\right)^2+\eta_1^2v^2\right]\times\frac{\left[\Omega\eta_2+\theta\left(1-\Omega^2v^2\right)\right]}{\left[\eta_1\left(1-\Omega^2v^2\right)+\Omega\eta_2\left(v^2-1\right)\right]};\\
\gamma_2(v) &= \left[\theta\left(v^2-1\right)-\eta_1\right]\times\frac{\left[\Omega^2\eta_2^2v^2+\left(1-\Omega^2v^2\right)^2\right]}{\left[\eta_1\left(1-\Omega^2v^2\right)+\Omega\eta^2\left(v^2-1\right)\right]}.
\end{aligned} \tag{29}$$

Here the parameter ν makes sense of the frequency of the periodic solution of the system of equations (23). From (26):

- the coordinate point $\gamma_1 = \gamma_2 = 0$ always belongs to the stability area;
- the intersection of the lines $a_3 = 0$ and $a_4 = 0$ lies outside this area.

Curve given by equations (29):

- when $\nu = 0$ passes the intersection of lines $a_4 = 0$ and $a_5 = 0$;
- when $\nu = \left(1+\eta_1\theta^{-1}\right)^{1/2}$ crossing the axis γ_1 at a point $\gamma_{10} = \eta_1(\theta + \theta^{-1} + \eta_1)$;
- when $\nu = \Omega^{-1}\left(1+\Omega\eta_2\theta^{-1}\right)^{1/2}$ crossing the axis γ_2 at the point $\gamma_{20} = \eta_2(\theta\Omega^{-1} + \theta^{-1}\Omega + \eta_2)$.

If part of curve (29) is above the line $\Delta_2 = 0$ (Fig. 7), this part describes unstable periodic solutions of equation system (7) and is not the boundary of the stability field.

Consider the effect of changes in the dynamic system parameters on the level of forced oscillations. When each of the four dimensionless parameters changes within the above-set limits, three others maintain constant averages. The numerical calculations according to precise formulas have established the following regularities of the change of the maximum value of the amplitude of forced oscillations (Fig. 8):

(a) inverse proportionality to η;
(b) maximum near $\theta = 1$;
(c) low dependence on the coefficient of connectivity γ away from the stability boundary and unlimited increase as one approaches the boundary;
(d) dependence on β, similar dependency of γ.

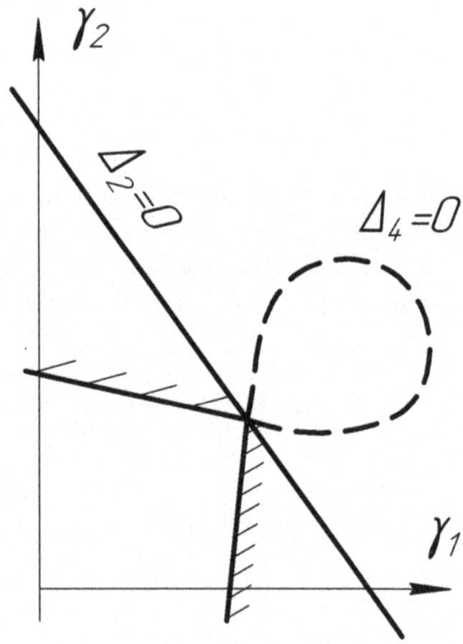

Fig. 7 Conditions for the emergence of unsustainable periodic decisions

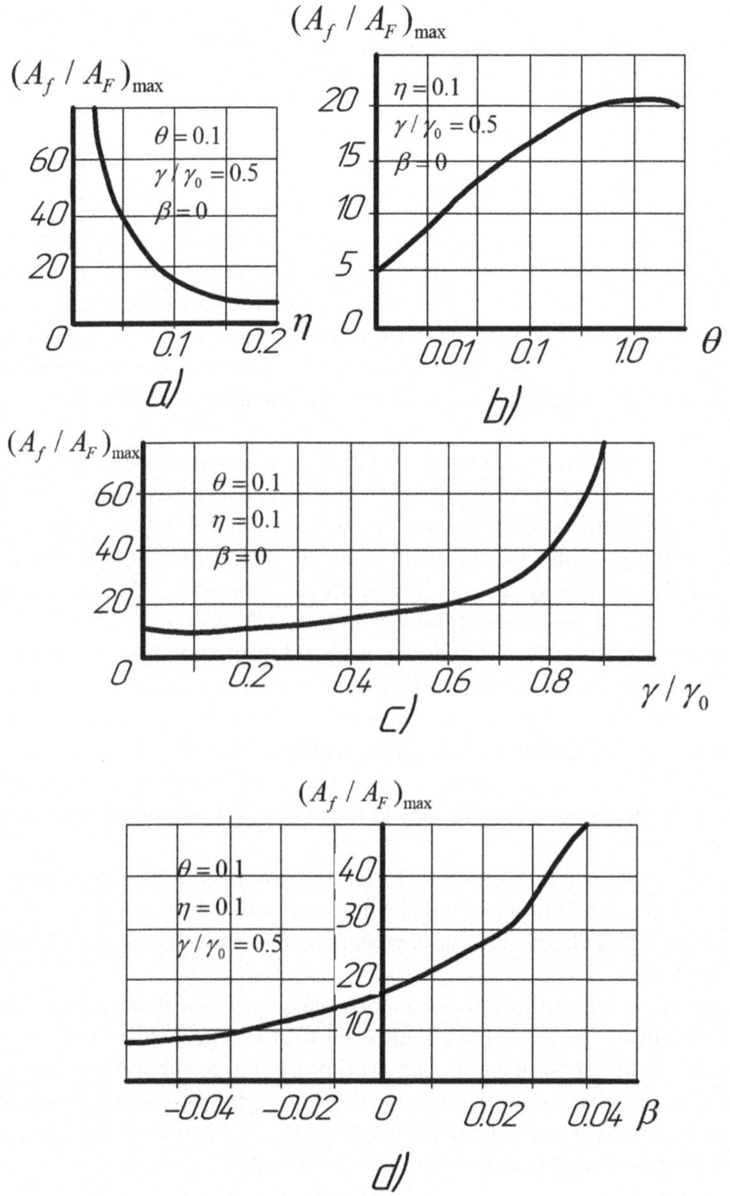

Fig. 8 Influence of changes in the dimensionless parameters of a closed spindle system on the amplitude of forced oscillations (A_f/A_F—the ratio of the amplitude of forced oscillations to the amplitude of the disturbing force)

The extreme properties of θ are related to the fact that the minimum boundary value of the coefficient of connectivity γ_0 is obtained with $\theta = 1$.

4 Conclusions

For the description of a closed dynamic system of machine is proposed a set of dimensionless parameters. These parameters, which characterize the relationship of the elastic system to the cutting process (γ, β), the dissipative properties of the elastic system (η) and the relationship of the constant time of particle formation to the period of free oscillation (θ), are used to solve differential motion equations. The use of these coefficients allows to simplify the calculations of vibration resistance criteria.

In dynamic systems, loaded with constant forces (gravity, cutting), the parameter change is a source of forced parameter fluctuations whose resonant values depend on: $\Delta\gamma$ at low system stability; from η—at high system stability.

Determining the influence of changes in the dimensionless parameters of the closed spindle-trunnion system on the amplitude of forced oscillations allows managing the reserve of resistance of the dynamic LDTW system.

The unlimited increase of amplitude of forced oscillations as the working point approaches the boundary of the stability field is a consequence of the linearization of the closed dynamic system. Non-linearity limits the level of forced oscillations and determines the amplitude of the auto oscillations during cutting. The excitation of auto-oscillations during cutting is most often determined by a combination of various factors that cause non-linearity of the cutting process and the elastic system of the lather.

The use of the stability criterion for a single-mass model allows us to study vibrations during fine boring of holes with rigid boring bars in the spindle-boring bar subsystem, calculating the parameters of the elastic system and adjusting the cutting modes.

The use of the stability criterion for a two-mass system with the proposed dimensionless parameters for finish boring allows us to study the dynamics of interactions between the spindle-boring bar and the workpiece-fixture subsystems under conditions of closure to the cutting process or under the combined action of bending and torsional vibrations when machining long holes. The features of the vibrations of non-stationary technological systems will be studied in the future.

Acknowledgments The research was partially supported by the International Association for Technological Development and Innovations.

References

Alammari Y, Sanati M, Freiheit T, Park SS (2015) Investigation of boring bar dynamics for chatter suppression. Proc Manufact 1:768–778. https://doi.org/10.1016/j.promfg.2015.09.059

Alifov AA (2020) About mixed forced, parametric and self-oscillations by limited excitation and delayed elasticity. PNRPU Mech Bull (3):12–19. https://doi.org/10.15593/perm.mech/2020.3.02

Barois T, Perisanu S, Vincent P, Purcell ST, Ayari A (2014) Frequency modulated self-oscillation and phase inertia in a synchronized nanowire mechanical resonator. New J Phys 16:083009. https://doi.org/10.1088/1367-2630/16/8/083009

Basovich S, Arogeti S (2021) Identification and robust control for regenerative chatter in internal turning with simultaneous compensation of machining error. Mech Syst Signal Process 149. https://doi.org/10.1016/j.ymssp.2020.107208

Belov EB, Leonov SL, Markov AM, Sitnikov AA, Khomenko VA (2017) Forecasting of machined surface waviness on the basis of self-oscillations analysis. IOP Conf Series Earth Environ Sci 50(1):art. no. 012053. https://doi.org/10.1088/1755-1315/50/1/012053

Butikov EI (2018) Analytical expressions for stability regions in the Ince–Strutt diagram of Mathieu equation. Am J Phys 86(4):257–267. https://doi.org/10.1119/1.5021895

Ftorek B, Oršanský P, Šamajová H (2018) Parametric oscillations of the mechanical systems. MATEC Web Conf 157:08002. https://doi.org/10.1051/matecconf/201815708002

Grigorian GA (2022) Oscillation and non-oscillation criteria for linear nonhomogeneous systems of two first-order ordinary differential equations. J Math Anal Appl 507:125734. https://doi.org/10.1016/j.jmaa.2021.125734

Ivanov V, Pavlenko I, Kuric I, Kosov M (2019) Mathematical modeling and numerical simulation of fixtures for fork-type parts manufacturing. In: Knapcikova L, Balog M (eds) Industry 4.0: trends in management of intelligent manufacturing systems, EAI/Springer innovations in communication and computing. Springer, Cham, pp 133–142. https://doi.org/10.1007/978-3-030-14011-3_12

Oborskyi G, Orgiyan A, Tonkonogyi V, Balaniuk A, Muraviova I (2021) Dynamics of fine boring with multicutting console drilling rods. In: Advanced manufacturing processes II, Lecture notes in mechanical engineering, pp 577–587. https://doi.org/10.1007/978-3-030-68014-5_56

Oborskyi G, Orgiyan A, Ivanov V, Balaniuk A, Pavlenko I, Trojanowska J (2023) Improvement of the dynamic quality of cantilever boring bars for fine boring. Machines 11(1):7. https://doi.org/10.3390/machines11010007

Ozlu E, Budak E (2007) Analytical modeling of chatter stability in turning and boring operations—Part I: model development. J Manufact Sci Eng 129(4). https://doi.org/10.1115/1.2738118

Parmar RK, Milovanović GV, Pogány TK (2023) Extension of Mathieu series and alternating Mathieu series involving the Neumann function Yν. Period Math Hung 86(1):191–209. https://doi.org/10.1007/s10998-022-00471-9

Ramesh K, Alwarsamy T, Jayabal S (2013) Investigation of chatter stability in boring tool and tool wear prediction using neural network. Int J Mater Prod Technol 46(1):47–70. https://doi.org/10.1504/IJMPT.2013.052789

Reith MJ, Bachrathy D, Stepan G (2016) Improving the stability of multi-cutter turning with detuned dynamics. Mach Sci Technol 20(3):440–459. https://doi.org/10.1080/10910344.2016.1191029

Smolenicki D, Boos J, Kuster F, Roelofs H, Wyen CF (2014) In-process measurement of friction coefficient in orthogonal cutting. CIRP Ann 63(1):97–100. https://doi.org/10.1016/j.cirp.2014.03.083

Vnukov Y, Dyadya SI, Kozlova OB, Trishin PR, Zubarev AE (2023) Influence of cutting time on types of oscillations during blade processing. Ukr J Mech Eng Mater Sci 9(1):53–66

Wu BS, Lim CW, Ma YF (2003) Analytical approximation to large-amplitude oscillation of a non-linear conservative system. Int J Non-Linear Mech 38(7):1037–1043. https://doi.org/10.1016/S0020-7462(02)00050-1

Dynamics of the Cutting Process

Yiriy Petrakov

Abstract Most machine parts are manufactured in the cutting process, which is carried out in a technological machining system and is always accompanied by vibrations. Vibrations (chatter) are the main restraining factor in increasing productivity. Among the known methods for eliminating chatter, the most promising is selecting the cutting mode according to the stability diagram, since this method does not require equipment modernization and is easily implemented on any CNC machine. It has been established that the main cause of chatter in the form of regenerative self-oscillations is machining along a trace. To compile a mathematical model, it is proposed to represent the cutting process in a closed system with negative feedback on elastic displacement and positive feedback through the function of the delay argument. As a result of the frequency-time approach to the stability study, a new stability criterion is established, which is based on the location of the Nyquist diagram on the complex plane. Such a stability criterion (unlike the known ones) allows for complete automation of the stability diagram construction. Methods for identifying dynamic parameters of machining systems using experimental modal analysis are presented. The results of experimental spectra obtained from the impulse response function after its processing by fast Fourier transforms are presented. The demonstrated results of experimental testing during turning and milling fully confirm the developed method of controlling dynamic phenomena during cutting.

Keywords Chatter in machining · Stability criterion · Stability lobes diagram

Y. Petrakov (✉)
National Technical University of Ukraine "Igor Sikorsky Kyiv Polytechnic Institute", Kyiv, Ukraine

V. Ivanov (ed.), *Emerging Technologies in Manufacturing and Materials Engineering*, Springer Tracts in Mechanical Engineering,
https://doi.org/10.1007/978-3-032-14914-5_15

1 Introduction

Most machine parts are manufactured using a cutting process, mainly on CNC machines. The main objective of the manufacturing process is the highest possible machining productivity while meeting all quality requirements for processing. The desire to increase productivity simply by increasing the cutting mode is constrained by the occurrence of vibrations, which lead to a significant deterioration in surface quality, increased wear of the tool, and the machine itself. This problem is significant when machining thin-walled parts widely used in the aerospace and automotive industries (Sun et al. 2023). Also, problems associated with vibration and machining dynamics during micro-milling often limit quality and productivity (Bai et al. 2024).

Any cutting process is performed in an elastic mechanical machining system and is necessarily accompanied by vibrations. The machining system has three types of vibrations: free, forced, and self-excited vibrations (Yue et al. 2019). The causes of the first two types of vibrations, inherent in any mechanical system, are relatively well studied, and various methods for their elimination have been proposed. Self-excited vibration is an undamped vibration of the machining system caused by the peculiarities of the interaction between the cutting force and the volume of the removed material, which occurs in an elastic system of a complex dynamic structure. These regenerative vibrations lead to chatter on the machined surface and constitute the main obstacle to stabilizing the cutting process and the controllability of the machining system.

In the stock removal process, the main disturbance acting on the machining system is the cutting force, which causes elastic displacements of the system from the initially specified position. Such displacements lead to a change in the volume of the material being removed, which in turn causes a change in the cutting force, which usually decreases the volume of interaction between the tool and the workpiece. Such interaction, in turn, provokes a decrease in the cutting force, which leads to a decrease in the volume of the material being removed, and so on. This is the main property of machining systems structurally closed through negative feedback.

In addition to such interaction, disturbances arise in the machining system in the form of a change in the allowance, which is removed during the pass and formed due to oscillations during the previous pass. This disturbance is the main cause of self-excited oscillations in the machining system, since its frequency coincides with the frequency of its main harmonic. This process is represented by another feedback in the machining system, which is positive and acts with a time delay equal to the time of one pass (Altintas 2012).

A large number of studies are devoted to the development of technologies that are capable of predicting and preventing vibration. Various methods are used to reduce the level of regenerative oscillations to the permissible amplitude value, which can be divided into three groups (Li et al. 2024). The first group includes passive chatter control (PCC) methods, consisting of changing the dynamic properties of the machining system due to different designs of dynamic vibration dampers. It is also proposed to change the geometry of the tool, for example, the pitch of the flutes of end mills or the application of the spindle speed variation (SSV), which helps to change the shape of the allowance in time and, as a result, the departure of the disturbance frequency from the frequency of the main harmonic of the machining system. The choice of reasonable disturbance parameters remains an unsolved problem for the SSV method (Li et al. 2024).

The second group consists of using active chatter control (ACC) methods, which consist of creating automatic systems that affect the machining system by excitation in antiphase to the main excitation of the cutting process. Such systems are complex to set up, and their use may be appropriate for rough milling operations with low cutting speeds and large allowances. The control strategy based on counter-excitation has limitations associated with the bandwidth limitations of real systems. The main disadvantage of these works is the lack of a generalized theoretical basis for a dynamic model of the machining system, which considers its structure and allows for unifying the stability criterion. In addition, all the methods considered are somehow related to the modernization of the machine's dynamic system and necessarily lead to significant material costs.

The method of cutting mode control based on stability lobes diagrams (SLD), which divide the entire area of the space "cutting depth—cutting speed" into stable and unstable, is considered separately. Obviously, such control does not require modernization of machining systems, but its effectiveness requires adequate representation of the dynamic model of the machining system and is associated with a reliable stability criterion.

The paper presents methods for constructing dynamic models of machining systems based on a system approach using the apparatus of transfer functions. A new stability criterion is also substantiated, and methods of experimental modal analysis for identifying dynamic machining systems are proposed. The theoretical results have been experimentally confirmed in modeling and full-scale testing of various machining systems.

2 Fundamentals and Challenges

A review of the work of scientists devoted to vibrations during the machining of parts by cutting shows that this phenomenon is generated by dynamic processes occurring in the machining system itself, and the use of a SLD to assign a chatter-free cutting mode allows for the highest possible quality of machining of various types of cutting—milling, turning, and grinding.

The best results in predicting stability are obtained by simultaneously representing the dynamic cutting process in the frequency and time domains (Altintas et al. 2020). Continuous analysis of turning in the time domain is represented as an equation, where the cutting force and rigidity are provided only in the normal direction. It is shown that the dynamics of the machining process must be simulated, considering the interaction between the machining process and the elastic system of both the machine tool and the workpiece. Although it is known that vibration occurring in different types of machine tools is essentially caused by the same basic physical phenomena, no general theory currently covers all particular cases. As a result, vibration occurring in each of the different types of machine tools must be considered as a separate problem, complicating theoretical developments and practical recommendations.

Most scientists believe that the main cause of vibrations is the recovery vibrations caused by machining on the trace. The model of the machining system is presented in the form of a structural diagram with positive feedback through a time-delay link. The proposed model of stable cutting is based on a stability diagram constructed from petals corresponding to the conditions of the solution of the differential equation of the processing system through the roots of its characteristic equation (Altintas et al. 2020).

Such a solution ignores the frequency properties of the system, is approximate and cannot be included in the numerical procedure of the automatic algorithm. Next, an algorithm for designing a diagram is proposed, which includes sequential calculations for each spindle speed. After determining the frequency transfer function of the processing system and the coefficients of cutting force, workpiece material and cutting width, an algorithm is used that contains 5 steps (Altintas 2012). However, among these steps, there are actions that involve the intervention of the designer, which makes it unsuitable for design automation.

Dynamic models of machining systems are constructed without considering their closed nature, natural feedback, and their behavior for assessing stability, which is replaced by an analysis of incoming signals in the form of cutting depth for the current and previous pass (Thévenot et al. 2006). According to the general theory of stability, the stability of systems cannot depend on external influences, but is determined by internal processes occurring in their structure.

In (Zhu and Liu 2020), it is noted that during machining on a machine tool, when one of the machine tool's operating modes is self-exciting, self-oscillations occur at frequencies close to the dominant frequencies of the machine tool, but not equal to them. The article examines the cross-sections of the workpiece, which, according to the authors, determine the thickness of the chips that will be removed in the next pass. Taking into account the phases between the waves, various phases are presented and it is stated that a cutting force will act in the system, which will cause oscillations of the dynamic processing system. Thus, the cutting force is presented as the main factor capable of causing a violation of the stability of the system. However, the stability of any system, including a mechanical one, depends on the internal processes occurring in its structure and should be determined by the analysis of its mathematical model, and not by external influences, which, in this case, is the cutting force.

Taking into account the complexity of the theoretical solution to the problem of designing a stability diagram and the need to adapt to the real dynamic parameters of the machining system for which the design is carried out, methods of experimental search for a vibration-free cutting mode are used. An improvement of these methods is proposed, which consists in the fact that the search is carried out using a specially created automatic control system with feedback through a vibration sensor (Brecher et al. 2018). When changing the cutting mode in accordance with a given program, online detection of chatter is performed and the cutting mode parameters are updated.

Similar studies were conducted for turning operations, which indicates a common nature of vibration occurrence during cutting (Turkes et al. 2011). The modeling of the dynamic machining system was limited to a single-mass model with one degree of freedom using an oriented transfer function. The SLD graph was constructed based on the solution of the characteristic equation and, since this equation for a system with a delay argument function is transcendental, the Taylor series expansion was also used here. To clarify the solutions, an experiment was conducted, during which an acceptable discrepancy between the results obtained during the experiment and the theoretical results using the Taylor series expansion was confirmed.

The paper (Law et al. 2020) shows the results of comparison of experimental and theoretical frequency response function (FRF) for a tool system consisting of a cutter and represented by a single-mass model. However, there is no information on how to adapt the obtained experimental results, which are usually multi-mass systems, to the adopted dynamic model of the system, which is used to construct SLD.

The algorithm for constructing the SLD assumes that the dynamic characteristics of the processing system are known. As one of the widely used methods for studying dynamic characteristics, operational modal analysis is more closely related to the actual working state of mechanical structures compared with experimental modal analysis. This is due to the fact that there is always a discrepancy between the excitation methods and the actual working state, which affects the accuracy of identifying the modal parameters. A new method of operational modal analysis for CNC machines is proposed based on the surface topography processed by side milling (Guan et al. 2024).

Self-excited regenerative vibration or chatter limits the basic requirements such as productivity, surface quality and dimensional accuracy of high-speed machining. It is the most critical factor that seriously affects the tool life and machine tool life. Among several methods used to suppress and prevent chatter, the machining conditions selected by the stability lobe diagram are the most reliable way. The stability lobe diagrams are usually generated by knowing the specific cutting force coefficients and the tool point frequency response functions (FRFs). A comparative study of the stability lobe diagrams generated by static modal analysis and operational modal analysis (OMA) showed that the cutting conditions selected from the SLD obtained by OMA are more realistic in preventing chatter during high-speed milling. To apply the OMA method, vibration signals were obtained at the cutter tip during machining using a non-contact laser vibrometer (Paliwal and Ramesh Babu 2020). In this case, the FRF of the machine is considered as a link between the dynamic force load and the vibration response (Guo et al. 2022).

It is noted that the search for stability using existing theoretical methods is time-consuming and recently it has been proposed to use deep learning methods using the

LSTM neural network (Wang et al. 2024). However, such methods do not yet guarantee an adequate solution for a specific dynamic machining system.

To construct the SLD, a cutting force model with deterministic coefficients of dependence on the cutting mode, material, tool, etc. is also required. An accurate local milling force prediction model is the key to reliably modeling the cutting force fluctuations and their effect on the machined surface shape (Liu and Jin 2020). However, accounting for such changes in the milling force remains a serious problem due to changes in the instantaneous chip thickness and non-constant cutting coefficients with changing milling conditions. Thus, a general milling force model with coefficients calculated based on average forces from experimental results may lose reliability when applied to different milling conditions. The cutting force coefficients are related to the specific cutting force, which is recommended for modeling under different milling process configurations (Janota et al. 2019). Different procedures used to predict the specific cutting force from experimental results are analyzed. A research method is proposed that allows estimating the specific cutting force based on information from the machine control system. The specific cutting force is calculated as the ratio of the material removal rate to the power measured on the machine spindle. This approach may be the only effective one when milling surfaces whose contour is outlined by variable curvature guides, since for such an operation there is no direct relationship between the feed, cutting speed and instantaneous thickness of the cut chips. Therefore, it can be expected that the proposed method of analyzing the material removal rate gives adequate results for such operations.

However, the creation of a model for predicting the milling force is quite a complex task, since the force is the result of multi-parameter relationships of the technological processing system. In general, based on the analysis of many works in the field of aviation engineering, studies of the force model are determined based on three methods: empirical dependencies, methods of finished elements and a model based on the instantaneous undeformed thickness of the chip (Duan et al. 2021).

It is known that the accuracy of machining in final milling is affected by elastic deviations of the machining system under the action of the cutting force. In (Kaneko et al. 2023), a new method for analyzing the elastic shift of the workpiece in final milling is proposed. The advantage of the method is the possibility of combining it with geometric modeling of milling based on voxels used to predict the cutting force. To predict the error in the shape of the machined surface, the rigidity of the tool is taken into account first of all with some corrections for additional elastic shifts of the entire machining system. To calculate the magnitude of the elastic shift of the tool, the force load is taken in the form of a point force. However, the feedback of the machining elastic system is not taken into account in the overall balance.

Most studies devoted to dynamic phenomena during cutting and vibration elimination tend to control the process by selecting the cutting mode in accordance with the stability diagram. Such control is easily implemented on a CNC machine and does not require its modernization. It is proposed to construct a stability diagram using a multi-step algorithm that allows designing a diagram based on an approximate method for solving a transcendental characteristic equation. This approach does not rely on a mathematically correct criterion and cannot be implemented in an automatically operating algorithm. Most mathematical models representing processing systems are based on single-mass systems. It is possible to increase the adequacy of dynamic models only by taking into account the interaction of at least two masses representing the workpiece and the tool and interacting through the cutting process. The creation of a digital modeling technique using a scientifically based stability criterion to solve the problem remains an urgent scientific and production task.

Methods for identifying the dynamic parameters of the machining system, necessary for designing the SLD in terms of adapting the experimentally obtained FRF to the adopted dynamic model, deserve development. In addition, it is important to determine the relationship between the cutting speed of the material and the cutting force. The purpose of this work is to create a new method for determining the vibration-free cutting mode based on the construction of SLD, which will provide the possibility of optimizing the process on CNC machines.

3 Ideas and Solutions

The cutting process always occurs in a closed machining system W_c, the stability of which can be determined by the Nyquist criterion as for a system with negative feedback. The practice of studying dynamic systems of metal-cutting machines shows that, as a rule, such systems are stable, that is, the Nyquist diagram of an open system on a complex plane does not cover a point with coordinates [−1,0]. In the diagram (Fig. 1): h_c is the commanded cutting depth, h_l is the cutting depth on the previous pass, h_a is the actual cutting depth, δy is the elastic displacement of the machining system, τ is the delay hour. Therefore, to study such systems, it is sufficient, without losing adequacy, to represent it as a system of no higher than the fourth, and most likely even the second order.

At the same time, an in-depth study of the cutting processes revealed that undamped oscillations caused by self-excitation necessarily arise in the system. The cause of such oscillations should be considered to be the machining on the trace, which in the structure of the machining system is represented by a link in the delay function with the transfer function $e^{-\tau s}$, where τ is the delay time, s is the Laplace operator. The delay time is determined depending on the type of machining, for example, for turning this is the time of one revolution of the workpiece, for milling—the cutting time between two adjacent teeth of the cutter, etc.

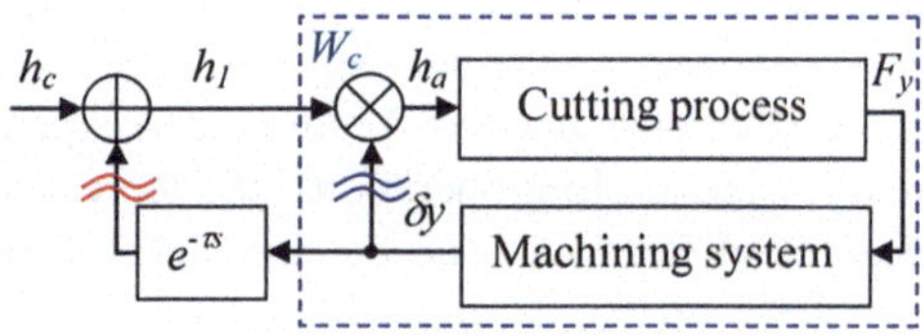

Fig. 1 Structure of the machining system

It is proposed to determine the stability of such systems using a new criterion based on the frequency approach using the Nyquist diagram. The diagram should be constructed similarly to the determination of stability in the W_c machining system, but with the transfer function replaced in accordance with the rules for transforming structural diagrams, and the disconnection should be carried out using positive feedback through the delay function.

Thus, it is proposed to formulate a new stability criterion for machining systems taking into account track machining as follows: a system with a delay function in positive feedback will be stable if its Nyquist diagram in the open state does not cover a point with coordinates [+1,0] on the complex plane. Such a criterion allows one to construct a stability diagram for a machining system in a fully automatic mode (Petrakov and Danylchenko 2022). The stability diagram is usually constructed as a dependence of the selected cutting mode parameter (depth for turning or feed for milling) on the cutting speed (spindle speed).

The first procedure of the created algorithm is the calculation of digital arrays of amplitudes $A(\varphi)$ of the frequency response for a given cutting mode—spindle speed N_{sp} and feed f_c. Then, all amplitudes $A_i(\varphi_i)$ of the frequency response are determined in the vicinity of the phase angle $\varphi_i = 2\pi i \pm \delta$, where $i = 0,1,2...$. Moreover, the value δ is taken equal to the step of the digital array. Such conditions guarantee rapid convergence of the algorithm procedures and finding the maximum amplitude A_{max}. Exiting the computational cycle according to condition 1 determines the values of the cutting mode components that correspond to one point of the desired SLD graph.

Otherwise, the feed value changes according to condition 2, then the model is calculated with this feed value and a new amplitude-frequency response determined. After that, the procedure for finding the maximum amplitude A_{max} is repeated again, also in the phase region $\varphi_i = 2\pi i \pm \delta$, where $i = 0,1,2...$etc. Thus, the algorithm provides for a return to the calculation of the system model after each iteration of determining the diagram point and moving along the cutting depth or spindle speed. Depending on the complexity of the model, such a cycle slows down the speed of calculations, however, as practice shows, the design time of the stability diagram is several tens of seconds.

As a result of such procedures, an array of $N_i(f_i)$ SLD values is calculated and output in a given range of spindle rotation frequency variation (highlighted in Fig. 2). The accuracy of determining SLD depends on the tolerance value and the step of changing both parameters.

Thus, the construction of the stability diagram assumes known mathematical models of the machining systems. Such models, which best reflect the processes operating in them, are presented in the form of structural diagrams, which are obtained from the analysis of dynamic models of the machining systems. Here, as always, it is necessary to seek a compromise between the complexity of the model, which depends on the number of masses and degrees of freedom, and the adequacy of the solutions obtained. Accumulated experience (Petrakov et al. 2024) has shown that for end milling operations, the representation by a two-mass model with two degrees of freedom is acceptable, and for turning operations—a single-mass system, but with two degrees of freedom.

Fig. 2 Block diagram of the algorithm for constructing SLD

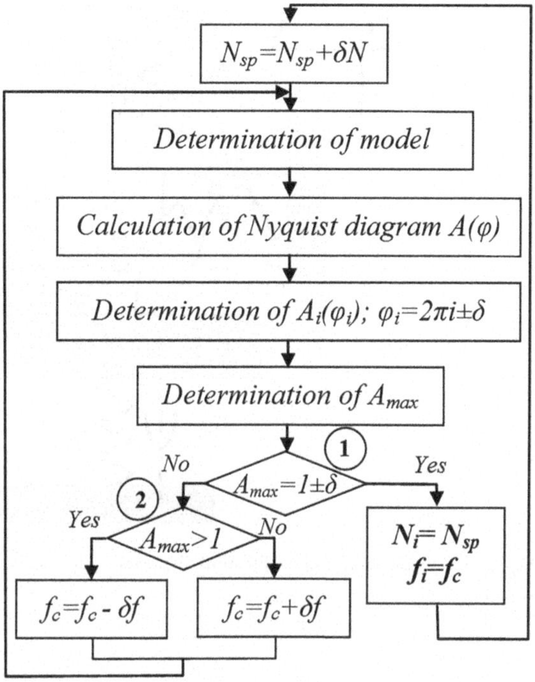

The end milling machining system is represented by two masses—a cutter (mill) and a part, each with two degrees of freedom (Fig. 3). The mill is connected to the bed in the direction of two coordinates X and Y by stiffness k_{mX}, k_{mY} with viscous friction λ_{mX}, λ_{mY}. The dynamic system of the workpiece is represented similarly—stiffness k_{wX}, k_{wY} with viscous friction λ_{wX}, λ_{wY}. Milling is performed with a commanded feed f_c at a commanded cutting depth H_c. Such set parameters of the cutting mode provoke a cutting force that can be represented on the basis of a mechanistic approach to the calculation by the geometric parameters of the undeformed chip cut by each tooth along its length (coordinate Z). This dependence is nonlinear and can be linearized with a sufficient degree of accuracy as a function of the cutting mode components:

$$F = k_f \left(f_t\right)_a + k_H H_a, \tag{1}$$

where $(f_t)_a$, H_a are actual feed and cutting depth.

Formula (1) represents the linear dependence of the cutting force on the feed and cutting depth with coefficients in the form of partial derivative values at the linearization point:

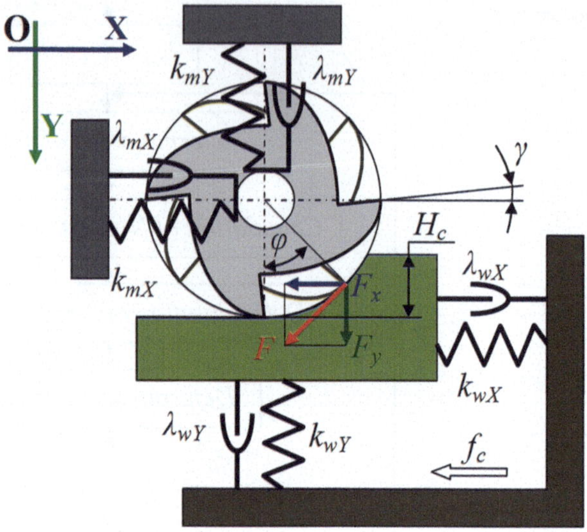

Fig. 3 Dynamic model of end milling machining system

$$k_f = \left(\frac{\partial F}{\partial f}\right)_0 = C_F \alpha \left(f_t\right)_a^{\alpha-1} H_a^{\alpha} B,$$
$$k_H = \left(\frac{\partial F}{\partial H}\right)_0 = C_F \left(f_t\right)_a^{\alpha} \alpha H_a^{\alpha-1} B, \tag{2}$$

where C_F, α are empirical coefficient and exponent, B is width of cutting.

The geometric arrangement of the nodes of the dynamic system changes under the action of elastic displacements (δ_{mX}, δ_{mY}, δ_{wX}, δ_{wY}) from the components of the cutting force in the direction of the coordinate axes:

$$H_a = H_c + \delta_{mY} + \delta_{wY},$$
$$\left(f_t\right)_a = \left(f_t\right)_c + \delta_{mX} + \delta_{mY}, \tag{3}$$

where H_c, $(f_t)_c$ are commanded feed and cutting depth.

This engagement is shown by a generalized mathematical model of the machining system, which is built using a systems approach and is presented as a connection of isolated elements with an indication of the connections between them (Fig. 4).

This model reflects the connections in the system well, taking into account formulas (2), and the machining on the trail is represented by the functions of the delay argument for each controlled coordinate. Since the dynamic system is represented as a two-mass system, the structural diagram shows two transfer functions in the direction of each coordinate axis:

along axis Y:

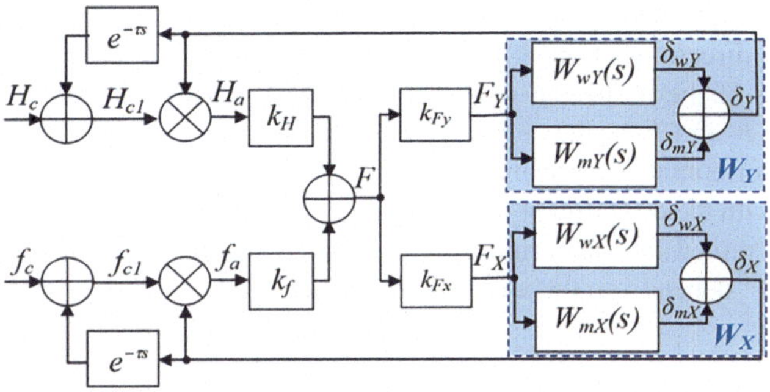

Fig. 4 Mathematical model of end milling machining system

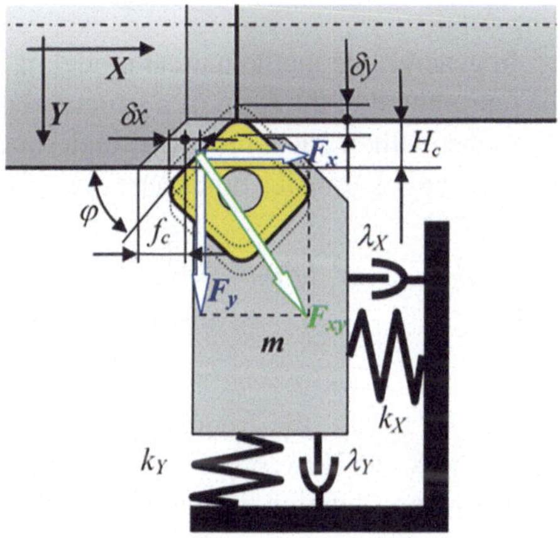

Fig. 5 Dynamic model of the turning machining system

$$W_{wY}(s) = \frac{1/k_{wY}}{\frac{s^2}{\omega_{wY}^2} + 2\xi\frac{s}{\omega_{wY}} + 1}, \quad W_{mY}(s) = \frac{1/k_{mY}}{\frac{s^2}{\omega_{mY}^2} + 2\xi\frac{s}{\omega_{mY}} + 1}, \tag{4}$$

along axis X:

$$W_{wX}(s) = \frac{1/k_{wX}}{\frac{s^2}{\omega_{wX}^2} + 2\xi\frac{s}{\omega_{wX}} + 1}, \quad W_{mX}(s) = \frac{1/k_{mX}}{\frac{s^2}{\omega_{mX}^2} + 2\xi\frac{s}{\omega_{mX}} + 1}, \tag{5}$$

where ω_{wX}, ω_{wY} are the frequency of natural oscillations of the workpiece system along the X and Y axes, ω_{mX}, ω_{mY} are the frequency of natural oscillations of the mill system along the X and Y axes, ξ is damping coefficient, s is the Laplace operator.

The turning system can be investigated in a similar way, but here it is sufficient to limit ourselves to a single-mass model with two degrees of freedom (Fig. 5).

The figure uses the same designations of elastic connections of the dynamic system as in Fig. 3.

Elastic displacements of the machining system change the area of undeformed chips in two directions at once: in the direction of the Y axis: $H_a = H_c \pm \delta y$ and in the direction of the X axis: $f_a = f_c \pm \delta x$. Elastic displacements are caused by the components F_x and F_y of the cutting force, which are the vectors of the decomposition of the horizontal component F_{xy}, acting along the front surface in the direction coinciding with the direction of the main angle in the plan φ of the cutting plate, and can be determined by the formula $F_{xy} = 0.6F_z$ (Machining handbook, n.d.). The component F_z depends on the area of undeformed chips:

$$F_z = f_a \left(k_f\right)_0 + H_a \left(k_H\right)_0, \tag{6}$$

where f_a, H_a are the current feed per revolution and cutting depth, $(k_f)_0$, $(k_H)_0$ are the coefficients of the linearized dependence of the cutting force on the feed and cutting depth.

In general, the mathematical model of the cutting process in an elastic TOS can be represented in the form of a structural diagram shown in Fig. 6.

In the mathematical model, a single-mass system with two degrees of freedom is represented by transfer functions:

$$W_Y(s) = \frac{1/k_Y}{\frac{s^2}{\omega_Y^2} + 2\xi\frac{s}{\omega_Y} + 1}, \quad W_Y(s) = \frac{1/k_X}{\frac{s^2}{\omega_X^2} + 2\xi\frac{s}{\omega_X} + 1}, \tag{7}$$

where ω_X, ω_Y are the frequency of natural oscillations of the machining system along the X and Y axes, ξ is damping coefficient, s is the Laplace operator.

This representation of mathematical models of dynamic systems allows us to obtain transfer functions of each processing system for the feed input and the cutting depth input, which makes it possible to evaluate the stability of the system using a new criterion for these effects. For the practical application of the new stability criterion, it is necessary to identify the system. The dynamic responses are subject to identification, primarily the frequencies of natural oscillations and the rigidity of the machining system in each case in accordance with the mathematical model used.

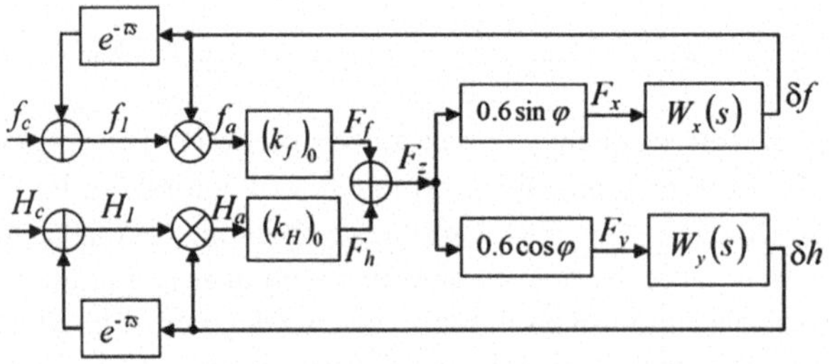

Fig. 6 Mathematical model of turning machining system

Identification of dynamic parameters is carried out by methods of experimental modal analysis, when the frequency response of the machining system is determined by the impulse response function (IRF) obtained by striking the system with a special impact hammer. To identify the dynamics of the milling machining system, it is necessary to obtain four IRF in the directions of two coordinate axes separately for the milling system and the workpiece system, and to identify the dynamics of the turning machining system, two experimental responses are sufficient. To eliminate random effects, it is recommended to perform a series of such experiments for subsequent averaging of the results. Figure 7 shows the scheme for performing an experiment for the milling system on a TM 0p CNC-machine from HAAS (Petrakov et al. 2024).

The amplitude-frequency responses of the dynamic system were determined through digital IRF files using the standard Fast Fourier Transform (FFT) procedure. Figure 8 shows a graph of the IRF file obtained during the experimental modal analysis of the "tool" system using a hammer impact and stored in a storage oscilloscope (Fig. 7).

Taking into account the representation of the dynamic model for each coordinate as a single-mass model, it is necessary to extract the natural frequency of the equivalent single-mass system from the signal spectrum obtained using the FFT (Fig. 9). For such an operation, a decision was made on the equivalence of the spectrum energy E and signals of the single-mass model. Therefore, the natural frequency of the single-mass model is:

$$(\omega_m)_Y = \sqrt{\frac{2E}{A_a^2}}, \tag{8}$$

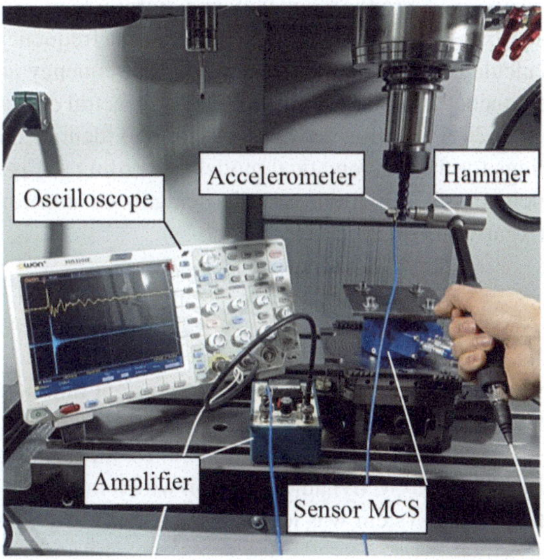

Fig. 7 Determination of the IRF of the "tool" system

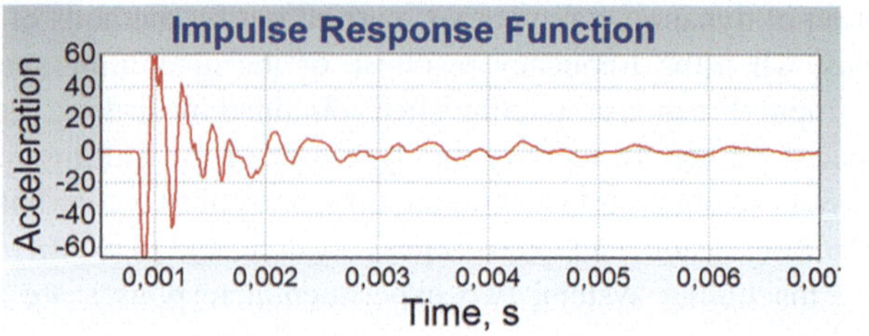

Fig. 8 IRF of the "tool" system

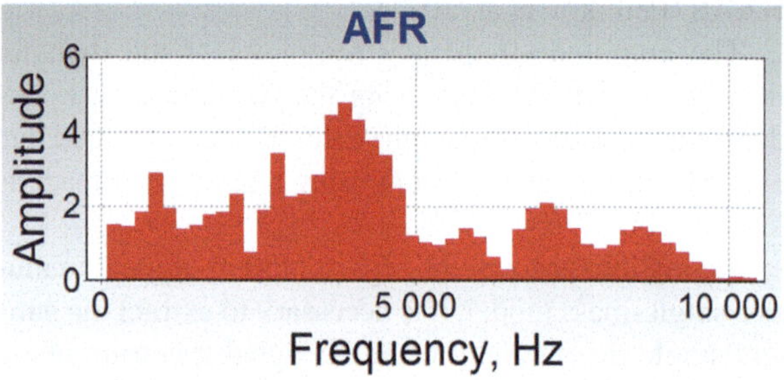

Fig. 9 Amplitude-frequency response

where A_a is the average amplitude of spectrum.

To calculate the energy of the spectrum obtained in the experimental modal test, one can use the formula:

$$E = \sum_{i=1}^{kn} \left(\omega_i^2 A_i^2 \right) / 2, \quad A_a = \left(\sum_{i=1}^{kn} A_i \right) / kn. \tag{9}$$

where ω_i, A_i are the frequency and amplitude of each selected harmonic.

To increase the adequacy, the natural frequency of the replacement mass was calculated as the average between the frequency according to formula (8) and the frequency calculated according to the spectral density of the signal energy (Agarwal et al. 2012). In this way, it is possible to identify the frequencies of natural oscillations for the mathematical models according to Figs. 4 and 6. The attenuation coefficients of oscillations are also calculated according to the well-known formula (Agarwal et al. 2012):

$$\xi = \frac{\ln\left(A(t) / A(t+T) \right)}{T}, \tag{10}$$

where $A(t)$ is the amplitude of the transient function at time t, $A(t + T)$ is the amplitude of the transient function at time $t + T$, where T is the period.

When machining on CNC machines, the rigidity was determined using a Sensor MCS 10-005-3C dynamometer, on which the workpiece was installed (Fig. 10). The machining system was loaded by the CNC machine drives using a manual encoder

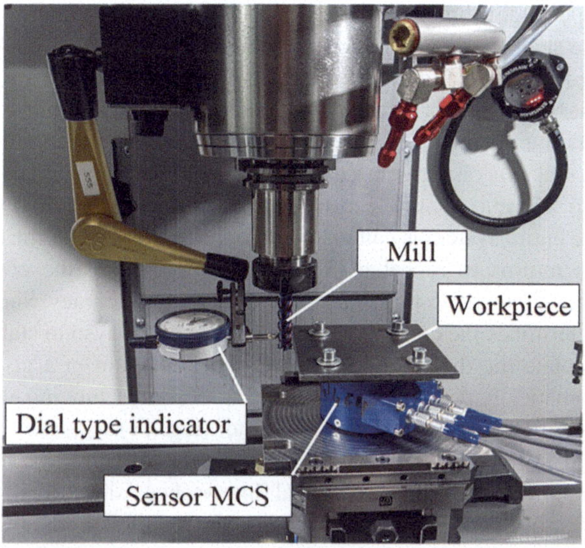

Fig. 10 Determination of stiffness in the direction of the axis Y

for moving the table along the coordinate axes. The workpiece was brought into contact with the mill, and the elastic displacement of each was controlled by a clock-type indicator. In this case, the force was controlled by the dynamometer, and the elastic displacement in the required directions was controlled by indicators. The stiffness was determined by a linearized dependence as the ratio of the force to the elastic displacement in the direction of the force.

According to the developed technology, to construct a stability diagram, it is necessary to use the created algorithm, which is based on the model of the processing system, built according to its structural diagram. It is this approach that makes it possible to obtain the transfer function of the system for the required input and calculate the frequency response of the system for this input. Frequency systems are determined by digital modeling using a special procedure that takes into account the function of the lagging argument. Transient characteristics are determined by the standard Runge-Kutta method of the 4th order, also taking into account the delay. Designing the Nyquist diagram of a system described by differential equations with delay is also performed by the standard procedure and designing the stability diagram according to the algorithm in Fig. 2.

All the above procedures are performed using numerical methods in a specially created application program, compiled in an object-oriented programming environment. To simulate the system using its mathematical model, the program should use the dynamic parameters identified during modal tests. The empirical coefficient and exponent in the cutting force formula for milling Steel 45 correspond to the data in (Machining handbook, n.d.), and the oscillation damping coefficient of 0.05 is determined from the experimental IRF.

The developed algorithm, which is based on the logistics of determining the digital array of the stability diagram in the specified components of the cutting mode (feed and speed) according to the new stability criterion of systems with a delay in positive feedback. The algorithm automatically performs iterations consisting of finding the maximum value of the frequency response amplitude in the phase angle range near the real axis on the complex plane. Then, if Amax>1, the feed is reduced by a given step, if Amax<1, the feed is automatically increased and the frequency response of the entire system is calculated with the new values of the cutting mode. The algorithm returns to repeated procedures according to the condition while $1 - \delta < \text{Amax} < 1 + \delta$, where δ is the specified calculation accuracy. Such a computational cycle determines the coordinates of one point of the system stability diagram graph. To calculate the digital model of the entire diagram in a given range, the algorithm automatically performs the described procedures, moving along the entire range of change in the cutting speed of the cutter (Fig. 11).

Above the SLD graph there is an unstable cutting zone, below—a stable one. The constructed diagram is of great practical importance, since it allows assign a vibration-free cutting mode taking into account the required productivity during the technological preparation of the operation. The created software allows simulate the transient and frequency characteristics of the system for the cutting modes selected in Fig. 11, which are indicated by circles. From Fig. 12 it is evident that the mode corresponding to points 1 (900 rpm) and 3 (1700 rpm) of the diagram ensures rapid damping of the system oscillations, and the mode corresponding to point 2 of the diagram (1300 rpm) causes an unstable process in the system.

At the same time, the property of the program to automatically construct Nyquist diagrams for the system for the specified cutting modes allows to fully verify the validity of the new stability criterion for systems with a retarded function in positive feedback. When the graph of the frequency transfer function of the system on the complex plane covers the point with coordinates [+1,0], the system loses stability, which is evident from the corresponding transient characteristics. It should be noted here that in practice, the phenomenon noted here as "loss of stability" does not lead

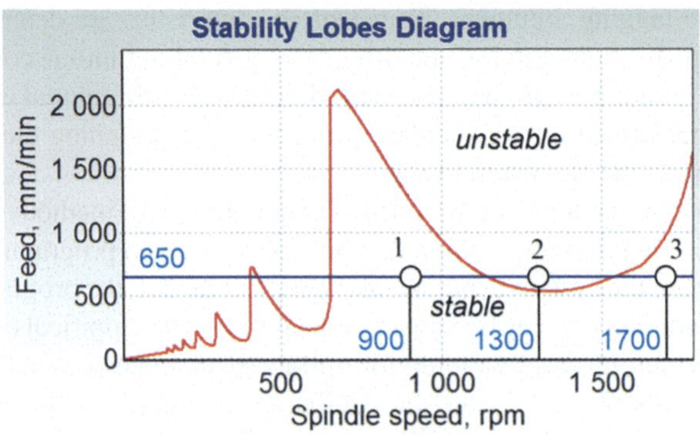

Fig. 11 Stability Lobes Diagram for end milling machining

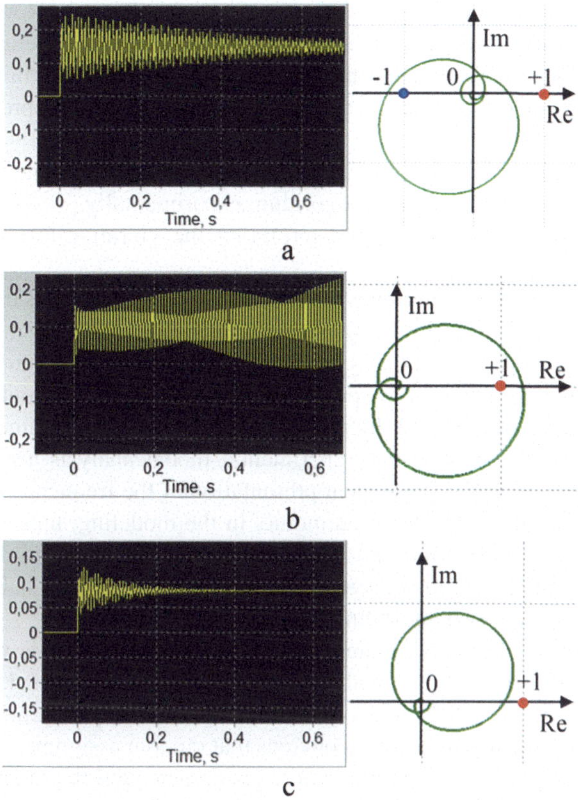

Fig. 12 Step Response and Nyquist Diagram: (**a**) $f = 650$ mm/min, $n = 900$ rpm; (**b**) $f = 650$ mm/min, $n = 1300$ rpm; (**c**) $f = 650$ mm/min, $n = 1700$ rpm

to catastrophic consequences, since it is obtained in relation to an idealized model, but is characterized by a significant increase in surface roughness.

To test the adequacy of the proposed solutions, full-scale experiments were conducted, during which a square-shaped workpiece was milled on a TM 0p HAAS CNC machine on a dynamometer table (Fig. 10). This arrangement of the workpiece allows for counter milling of all four sides of the workpiece during the experiment without stopping according to a single CNC-program. Milling was performed in cutting modes in accordance with points "1", "2" and "3" of the SLD (Fig. 11).

The stability of the machining system was assessed by measuring the roughness of the machined surface. The roughness profile of the surface machined under the cutting conditions corresponding to points 1 and 3 has an amplitude of 1.38 and 0.64 μm, respectively, and the surface machined under the cutting conditions of point 2 has Ra = 3.17 μm.

Under the condition of geometric roughness formation, without taking into account the random component, there is a linear dependence of the roughness parameter Ra on the spindle speed. Therefore, when machining under the conditions of point 2, Ra = 1.45 μm is expected (line 2 in Fig. 13). However, according to the results of profile measurements, Ra under these cutting conditions is 2.2 times greater. This indirectly confirms the chatter cutting mode in this case, which indicates a violation of the process stability in accordance with the prediction of the process stability. Similar results were observed in all experiments.

Thus, the graph of the stability diagram, constructed using the new stability criterion, which is used in the created algorithm, experimentally proved the effectiveness of the developed method for determining the vibration-free cutting mode during end milling.

Since the new stability criterion is based on the general principles of representing the cutting process, it will also be valid for turning. To conduct such experiments, following the developed methodology, an application program was created, with the help of which the modeling of the processing system was carried out according to its mathematical model (Fig. 6). Identification of the dynamic parameters was also carried out using the methods of experimental modal analysis according to the impulse response, with the subsequent presentation of the frequency transfer function and the selection of the main harmonic. In the modeling, numerical methods were used for constructing the transient and frequency response of the processing system in the form of a Nyquist diagram, as well as constructing a stability diagram according to the developed algorithm (Fig. 14).

The diagram highlights two points 1 and 2 of the cutting mode with the same cutting depth of 2.3 mm, but with different spindle speeds of 1550 rpm and 1650 rpm, respectively. The results of modeling and field experiments are shown in Figs. 15 and 16. Here it also becomes obvious that random assignment of the cutting mode without taking into account dynamic phenomena and the obtained patterns, as, incidentally, happens in production, can lead to unexpected negative consequences.

It should be noted that the terms "stability" or "not stability" should be adapted to the model, not to the real system, since the created models are always a certain approximation to full adequacy. Nevertheless, a correlation has been experimentally recorded, manifesting in an anomalous increase in the roughness of the machined surface compared to the cutting mode corresponding to the mode stable.

The greatest efficiency of application of the results of the work should be expected from the creation of systems using automatic control by feedback signals

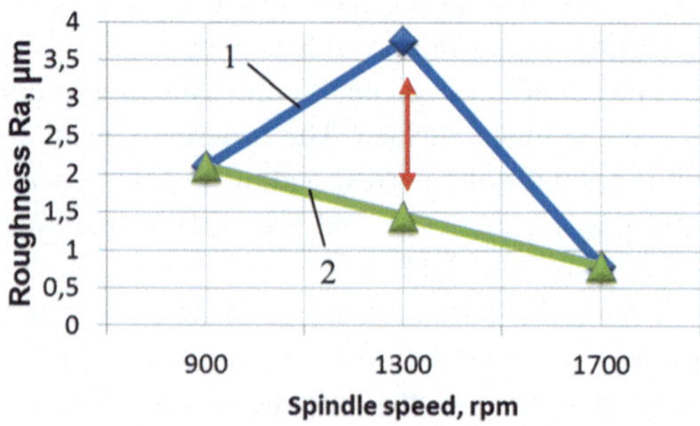

Fig. 13 Results of roughness measured

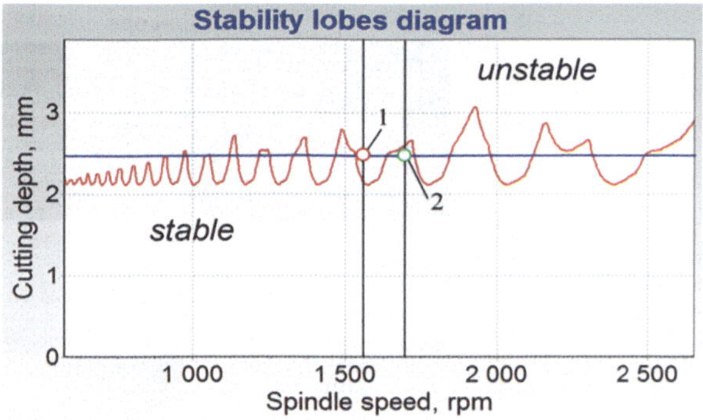

Fig. 14 Stability Lobes Diagram for turning machining

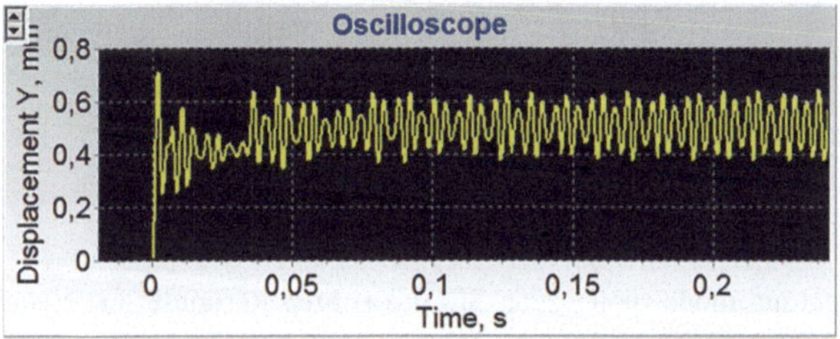

a

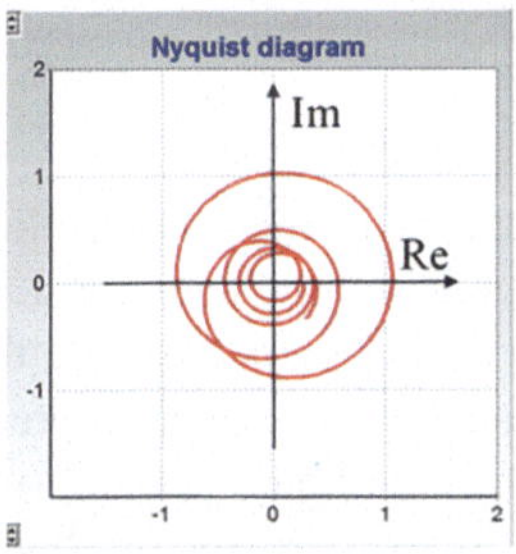

b

c

Fig. 15 Unstable cutting mode (point 1 on SLD): (**a**) Step Response; (**b**) Nyquist diagram; (**c**) machined surface

from vibration sensors, with subsequent correction of the control program with automatic change of the cutting mode specified in the control program in accordance with the stability diagram. A similar system is advertised in the NAVI technology of OKUMA (Machining Navi, n.d.). With the increased vibration level, it is considered appropriate to search for a new value of the cutting speed in the direction of the most significant productivity using the developed technology.

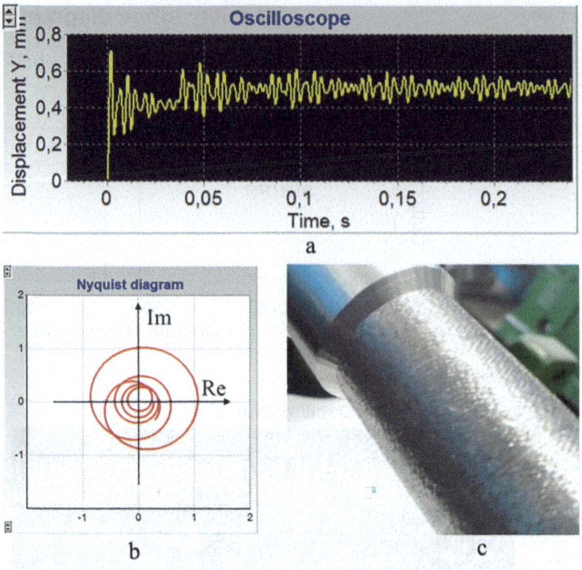

Fig. 16 Stable cutting mode (point 2 on SLD): (**a**) Step Response; (**a**) Nyquist diagram; (**a**) machined surface

4 Conclusions

The stability assessment of machining systems with the construction of stability diagrams in the coordinates of the cutting mode components must be performed using a new stability criterion. Such a stability criterion is based on the analysis of the frequency characteristics of the machining system, which is presented with a lagging function in positive feedback along the coordinate normal to the machined surface. The machining system will be stable in assessing the oscillation amplitude if the graph of its frequency characteristic does not cover the point with coordinates [+1,0] on the complex plane.

Acknowledgments The research was partially supported by the International Association for Technological Development and Innovations.

References

Agarwal RP, Berezansky L, Braverman E, Domoshnitsky A (2012) Introduction to oscillation theory. In: Nonoscillation theory of functional differential equations with applications. Springer, New York. https://doi.org/10.1007/978-1-4614-3455-9_1

Altintas Y (2012) Manufacturing automation: metal cutting mechanics, machine tool vibrations, and CNC design. Cambridge University Press, New York. https://doi.org/10.1017/CBO9780511843723

Altintas Y et al (2020) Chatter stability of machining operations. J Manuf Sci Eng 142:110801-1-19. https://doi.org/10.1115/1.4047391

Bai Q et al (2024) Machining dynamics and chatters in micro-milling: a critical review on the state-of-the-art and future perspectives. Chin J Aeronaut 37(7):59–80., ISSN 1000-9361. https://doi.org/10.1016/j.cja.2024.02.022

Brecher C. et al. Efficient determination of stability lobe diagrams by in-process varying of spindle speed and cutting depth. J Adv Manufact, 6, 272–279 (2018) Springer. https://doi.org/10.1007/s40436-018-0225-x

Duan Z, Li C, Ding W et al (2021) Milling force model for aviation aluminum alloy: academic insight and perspective analysis. Chin J Mech Eng 34:18. https://doi.org/10.1186/s10033-021-00536-9

Guan L, Chen Y, Wang Z (2024) Operational modal analysis of CNC machine tools based on flank-milled surface topography and spectrum. Vibration 7:738–763. https://doi.org/10.3390/vibration7030039

Guo W et al (2022) The experimental study on interaction of vibration and dynamic force in precision milling process. Int J Adv Manuf Technol. https://doi.org/10.1007/s00170-021-08568-0

Janota M, Kolář P, Sulitka M (2019. 2019 15th international conference on high speed machining) Operational method for identification o specific cutting force during milling. MM Sci J HSM. https://doi.org/10.17973/MMSJ.2019_11_2019078

Kaneko K, Shimizu J, Shirase K (2023) A voxel-based end milling simulation method to analyze the elastic deformation of a workpiece. J Manuf Sci Eng 145(1):011005. https://doi.org/10.1115/1.4055794

Law M et al (2020) Modal analysis of machine tools using visual vibrometry and output-only methods. CIRP Ann Manuf Technol. https://doi.org/10.1016/j.cirp.2020.04.043

Li Z et al (2024) Chatter suppression techniques in milling processes: a state of the art review. Chin J Aeronaut 37(7):1–23. https://doi.org/10.1016/j.cja.2023.10.001

Liu S, Jin S (2020) Predicting milling force variation in time and space domain for multi-toothed face milling. Int J Adv Manuf Technol 108:2269–2283. https://doi.org/10.1007/s00170-020-05319-5

Machining handbook - 2-part (EN) (n.d.). https://www.hoffmann-group.com/GB/en/houk/know-how/machining-handbook/e/61245/

Machining Navi—less chatter leads to greater output (n.d.). www.okuma.com/machining-navi

Paliwal V, Ramesh Babu N (2020) Prediction of stability lobe diagrams in high-speed milling by operational modal analysis. Proc Manuf 48:283–293. https://doi.org/10.1016/j.promfg.2020.05.049

Petrakov Y, Danylchenko M (2022) A time-frequency approach to ensuring stability of machining by turning. East Eur J Enterprise Technol 6(2 (120)):85–92. https://doi.org/10.15587/1729-4061.2022.268637

Petrakov YV, Ohrimenko OA, Sapon SP, Sikailo MO, Fedorynenko DY (2024) Determination of chatter-free cutting mode in end milling. J Eng Sci 11(2):A1–A11. https://doi.org/10.21272/jes.2024.11(2).a1

Sun Y et al (2023) A state-of-the-art review on chatter stability in machining thin walled parts. Machines 11:359. https://doi.org/10.3390/machines11030359

Thévenot V et al (2006) Integration of dynamic behavior variations in the stability lobes method: 3D lobes construction and application to thin walled structure milling. Int J Adv Manuf Technol 27:638–644. https://doi.org/10.1007/s00170-004-2241-1

Turkes E et al (2011) Linear analysis of chatter vibration and stability for orthogonal cutting in turning. Int J Refract Met Hard Mater 29:163–169. https://doi.org/10.1016/j.ijrmhm.2010.10.002

Wang G et al (2024) Prediction of stability lobe diagrams in milling based on deep learning method. In: Third international conference on advanced manufacturing technology and manufacturing systems (ICAMTMS 2024), vol 13226, p 1322617. https://doi.org/10.1117/12.3039174

Yue C et al (2019) A review of chatter vibration research in milling. Chin J Aeronaut 32(2):215–242. https://doi.org/10.1016/j.cja.2018.11.007

Zhu L, Liu C (2020) Recent progress of chatter prediction, detection and suppression in milling. Mech Syst Signal Process 143:106840. https://doi.org/10.1016/j.ymssp.2020.106840

Using Markov Chains in Manufacturing Engineering

Yaroslav Kusyi and Michal Hatala

Abstract The product life cycle, which includes developing and manufacturing a product and its future operation, is vital in scientific research for mechanical engineering. Object-oriented technologies, with stepwise implementation of interconnected phases and stages based on the prototyping process, are realized in traditional automated systems for the technological preparation of modern manufacturing. A distinguishing aspect of functionally-oriented technologies, on the one hand, and smart manufacturing, on the other hand, is the technological assurance of operational characteristics, reliability metrics, environmental protection, assurance of economic criteria, and social aspects at product manufacturing according to the concept of Industry 4.0/5.0. Technological defects formed from structural stress concentrators at parts manufacturing evolve into damage due to unfavourable operating conditions, which can lead to machine failures. Thus, a product behaviour prediction during important phases of its life cycle is an essential task for current manufacturing engineering. An analysis methodology for the effect of components of the technological system during part machining on the assurance of product quality using Markov chains is developed. Optimization of cutting modes during item machining according to economic criteria is presented. An original technique for determining the failure rate of technological system components by cutting is suggested. The influence of a machine tool, a fixture, and a cutting tool on ensuring the regulated quality parameters of the processed product is determined. Ways of further research in this direction are outlined.

Keywords Reliability engineering · Product life cycle · Technological system

Y. Kusyi (✉)
Lviv Polytechnic National University, Lviv, Ukraine

Technical University of Košice, Prešov, Slovak Republic
e-mail: yaroslav.m.kusyi@lpnu.ua

M. Hatala
Technical University of Košice, Prešov, Slovak Republic

V. Ivanov (ed.), *Emerging Technologies in Manufacturing and Materials Engineering*, Springer Tracts in Mechanical Engineering,
https://doi.org/10.1007/978-3-032-14914-5_16

1 Introduction

Most machine parts must be considered a potential source of increased danger to people and/or the environment. This requires comprehensive assurance of the regulated performance of products in their life cycles based on the important demands of Industry 4.0/5.0 and the concept of sustainable machining (Davim 2017; Grzesik 2016). Speeds on all modes of transport are steadily increasing, energy consumption in industry is improving, unique complexes for electricity production, exploration, extraction, and transportation of minerals, and, in particular, various types of gas and oil are developing. All these factors present the problem of ensuring product safety in a new way and require systematicity in solving scientific, engineering, and technical tasks (Tran 2024; Rigdon et al. 2022).

This problem is exacerbated for technical facilities, the operation of which is planned until they reach the limit conditions or states. The term "boundary (marginal) state" symbolizes the exhaustion of functioning possibilities and can be interpreted differently. In some cases, moral deterioration causes the machine to be decommissioned; in others, it is a sharp reduction in efficiency, which makes it impossible to further operate from an economic perspective or a drop in safety indicators beyond the permissible level. In most cases, it is difficult to establish the true causes and signs of indicators that determine the onset of the limit state. Therefore, manufacturing products, which are restored once, are characterized by a resource for medium or overhaul repair and a complete warranty resource for write-off or disposal. The inter-repair service also serves as an additional factor for assessing the performance of objects restored many times. Therefore, the boundary conditions or states of products differ for different operating conditions by the specified indicators. On the other hand, many technical problems are closely related to the product life cycle. This requires analysing the main relationships during product manufacture, which are very difficult and unpredictable (Davim 2017; Modarres and Groth 2023).

The problems described require more attention to the technological assurance of product quality.

Technological defects that are formed from different types of concentrators in parts machining turn into damage due to intensive and challenging operating conditions. This process can fail the machines and their decommissioning (Fig. 1) (Liao et al. 2021).

Thus, predicting product behaviour during important phases of its life cycle is an urgent problem in modern manufacturing engineering.

The incessant increase in the essential requirements for product quality has required fast changes in the key directions of industry and basic criteria for the effectiveness of product manufacturing for more than 100 years. Now we live in a digital technology period, and priorities are shifting faster than a few decades ago (Kusyi et al. 2022).

In general, we can determine two main approaches for technological process planning in product manufacturing. Object-oriented technologies are closely related

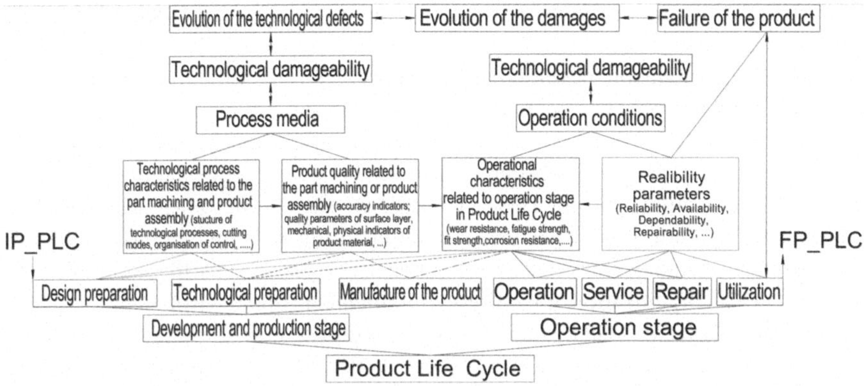

Fig. 1 Simplified scheme of relationships in the Product Life Cycle

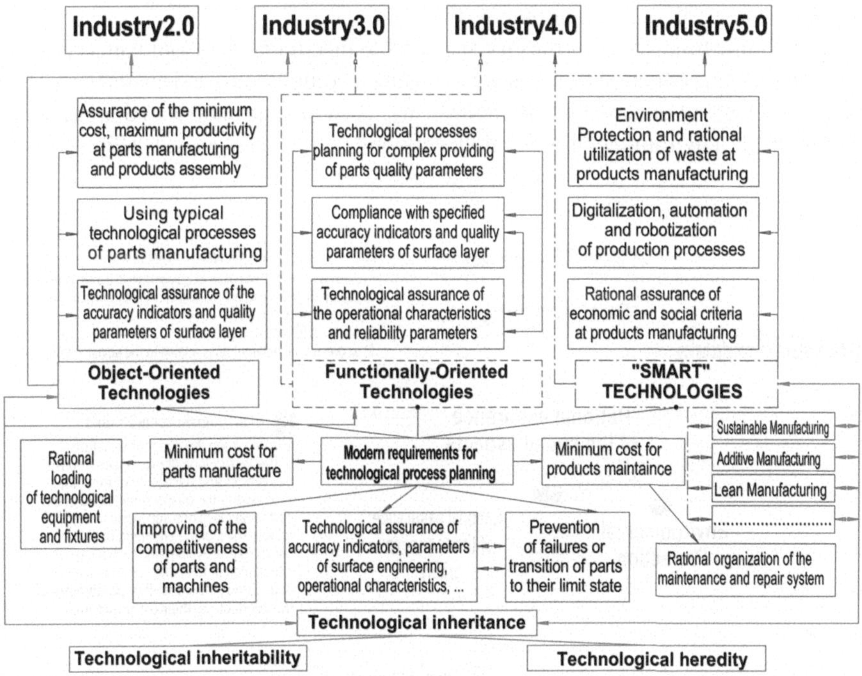

Fig. 2 Different approaches for the technological process planning at product manufacturing

to the stepwise production structure for technological assurance of product quality (see Fig. 2). The main peculiarity of the functionally-oriented technologies is the technological assurance of reliability metrics and operational parameters of products (see Fig. 2) (Liao et al. 2021; Kusyi et al. 2022). "SMART" technologies, including Lean Manufacturing, Additive Manufacturing, Sustainable Manufacturing, etc., are the highest level of functionally-oriented technologies considering environmental protection, economic criteria, and ensuring social aspects (Akhilesh and Möller 2020).

Nowadays, sustainable development has a priority role in the industry, ensuring a "golden" balance between economic growth, material production, and consumption in conditions of optimal restoration of ecosystems, rational disposal of production waste and carcinogenic substances, and effective support of the life of present and future generations (Davim 2017; Grzesik 2016). On the other hand, product quality is formed not only by design for manufacturing and manufacturing technology, different technological processes and resources, but *also* by the verifiable sustainability of an organization, ensuring social, financial, and environmental aspects. Thus, support of sustainable manufacturing requires new paradigms, including Zero-Defect Manufacturing (ZDM) (related to Reliability Engineering), Quality 4.0, and Digital Lean in Quality Management (QM) (Psarommatis and Azamfireic 2024).

Therefore, sustainable machining and manufacturing are important areas of "SMART" technologies development (Fig. 3).

At the same time, the priorities for sustainable machining development as part of the sustainable production of items are to ensure a comprehensive decision of scientific problems and the solution of engineering tasks in four interrelated areas: production itself, environmental protection, economic, and social aspects (Fig. 3).

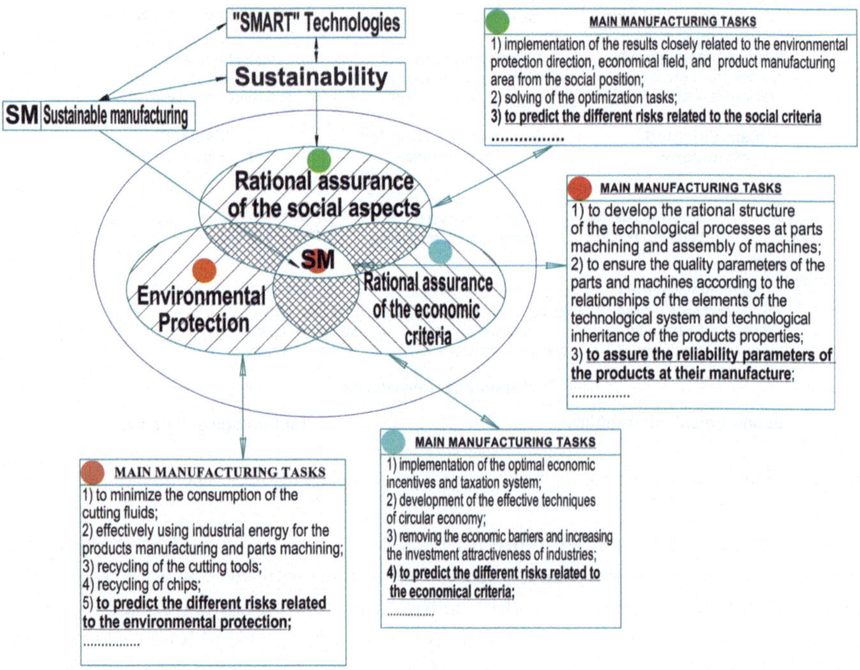

Fig. 3 General characteristics of "SMART" technologies

Predicting the risks of the parts and machines' failures during their life cycles, considering environmental protection, economic criteria, and social aspects (Tran 2024; Modarres and Groth 2023), is an important task for modern manufacturing engineering (Fig. 3).

Thus, searching for effective approaches for technological process planning of parts and product manufacturing is an urgent task of modern production.

2 Fundamentals and Challenges

QM consists of some techniques: quality enhancement, quality inspection, and quality assurance to enhance the overall merit of the product (Romero et al. 2019). ZDM, as a new technique to QM, covers the manufacturing system as a whole, rather than focusing on production (Psarommatis and Azamfireic 2024). Nowadays, three paired strategies: "detect-maintenance (repair)," "detect-prevent (block)", and "predict-prevent (block)" are being developed very intensively according to the ZDM concept (Psarommatis and Bravos 2022; Psarommatis et al. 2020).

Reliability engineering is one of the fundamentals for developing the ZDM concept (Psarommatis et al. 2020; Selech et al. 2025). Reliability engineering is a combination of physics, statistics, and engineering (see Fig. 4). An engineer must understand the physical nature of the analysed process, on the one hand. Still, he must choose the mathematical apparatus to model a specific physical process. Additionally, an engineer must adapt the obtained results to the real process, such as machining, assembly, etc. (Wanga et al. 2024; Melchers and Beck 2018).

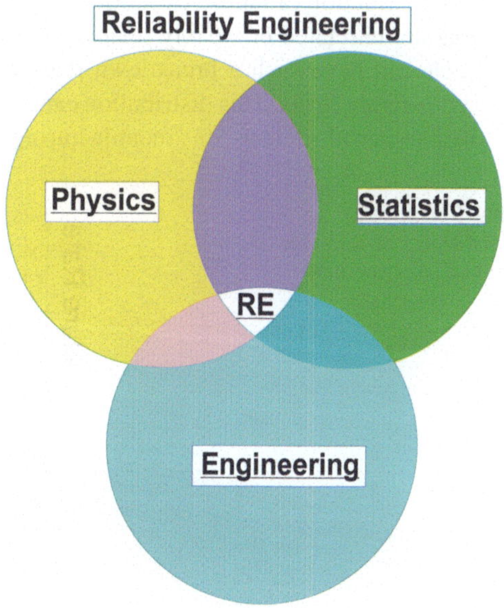

Fig. 4 Description of the reliability engineering

Modern reliability engineering concentrates on the following paradigms: reliability must be designed into products and processes according to the best available scientific and research methods; reliability practices must start early in the design and technological preparation of production and be well integrated into the overall product life cycle; reliability calculation and reliability achievement are both necessary factors of production success (Grzesik 2016; Melchers and Beck 2018).

Reliability engineering consists of some basic principles.

Physics. A defined physical process is characterized by an initial task to decide a scientific problem or solve an engineering task. It is usually characterized by the customer and described in the initial data for project development.

Statistics. There are different life distribution models for describing physical processes in manufacturing engineering: exponential, Weibull, gamma, normal, lognormal, etc. (Zhao and Lu 2021; Dong and Nassif 2019).

Failure rate, as an essential reliability parameter, is the chance of failing in the next small unit of time, given that the item has operated for that long. There are three possible variants of its changes: increasing failure rate implies wear-out failure; decreasing failure rate implies infant mortality; constant failure rate implies random failure (see Fig. 5). In this case, the exponential distribution is characterized by a constant failure rate, the gamma distribution provides an increasing failure rate, and the normal distribution provides a decreasing failure rate. Weibull distribution can increase, decrease, or have a constant failure rate, and it can be monotonic (see Fig. 5) (Zhao and Lu 2021).

Exponential distribution. Number of parameters is 1. It characterizes the processes in which the failure of an element occurs suddenly, that is, regardless of how long the object has been operated and in what condition it is. First of all, the exponential distribution of failure time occurs in situations of sudden, accidental failures that are unrelated or weakly related to wear and aging processes. Good quality tools can have such a breakdown if they operate under extreme dynamic and temperature loads. This distribution can also be used in cases of poor tool quality, technological defects in manufacturing, unfavourable operating conditions

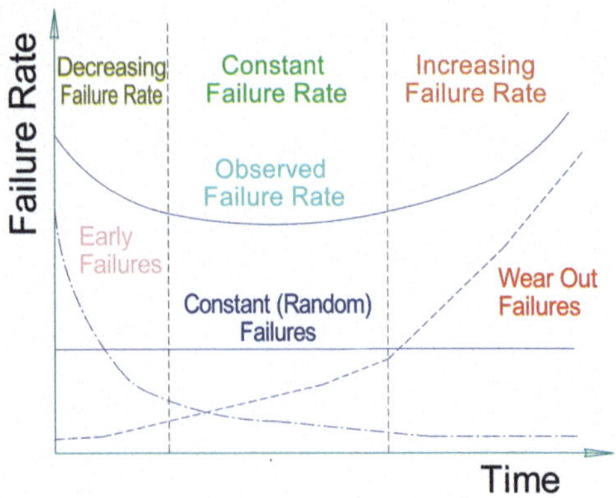

Fig. 5 Different interdependencies of the failure rate over time

(inexperienced operator, manual feeding, poor technical condition of the machine, defects in the workpiece material). The failure rate is constant over time (Dong and Nassif 2019).

Weibull distribution. The number of parameters is 2. It is used to describe the reliability of products that have hidden defects, but which do not fail for a long time. In this case, the failure rate is greatest in the initial period, and then rapidly decreases. This distribution is also widely used to describe the reliability of machine parts and mechanisms during their run-in period. The exponential law is a special pleading of the Weibull-Gnedenko distribution. It approaches the normal and gamma distribution under some conditions. The intensity of failures increases over time and, in some cases, is a constant value or decreases (Melchers and Beck 2018; Zhao and Lu 2021).

Gamma distribution. The number of parameters is 2. It characterizes the tool wear under the following conditions: the average wear rate is constant; the original quality of the tool is homogeneous, and the wear rate is subject to random changes. Physically, the transition from gamma distribution to normal is justified if wear phenomena are intertwined with each other for a long time before failures begin to occur. It follows a pattern of damage accumulation. The failure rate increases over time (Zhao and Lu 2021; Dong and Nassif 2019).

Normal distribution. The number of parameters is 2. It is widely used to describe wear and aging processes in a variety of technical facilities. Normal distribution can be due to the uniformity of the quality of the tool manufactured according to a stable technological process, constant average wear intensity, with a gradual change in the tools parameters during their operation, when the proportion of sudden failures is very small (Tran 2024; Dong and Nassif 2019).

Sudden failures of parts and machines in manufacturing engineering are most dangerous during their manufacturing and lifespan, as they can have catastrophic consequences and lead to human casualties. The Weibull-Gnedenko division or its special case—an exponential law is used to evaluate sudden product failures, in particular, destruction and failure of metal-cutting tools. The advantage of the Weibull-Gnedenko distribution is its versatility to describe both sudden and gradual failures. However, the Weibull-Gnedenko distribution is a two-parameter, and mathematical models based on it do not have an analytical solution. The exponential law is one parameter, which simplifies the formation of initial data and the search for solutions. Models based on the exponential distribution law have an analytical solution and well describe many real processes. Along with, the probability of trouble-free operation when using the exponential distribution law is always underestimated. Thus, if reliability indicators are provided under the exponential law, then they are provided unambiguously for other distribution laws (Grzesik 2016; Kusyi et al. 2022).

Engineering. Adaptation of the results obtained to the real process in production or other industries is the final stage of reliability engineering. Intensive production development in the digital age requires a comprehensive solution of tasks and the search for rational solutions in several technological or production areas. In particular, the analysis and estimation of the technical condition of the technological system components, the optimization of cutting modes, and the *rummage for the excellent* ratio of process productivity/product quality can be provided at parts machining according to the reliability engineering concept. This approach optimizes work and reduces the cost of time and resources, first of all, at the design for manufacturing and manufacturing technology of products, as well as during product manufacturing and their further operation (Tran 2024; Modarres and Groth 2023).

One of the optimal options for solving these problems is using the mathematical apparatus of Markov chains to analyse the parts machining, assembly, etc., in manufacturing engineering.

3 Ideas and Solutions

A stepwise description of the algorithm of modern engineering design that includes a rational combination of object-oriented and functionally-oriented principles for technological design during parts machining is given below.

3.1 Analysis of initial data or terms of reference for design (object-oriented principle for technological design)

This section is closely associated with the "physics"-principle of reliability engineering. The customer generates initial design data, according to which the terms of reference are developed. The designer develops an assembly drawing of the machine, designs working drawings of parts, selects the material for their manufacture, and establishes technical requirements. The work of the designer and technologist is coordinated by the accepted design principle (object-oriented or functionally-oriented) and the selected optimization criteria. The technological processes of planning parts machining and assembling the machine are also assured to the object-oriented or functionally-oriented design principle. Calculation of allowances, dimensional analysis of the technological process, cutting modes calculation and selection of certain metal-cutting machines, and technical standardization of the technological process are carried out with a detailed development of specified technology (Davim 2017).

3.2 Analysis of the selected technological system of machining the product (functionally-oriented principle for technological design)

This section is closely associated with the "statistics"-principle of reliability engineering. The technological operation is the object of research at this phase of technological design. Since it is a component of the technological process and contains one or more technological steps, moreover the core concepts of technical systems theory adapt to it. Each technological operation is characterized by a defined technological system: machine tool—fixture—cutting tool—workpiece (part) and process medium: machine tool—fixture—cutting tool. A variety of technological systems and process medium for selected cutting diagrams is shown in Fig. 6. The analysis of the relationships between the technological system components and the conditions for technological assurance of the product quality is the main subject of research at this stage of technological design. Calculation of cutting modes and selection of models of metal-cutting machines, type of fixtures, and metal-cutting tools (see the previous stage) are insufficient to predict the technological assurance of product quality. Operational indicators of parts and machines, which are described by their operational conditions and are established according to the regulated metrics of product reliability, are not taken into account in traditional approaches to technological design. The development of new approaches and methods in the technological process planning for product manufacture, in particular, using Markov chains, is one of the priority areas for the development of modern manufacturing engineering (Kusyi et al. 2022).

Markov chains are mathematical models that use concepts from probability to describe how a system changes from one state to another (Kozłowski et al. 2023; Yakovyna et al. 2020). Mathematical apparatus of Markov chains allows the analysis of physical processes in different branches of industry, mainly in radio electronics, electro-mechanical engineering, IT technology, transport, etc. to evaluate the reliability metrics of production systems that include the typical components with

a b

Fig. 6 Different variants of the technological systems (**a**—for milling; **b**—for turning) (https://www.istockphoto.com/uk/search/2/image-film?phrase=metal+cutting+machine)

well-known reliability parameters, especially the failure rates (Yakovyna and Symets 2021; Borucka et al. 2023). This approach can't realized in manufacturing engineering because standard components of manufacturing systems (axes, gears, shafts, pipes, bearings, etc.) work in production machines in difficult operational conditions (Dehtiarov et al. 2024; Ivanov and Zajac 2017). Thus, we need to find an approach (idea, technique, etc.) for calculating the failure rates of the elements of the technological systems for introducing the Markov chains technique for machining parts, assembly of machines, etc. (see Fig. 6) (Kusyi et al. 2022).

The designation in Fig. 6 is as follows: P_1 (t) is the probability of technological assurance of the regulated product quality in the i-th cutting operation during part machining in case of failure to reach the limit state of the machine tool, fixture located on the machine tool and cutting tool (or tools) in general and their components in particular; F_2 (t), F_3 (t), F_4 (t) are the probability of failure to assure regulated product quality in the i-th cutting operation during part machining when the limit state is reached by the machine tool, fixture, and cutting tool (or tools) and their components, respectively (Kusyi et al. 2022).

Technological system, process medium, and technological graph of the reliability for the *i*-th cutting operation are presented in Fig. 7, a and b, respectively (Kusyi et al. 2022).

A limit state is reached when an object's operation becomes unsafe, impractical, or when further restoration of its functional parameters is impractical or impossible (Kusyi et al. 2022).

The system of Chapman-Kolmogorov differential equations for the reliability graph (see Fig. 7b) with constant failure rates of the components of the process medium, taking into account the impact of the components of a certain technological system by cutting on technological assurance of the product quality for the i-th operation is as follows (Kusyi et al. 2022):

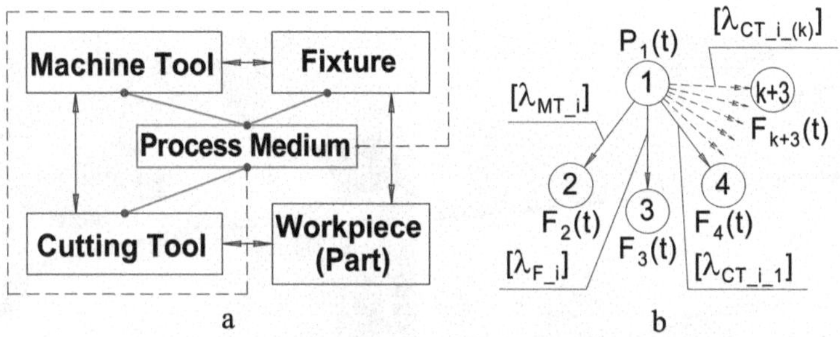

Fig. 7 Technological system/process medium (**a**) and technological graph of the reliability for description of the i-th technological operation using Markov chains to prognosticate the probability of assurance of the workpieces quality, taking into account the impact of the components of a certain technological system (**b**)

$$\begin{aligned}
&\frac{dP_1(t)}{dt} = -\lambda_{MT_V} \cdot P_1(t) - \lambda_{F_V} \cdot P_1(t) - \sum_{k=1}^{z} \lambda_{CT_(k)i} \cdot P_1(t);\\
&\frac{dF_2(t)}{dt} = \lambda_{MT_V} \cdot P_1(t);\ \frac{dF_3(t)}{dt} = \lambda_{F_V} \cdot P_1(t);\ \frac{dF_4(t)}{dt} = \lambda_{CT_i_1} \cdot P_1(t);\\
&\ldots\ldots\ldots\ldots\ldots\ldots\ldots\ldots\ldots\ldots\\
&\frac{dP_{z+3}(t)}{dt} = \lambda_{CT_i_z} \cdot P_1(t);\\
&P_1(t) + F_2(t) + F_3(t) + F_4(t) + \ldots\ldots\ldots F_{z+3}(t) = 1, \quad t \in [0;t].
\end{aligned} \tag{1}$$

where λ_{MT_i}, λ_{F_i}, $\lambda_{CT\_i\_1}$, $\lambda_{CT\_i\_z}$—failure rates for the machine tool, fixture located on the machine tool, and 1….z cutting tools, respectively.

The failure rate is the conditional probability of an object failing at a specific moment in time, given that it has not failed before (Tran 2024; Modarres and Groth 2023).

Generally (Melchers and Beck 2018):

$$\lambda(t) = f(t)/P(t). \tag{2}$$

Equation (2) shows that the failure rate λ(t) is the quotient of the density of the fault-free operation time distribution f(t) and the probability of failure-free operation P(t).

A product life cycle is an important characteristic for the analysis of physical processes related to the sustainable machining and manufacturing, and the requirements of Industry 4.0/5.0. In this regard, it is advisable to take the failure rates of the process medium components inversely proportional to the values that describe the duration of their respective life cycles.

The failure rates of specific technological system components are determined as follows (Kusyi et al. 2022):

- failure rate of a certain machine tool λ_{MT}

$$\lambda_{MT} = 1/T_{MRC\_MT} \tag{3}$$

- where T_{MRC_MT} is the average time between overhauls in a machine's repair cycle ("machine tool life"), [months → min];
- failure rate of the fixture located on the the machine tool λ_F

$$\lambda_F = 1/T_F \tag{4}$$

- where T_F—regulated fail-safe time for failure of the fixture ("fixture life"), [hours → min];
- failure rate of the cutting tool (s) T_{CT}

$$\lambda_{CT} = k_f / T_{CT} \tag{5}$$

- where k_f is a coefficient related to the machined material and type of the cutting tool(s); T_{CT} is the resistance (durability) of the cutting tool(s) ("tool life"), [min].

The technological reliability graphs are developed for each technological step of a certain operation during part machining. Failure rates for each element of the process medium are calculated by equations (3)–(5) and a system of some differential equation (1) is developed. These sequences are repeated for all operations of the technological process of parts machining.

3.3 Development of optimization criteria from the standpoint of reliability of technical systems and calculation of their critical values

This initial stage of the "engineering" principle enables the search and description of optimization criteria for the solution of technological tasks in reliability engineering—"engineering" principle.

The technological cost of products as the main indicator of the efficiency of manufacturing engineering determines the degree of perfection of design and technological solutions, the level of competitiveness of products, the potential for the development of enterprises, etc.

The technological cost of machining $C_{techn.i}$ at the i-th technological operation is determined as follows (Kusyi et al. 2023):

$$C_{techn.i} = C_{stand.i} \cdot t_{total\,i}, \text{or}\, C_{techn.i} = C_{stand.i} \cdot t_{maAh.i}, \tag{6}$$

where $C_{stand.i}$ is the standard (regulated) cost at the i-th (specific) technological step or operation, [€/min], etc.; $t_{total\ i}$, $t_{mach.i}$ are normative total unit time and machining time at the i-th technological step/operation, respectively, [min], etc.

The standard (regulated) cost at the i-th technological operation during part machining is calculated as follows (Kusyi et al. 2023):

$$C_{stand.i} = C_{EOi} + C_{ACi} + C_{SCi}, \tag{7}$$

where C_{EOi} is a price standard cost per 1 min of device or equipment operation, [€/min], etc.; C_{ACi} is an additional cost due to capital investments, [€/min], etc.; C_{SCi} is a cost of live labour for society, [€/min], etc.

Total unit time $t_{total\ i}$ is determined as follows (Yakovyna and Symets 2021):

$$t_{total\,i} = t_{mach.i} + t_{p.i} / n, \tag{8}$$

where $t_{p.i}$ is the normative preparatory time, [min]; n is the calculated batch size, [items].

Machining time includes four components: cutting time $t_{cut.i}$, auxiliary time $t_{aux.i}$, additional time for machining tending $t_{add.i}$, and time for personal needs $t_{p.n.i}$ (Kusyi et al. 2023):

$$t_{mach.i} = t_{cut.i} + t_{aux.i} + t_{add.i} + t_{p.n.i} \tag{9}$$

The cutting time $t_{cut.i}$ is the time during which chips are removed from the workpiece as a consequence of the contact interaction of the machine tool with the part surface to be processed without the participation of the worker. The cutting time during parts machining $t_{cut.i}$ is in direct proportion to the cutting path and inverse variation to the ratio of the cutting modes elements (Kusyi et al. 2022, 2023):

$$t_{cut.i} = L_{mach.st.i} / S_{\min pasp.} \tag{10}$$

$$t_{cut.i} = L_{mach.st.i} / \left(n_{pasp.} \cdot S_{0\,pasp.}\right) \tag{11}$$

where $L_{mach.st.\,i}$ is the estimated cut length characterized by the specific machining using certain metal-cutting tools on the concrete technological step or operation; $S_{min.pasp.}$, $S_{0pasp.}$, $n_{pasp.}$ Are passport (regulated according to the normative indicators of the machine tool) values of the cutting modes elements: feed per minute, feed per spindle revolution, and rotary speed, respectively.

The important chore of object-oriented technologies is to ensure item or machine quality parameters at the development and production stage of a product in its life cycle by achieving only the requirements of design documentation. The optimization criterion during the choice of a rational variant of the technological process at part machining and assembly of products is the minimum justified value of the technological at the maximum productivity of machining or assembly and the rational load factor of the main technological equipment without a functional analysis of the operational characteristics of the product. The objective function for this optimization and the function of the synthesized parameters are presented below (Davim 2017; Kusyi et al. 2022):

$$t_{Aut} \in f\left(t, S, V_{cut.}, n\right) \to \min, \tag{12}$$

where t, S, n, $V_{cut.}$ are the cutting modes elements: cutting depth, feed (feed per minute S_{min} or feed per spindle revolution S_0), cutting speed, and rotary speed, respectively.

The optimization criterion for functionally-oriented technologies, in contrast to object-oriented ones, is the maximum justified value of the technological cost of a product manufacturing from the point of view of technological assurance of reliability metrics of production processes, components of the technological system, and process medium. Thus, the objective function for this optimization and the function of the synthesized parameters are presented below (Kusyi et al. 2023):

$$t_{Aut} \in f\left(t, S, V_{cut.}, n\right) \to \max, \tag{13}$$

Condition (13) is provided by the condition of rational design of the technological routes structure during surface machining of the product and the optimal choice of the method of machining by cutting, specific metal-cutting machine, fixtures, and metal-cutting tools. Solutions of the system of differential equations (1) taking into account (2)–(4) are obtained in the form P = f(t), where P is the regulated reliability metric, P = f(t) ∈ [0; 1] (Kusyi et al. 2023).

If P = f(t) is coordinated with the regulated reliability metrics, which provide the operational characteristics of the product, its accuracy parameters, and surface quality, and the time indicator is adapted to the cutting time of surface machining at part machining, then with the combined solution of the system (1) and the limit regulated values of reliability indicators, the desired solutions of the condition (13) are obtained.

3.4 Implementation of results of computer modelling related to the technological processes planning during items machining

The final stage of the "engineering"-principle is the adaptation of the obtained results using the Markov chains technique to the technological processes planning at parts machining.

There are different ways to implement the obtained results of computer modelling for the technological process planning during items machining.

Prediction of the technical condition of the process medium components by the current states in their life cycles allows us to determine the suitability of a machine tool, fixture, and cutting tool for technological assurance of product quality parameters. The resource of the process medium components or the limit value of the cutting time is calculated for the solution of necessary tasks in this research area.

The optimization of cutting modes with respect to regulated reliability metrics provides a framework for defining the technological conditions necessary to maintain product quality parameters. It also facilitates the selection of rational cutting mode elements, ensuring the reliability of the technological system.

3.5 Prediction of the technical condition of the process medium components according to the regulated reliability metrics

We will consider this approach using the example of operation 005 Vertical milling at machining of the base OSHPV 25.001 (see Fig. 8).

3.5.1 Analysis of initial data or terms of reference for design (object-oriented principle for technological design)

We analyse the operation **005 Vertical milling** at machining of the base OSHPV 25.001:

005 Vertical-milling

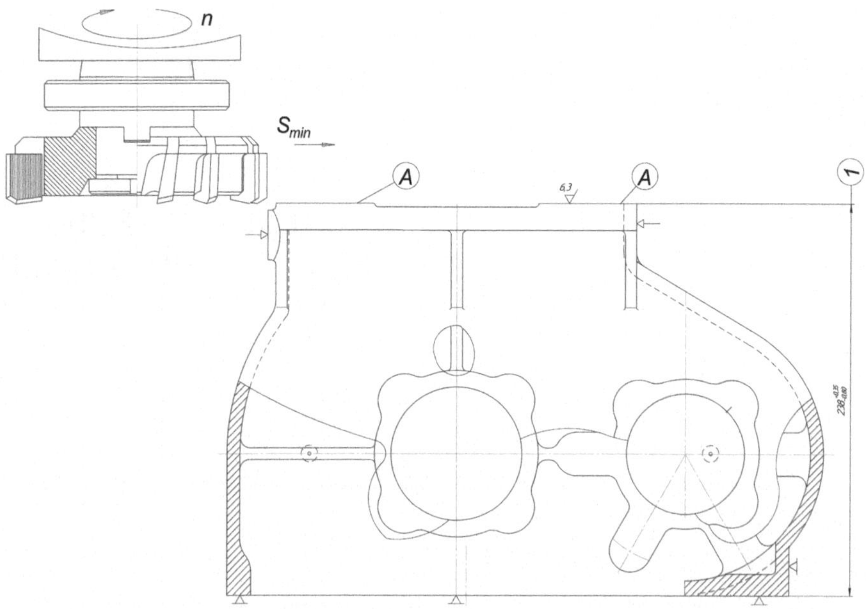

Fig. 8 Operational sketch for machining the base of the OSHPV 25.001 (operation 005)

Mill reference plane A to the regulated dimension 1 (see Fig. 8).

The base OSHPV 25.001 is machined on operation **005 Vertical milling** using an universal milling machine, automated fixture, and face milling cutter ∅ 250 mm.

Elements of the cutting modes are cutting depth t = 2.0 mm; feed per minute $S_{min\ pasp.} = 200$ mm/min; cutting speed $V_{cut.} = 125.66$ m/min; rotary speed $n_{pasp.} = 160\ min^{-1}$.

The estimated cut length $L_{mach.st.} = 490$ mm according to the cutting process diagram.

The cutting time at parts machining on the operation **005 Vertical milling** is calculated by (10):

$$t_{cut.005} = 490\ /\ 200 = 2.45\left[\text{min}\right].$$

3.5.2 Analysis of the selected technological system of the product machining (functionally-oriented principle for technological design)

The technological graph of reliability developed for operation **005 Vertical milling** is presented in Fig. 9.

A system of Chapman-Kolmogorov differential equations which describes the relationships between the elements of the technological system was developed for operation 005 Vertical milling at machining of the base OSHPV 25.001 using the technological graph of the reliability (Kusyi et al. 2022, 2023):

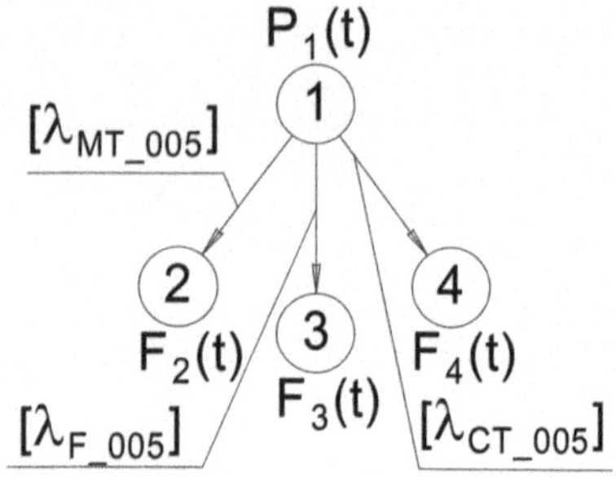

Fig. 9 Technological reliability graph for the Vertical milling operation (005) on the OSHPV 25.001 base

$$
\begin{aligned}
&\frac{dP_1(t)}{dt} = -\lambda_{T\_005} \cdot P_1(t) - \lambda_{F\_005} \cdot P_1(t) - \lambda_{CT\_005} \cdot P_1(t);\\
&\frac{dF_2(t)}{dt} = \lambda_{T\_005} \cdot P_1(t); \quad \frac{dF_3(t)}{dt} = \lambda_{F\_005} \cdot P_1(t);\\
&\frac{dF_4(t)}{dt} = \lambda_{CT\_005} \cdot P_1(t);\\
&{}_1(t) + F_2(t) + F_3(t) + F_4(t) = 1, \quad \mathrm{t} \in [0;\mathrm{t}]
\end{aligned}
\tag{14}
$$

An important step is the calculation of the failure rate for the technological system components using (3)–(5):

– the failure rate for the machine tool is calculated using (3) and (15):

$$T_{MRC\_MT\_005} = \left(\left[F_{act.}\right] \cdot \left[T_{MRC\_MT\_005}\right] \cdot 60\right) / 12. \tag{15}$$

In (15) [$T_{MRC\text{-}MT}$] represents the average overhaul period for the machine's repair cycle, [months]. The actual annual operating time of the equipment is denoted by [$F_{act.}$], [hours]; [$F_{act.}$] = 4055 hours for a two-shift operation as cited in (Kusyi et al. 2022). The number 12 simply indicates the number of months in a year.

$$T_{MRC\_MT\_005} = (4055 \cdot 9 \cdot 60) / 12 = 182475\,\text{min},$$

$$\lambda_{MT\_005} = 1/182475 = 5.48 \cdot 10^{-6}\ \text{min}^{-1},$$

– the failure rate for the fixture located on the machine tool is determined by (4):

$$\lambda_{F\_005} = 1/(1500 \cdot 60) = 1.11 \cdot 10^{-5}\ \text{min}^{-1},$$

– where T_{F_005} = 1500 hours,
– the failure rate for the cutting tool (tools) is calculated by (5):

$$\lambda_{CT} = 0.75/300 = 2.5 \cdot 10^{-3}\ \text{min}^{-1}$$

– for the face milling cutter k_f = 0.75; T_{CT_005} = 300 hours.

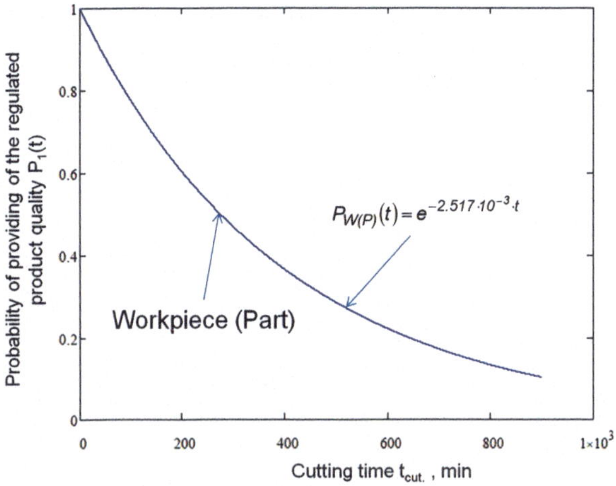

Fig. 10 The probability of technological assurance the normative workpiece quality for the base OSHPV 25.001 during its machining at operation 005 if the limit states of the process medium components are not reached

The solutions of the system (14) after substituting the calculated values of the failure rates will receive the specified form and present using graphical dependencies of changes in functions $P_1(t)$, $F_2(t)$, $F_3(t)$, $F_4(t)$ over time in Figs. 10, 11 and 12.

3.5.3 Development of optimization criteria from the standpoint of reliability of technical systems and calculation of their critical values

There are different reliability metrics used in manufacturing engineering: uptime percentage, mean time to failure, mean time to repair, etc. (Garg and Mangey Ram 2022; Friederich and Lazarova-Molnar 2024; Berladir et al. 2024). Gamma-percentile operating time to failure $P(t_\gamma)$ characterizes the operating time, during which the failure of the object will not occur with a probability $P(t_\gamma) = 1.0$; 0.99; 0.95; 0.90; 0.80; 0.50 and is regulated by certain standards. The values $P(t_\gamma) = 1.0$; 0.99; 0.95; 0.90 are used to analyse the parameters of new parts and machines, but the values $P(t_\gamma) = 0.80$; 0.50 allow to analyse the parameters of repaired products (Kusyi et al. 2022; Ivanov and Zajac 2017).

A system of mathematical dependencies, which is developed to solve an optimization task in the general case, for regulated values $P(t_{\gamma1})$...., $P(t_{\gamma6})$, are presented as follows (Kusyi et al. 2022):

$$_1(t) = e^{-(A \cdot t)}; \begin{cases} [P_2(t)] = [P(t_{\gamma_1})]; [P_3(t)] = [P(t_{\gamma_2})]; [P_4(t)] = [P(t_{\gamma_3})]; \\ [P_5(t)] = [P(t_{\gamma_4})]; [P_6(t)] = [P(t_{\gamma_5})]; [P_7(t)] = [P(t_{\gamma_6})], \end{cases} \quad (16)$$

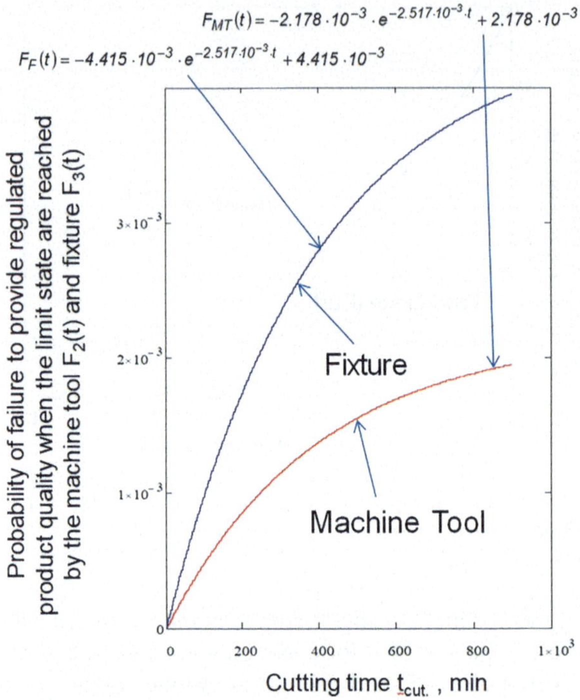

Fig. 11 The probabilities of failure to ensure the normative workpiece quality for the base OSHPV 25.001 during its machining at operation 005 when the limit state is reached for machine tool and fixture

where $P_1(t) = e^{-(A \cdot t)}$ describes the specific reliability function for a defined technological operation/step of the certain operation, obtained using the Markov chains technique and solution of the system (16); $P(t_{\gamma 1})$...., $P(t_{\gamma 6})$—regulated values of the gamma-percentile operating time to failure (see above).

The system (16) for Vertical milling operation (005) during machining of the base OSHPV 25.001 is presented as follows (Kusyi et al. 2022):

$$_1(t) = e^{-2.517 \cdot 10^{-3} \cdot t}; \begin{cases} \left[P_2\left(t_{\gamma_1}\right)\right] = 1.00; \left[P_3\left(t_{\gamma_2}\right)\right] = 0.99; \left[P_4\left(t_{\gamma_3}\right)\right] = 0.95; \\ \left[P_5\left(t_{\gamma_4}\right)\right] = 0.90; \left[P_6\left(t_{\gamma_5}\right)\right] = 0.80; \left[P_7\left(t_{\gamma_6}\right)\right] = 0.50. \end{cases} \tag{17}$$

The rational value of cutting time at parts machining will be determined taking into account its limit values associated with the object-oriented and functionally-oriented approaches in technological processes planning at product manufacture:

$$t_{cut.\min}\left[(12)\right] \le \left[t_{cut.optimal} \in f\left(t, S, n, V_{cut.}\right)\right] < t_{cut.\max}\left[(13)\right] \tag{18}$$

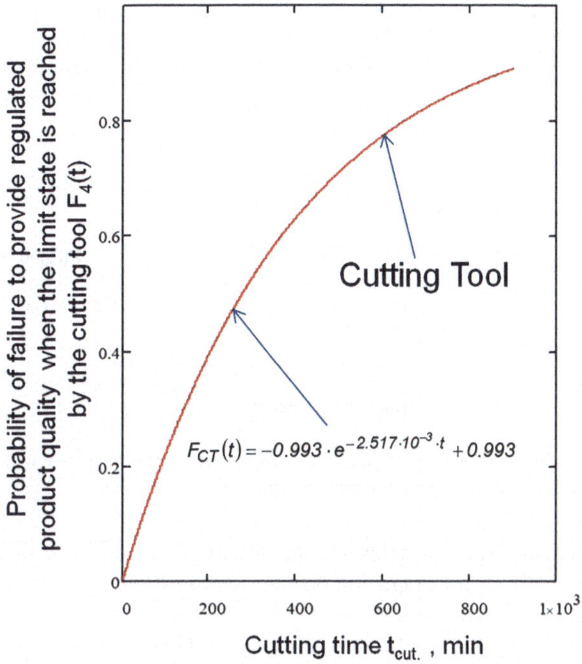

Fig. 12 The probability of failure to ensure the normative workpiece quality for the base OSHPV 25.001 during its machining at operation 005 when the limit state is reached for cutting tool

The joint solution of equations (17) to check the condition (13) and condition on the right (18) can be performed graphically (see Fig. 13) and analytically.

The analytical solutions of equation (17) give the following results:

$[t_{cut.max}]$=0 for $[P_2(t_{\gamma 1})]$=1.00; $[t_{cut.max}]$=3.99 for $[P_3(t_{\gamma 2})]$=0.99; $[t_{cut.max}]$=20.38 for $[P_4(t_{\gamma 3})]$=0.95; $[t_{cut.max}]$=41.87 for $[P_5(t_{\gamma 4})]$=0.90; $[t_{cut.max}]$=88.67 for $[P_6(t_{\gamma 5})]$=0.80; $[t_{cut.max}]$=275.43 for $[P_7(t_{\gamma 6})]$=0.50.

3.5.4 Implementation of results of computer modelling related to the technological processes planning at parts machining

The obtained results of the solution of the system (17) allow us to check the conditions (13) and the condition on the right (18), for example, for $P(t_{\gamma 1}) = 0.99$:

$$t_{cut.} = 2.45\,\text{min} < \left[t_{cut.\max}\right] = 3.99\,\text{min},$$

since the condition is met, the product parameters are ensured from the reliability standpoint of the components of the process medium for 005 Vertical milling at machining of the base OSHPV 25.001.

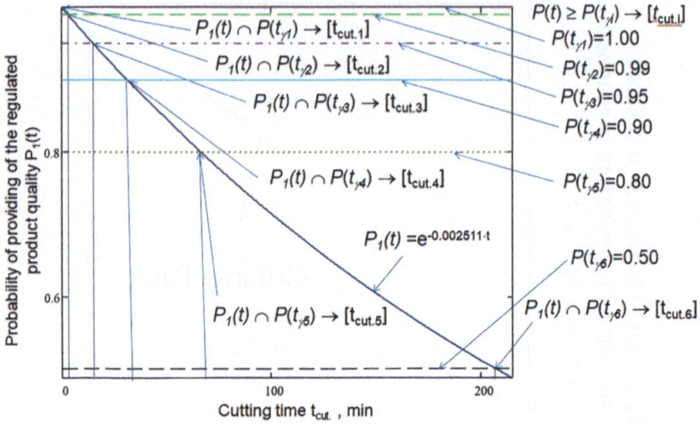

Fig. 13 Checking the workpiece's ability to be machined during technological operation 005, based on the gamma-percentile operation time to failure (t_γ)

On the other hand, we can calculate the actual values of $P_1(t)$; $F_2(t)$; $F_3(t)$; $F_4(t)$ for actual $t_{cut.}$ =*2.45* min for operation 005 Vertical milling:

$$P_1(t) = 0.993852; F_2(t) = 0.000013; F_3(t) = 0.000027; F_4(t) = 0.006105,$$

or calculate the limit values of the probabilities of the elements of the technological system for $[t_{cut.\,max}]$ = 3.99 min:

$$\begin{aligned}\left[P_1(t)\right] = \left[P_3(t_{\gamma 2})\right] = 0.990000; \left[F_2(t)\right] \\ = 0.000022; \left[F_3(t)\right] = 0.000044; \left[F_4(t)\right] = 0.009934,\end{aligned}$$

or check the working conditions of elements of the technological system:

$$P_1(t) \geq \left[P_1(t)\right]; F_2(t) \leq \left[F_2(t)\right]; F_3(t) \leq \left[F_3(t)\right]; F_4(t) \leq \left[F_4(t)\right] \quad (19)$$

Therefore, $P_1(t)$ = 0.993582 ≥ $[P_1(t)]$ = 0.990000; $F_2(t)$ =0.000013 ≤ $[F_2(t)]$ = 0.000022; $F_3(t)$ = 0.000027 ≤ $[F_3(t)]$ = 0.000044; $F_4(t)$ = 0,006105 = ≤ $[F_4(t)]$ = 0.009934, and the condition (19) is met.

3.6 Optimization of the cutting modes at parts machining

We will also consider this approach using the example of operation 005 Vertical milling at machining of the base OSHPV 25.001 (see Fig. 8).

3.6.1 Analysis of initial data or terms of reference for design (object-oriented principle for technological design)

The subparagraph 3.6.1 is identical to the subparagraph 3.5.1, thus we will not focus on it.

3.6.2 Analysis of the selected technological system of machining the product (functionally-oriented principle for technological design)

The subparagraph 3.6.2 is identical to the subparagraph 3.5.2, thus we will not focus on it.

3.6.3 Development of optimization criteria from the standpoint of reliability of technical systems and calculation of their critical values.

The subparagraph 3.6.3 is identical to the subparagraph 3.5.3, thus we will not focus on it.

3.6.4 Implementation of results of computer modelling related to the technological processes planning at parts machining

Initial data were [$t_{cut.\max}$] for [$P_k(t_{\gamma z})$], equations (10), and (11).

The limit values of the cutting modes elements are determined as follows:

$$\left[S_{\min pasp.}\right] = L_{mach.st.} / \left[t_{cut.\max}\right], \quad (20)$$

$$\left[n_{pasp.}\right] = \left[S_{\min pasp.}\right] / S_Z \cdot z, \quad (21)$$

$$\left[V_{cut\,pasp.}\right] = \left(\pi \cdot D_{cut.} \cdot \left[n_{pasp.}\right]\right) / 1000. \quad (22)$$

If $P(t_\gamma) = 0.99$ than $[t_{cut.}] = 3.99$ min:

$$\left[S_{\min pasp.}\right] = 490 / 3.99 = 122.8 \approx 123\,\text{mm} / \text{min},$$

$$\left[n_{pasp}\right] = 123 / 0.1 \cdot 14 = 87.86 \approx 88\,\text{min}^{-1},$$

$$\left[V_{cut\,pasp}\right] = \left(\pi \cdot 250 \cdot 88\right) / 1000 = 69.12\,\text{m} / \text{min}.$$

The limit conditions are presented as follows:

$$[S_{\min pasp} \geq \left[S_{\min pasp}\right]; n_{pasp.} \geq \left[n_{pasp}\right]; \left[V_{cut\,pasp.}\right] \geq \left[V_{cut\,pasp.}\right]. \quad (23)$$

$$S_{\min pasp.} = 200\,\text{mm}/\text{min} > \left[S_{\min pasp}\right] = 123\,\text{mm}/\text{min},$$

$$n_{pasp.} = 160\,\text{min}^{-1} > \left[n_{pasp}\right] = 88\,\text{min}^{-1},$$

$$V_{cut\,pasp.} = 125.66\,\text{m}/\text{min} > \left[V_{cut\,pasp}\right] = 69.12\,\text{m}/\text{min}.$$

Since all conditions related to (23) are met, machining by milling ensures the parameters of the product using the regulated cutting modes from the reliability standpoint of the components of the process medium for 005 Vertical milling at machining of the base OSHPV 25.001.

4 Conclusions

Reliability engineering requires the development of new approaches and techniques for the comprehensive solution of tasks at essential stages and stages in the life cycles of mechanical engineering products in the era of smart technologies and defect-free production. The systemic nature of such approaches lies in forming interconnected algorithms related to technological chains in the research media of physics-statistics-engineering during the analysis of physical processes at parts machining and assembly of machines.

The mathematical apparatus of Markov chains is proposed to be used at the design stage and technological production preparation to analyze the processes at parts machining. The influence of the elements of the technological system: metal-cutting machine- fixture (clamping device)- metal-cutting tool (s)-workpiece (part), and the optimization of cutting modes are realized using the developed Markov chain technique according to step-by-step algorithms at parts machining by cutting. In particular, the technological assurance of normative workpiece quality at the operation 005 Vertical milling during machining of the base OSHPV 25.001 is achieved according to the gamma-percentile operation time to failure $[P(t_\gamma)] = 0.99$, while the total probability of failure to achieve quality parameters due to failures of elements of the process medium is $6.15 \cdot 10^{-3}$.

Further research in this direction will include the analysis of the reliability metrics with different operating times of the elements of certain technological mediums: metal-cutting machines, clamping devices (fixtures), and metal-cutting tools; realization of the experimental studies to check the theoretical and modelling results; comparison with the results using Bayesian networks, Monte Carlo simulations, AI-driven predictive maintenance, etc.; and adaptation the obtained mathematical dependencies to the operation stage of products life cycles.

Acknowledgments The research was partially supported by the International Association for Technological Development and Innovations. This work was supported by the projects VEGA 1/0391/22, KEGA 017TUKE-4/2023 granted by the Ministry of Education, Research, Development and Youth of the Slovak Republic. This publication is the result of the Project implementation:

Development of excellent research capacities in the field of additive technologies for the Industry of the twenty-first century, ITMS: 313011BWN5, supported by the Operational Program Integrated Infrastructure funded by the ERDF. This work was supported by the Slovak Research and Development Agency under the Contract no. APVV-21-0228.

References

Akhilesh KB, Möller D (2020) Smart technologies. Scope and applications, 1st edn. Springer Nature Singapore Pte Ltd. https://doi.org/10.1007/978-981-13-7139-4

Berladir K, Hovorun T, Trojanowska J, Ivanov V, Iakovets A (2024) Failure analytics of defects in mechanical engineering parts using root cause analysis: case study. In: Trojanowska J, Kujawińska A, Pavlenko I, Husar J (eds) Advances in manufacturing IV. MANUFACTURING 2024, Lecture notes in mechanical engineering. Springer, Cham. https://doi.org/10.1007/978-3-031-56444-4_26

Borucka A, Kozłowski E, Parczewski R, Antosz K, Gil L, Pieniak D (2023) Supply sequence modelling using hidden markov models. Appl Sci (Switz) 13(1):231. https://doi.org/10.3390/app13010231

Davim JP (ed) (2017) Sustainable machining. Materials forming, machining and tribology, 1st edn. Springer, Cham. https://doi.org/10.1007/978-3-319-51961-6

Dehtiarov I, Ivanov V, Evtuhov A, Ostroha R, Ciszak O (2024) Research of static stiffness of flexible fixtures for Y-type fork parts. In: Ivanov V, Trojanowska J, Pavlenko I, Rauch E, Piteľ J (eds) Advances in design, simulation and manufacturing VII. DSMIE 2024, Lecture Notes in Mechanical Engineering. Springer, Cham. https://doi.org/10.1007/978-3-031-61797-3_14

Dong M, Nassif AB (2019) Combining modified Weibull distribution models for power system reliability forecast. IEEE Trans Power Syst 34(2):8506442. https://doi.org/10.1109/TPWRS.2018.2877743

Friederich J, Lazarova-Molnar S (2024) Reliability assessment of manufacturing systems: a comprehensive overview, challenges and opportunities. J Manuf Syst 72:38–58. https://doi.org/10.1016/j.jmsy.2023.11.001

Garg H, Mangey Ram M (2022) Reliability management and engineering. Challenges and future trends, 1st edn. CRC Press, Boca Raton, FL. https://doi.org/10.1201/9780429268922

Grzesik W (2016) Advanced machining processes of metallic materials: theory, modelling, and applications, 2nd edn. Elsevier

Ivanov V, Zajac J (2017) Flexible fixtures for CNC machining centers in multiproduct manufacturing. EAI Endor Trans Indus Netw Intell Syst 4(12):e4. https://doi.org/10.4108/10.4108/eai.10-1-2018.153552

Kozłowski E, Borucka A, Oleszczuk P, Jałowiec T (2023) Evaluation of the maintenance system readiness using the semi-Markov model taking into account hidden factors. Eksploat Niezawodn--Mainten Reliab 25(4):172857. https://doi.org/10.17531/ein/172857

Kusyi Y, Stupnytskyy V, Onysko O, Dragašius E, Baskutis S, Chatys R (2022) Optimization synthesis of technological parameters during manufacturing of the parts. Eksploatacja i Niezawodnosc-Maint Reliab 24(4):655–667. https://doi.org/10.17531/ein.2022.4.6

Kusyi Y, Kostiuk O, Kuk A, Attanasio A, Cocca P (2023) Optimization of cutting modes during sustainable machining of products based on economic criteria. In: Tonkonogyi V, Ivanov V, Trojanowska J, Oborskyi G, Pavlenko I (eds) Advanced manufacturing processes V. InterPartner 2023, Lecture notes in mechanical engineering. Springer, Cham. https://doi.org/10.1007/978-3-031-42778-7_16

Liao Z, la Monaca A, Murray J, Speidel A, Ushmaev D, Clare A, Axinte D, M'Saoubi R (2021) Surface integrity in metal machining - Part I: Fundamentals of surface characteristics and

formation mechanisms. Int J Mach Tools Manuf 162:103687. https://doi.org/10.1016/j.ijmachtools.2020.103687
Melchers RE, Beck AT (2018) Structural reliability analysis and prediction, 1st edn. Wiley. https://doi.org/10.1002/9781119266105
Modarres M, Groth K (2023) Reliability and risk analysis, 2nd edn, Boca Raton, FL. https://doi.org/10.1201/9781003307495
Psarommatis F, Azamfireic V (2024) Zero defect manufacturing: a complete guide for advanced and sustainable quality management. J Manuf Syst 77:764–779. https://doi.org/10.1016/j.jmsy.2024.10.022
Psarommatis F, Bravos G (2022) A holistic approach for achieving sustainable manufacturing using zero defect manufacturing: a conceptual framework. Proc CIRP 107:107–112. https://doi.org/10.1016/j.procir.2022.04.018
Psarommatis F, May G, Dreyfus P-A, Kiritsis D (2020) Zero defect manufacturing: state-of- the-art review, shortcomings and future directions in research. Int J Prod Res 58(1):1–17. https://doi.org/10.1080/00207543.2019.1605228
Rigdon SE, Pan R, Montgomery DC, Freeman LJ (2022) Design of experiments for reliability achievement, 1st edn. Wiley, Hoboken, NJ. https://doi.org/10.1002/9781119237754
Romero D, Gaiardelli P, Powell D, Wuest T, Thürer M (2019) Total quality management and quality circles in the digital lean manufacturing world. In: Ameri F, Stecke KE, von Cieminski G, Kiritsis D (eds) Advances in production management systems. production management for the factory of the future. Springer, Cham, pp 3–11. https://doi.org/10.1007/978-3-030-30000-5_1
Selech J, Matijosius J, Kilikevicius A, Marinkovic D (2025) Reliability analysis and the costs of corrective maintenance for a component of a fleet of trams. Tehn Vjesn 32(1):205–216. https://doi.org/10.17559/TV-20241121002144
Tran KP (2024) Artificial intelligence for safety and reliability engineering. Methods, applications, and challenges, 1st edn. Springer, Cham. https://doi.org/10.1007/978-3-031-71495-5
Wanga W, Liua W, Fanga Y, Zhengb Y, Linc C, Jiangd Y, Liuc D (2024) Reliability analysis of subway sliding plug doors based on improved FMECA and Weibull distribution. Eksploatacja i Niezawodnosc—Maint Reliab 26(2):178275. https://doi.org/10.17531/ein/178275
Yakovyna V, Symets I (2021) Reliability assessment of CubeSat nanosatellites flight software by high-order Markov chains. Proc Comput Sci 192:447–456. https://doi.org/10.1016/j.procs.2021.08.046
Yakovyna V, Seniv M, Symets I, Sambir N (2020) Algorithms and software suite for reliability assessment of complex technical systems. Radio Electron Comput Sci Contr (4):163–177. https://doi.org/10.15588/1607-3274-2020-4-16
Zhao, Y.-G., Lu, Z.-H.: Structural reliability: approaches from perspectives of statistical moments. 1, Wiley-Blackwell, Hoboken, NJ (2021). https://doi.org/10.1002/9781119620754

Computational Methods for Predicting Material Properties in Industry 4.0

Khrystyna Berladir

Abstract This chapter provides an overview of modern computational methods employed in materials science for predicting the properties of multicomponent materials, addressing the increasing demand for tailored materials in Industry 4.0 and the emerging Industry 5.0. It highlights the limitations of traditional experimental techniques and positions computational modeling as a crucial tool for accelerating materials discovery and reducing development costs. The chapter reviews several key computational methods, including Density Functional Theory for electronic structure calculations, Molecular Dynamics simulations for atomic motion analysis, Finite Element Modeling for stress and deformation analysis, Phase-Field Modeling for simulating phase boundary evolution, Monte Carlo Simulation for modeling systems with uncertainty, Machine Learning for identifying patterns in materials data, Computational Fluid Dynamics for fluid flow analysis and Peridynamics for failure modeling. The chapter outlines each method's fundamental principles, applications, and frequently utilized software packages. Although there are many methods, none of them is universal. The choice of methods depends on the specific task, the availability of data, and the requirements for the accuracy of the forecast. This chapter also highlights a practical experience of applying FEM for predicting composite materials' properties, which are complex multicomponent systems, and this is why many factors must be considered to predict their properties. While the study successfully modeled composite behavior, some limitations include the ideal contact assumption, friction considerations, and alternative filler shapes.

Keywords Industry 4.0 · Density functional theory · Molecular dynamics · Finite element modeling · Phase-field modeling · Monte Carlo simulation · Computational fluid dynamics · Machine learning · Peridynamics · Multicomponent materials · Process innovation

---

K. Berladir (✉)
Sumy State University, Sumy, Ukraine

Technical University of Košice, Presov, Slovak Republic

V. Ivanov (ed.), *Emerging Technologies in Manufacturing and Materials Engineering*, Springer Tracts in Mechanical Engineering,
https://doi.org/10.1007/978-3-032-14914-5_17

1 Introduction

Materials science is coming to the forefront more than ever in the modern world. It plays a key role in scientific and applied sciences, as it directly ensures the development of innovations in industry, energy, medicine, aerospace technologies, electronics, etc. The properties of materials determine the efficiency of various structures, the durability of products, and their environmental safety (Jasiulewicz-Kaczmarek et al. 2023). At the same time, the process of creating new materials is complex and resource-intensive, which makes the issue of predicting their properties particularly relevant in the context of the already existing Industry 4.0 and the emerging digital era of Industry 5.0 (Adel 2022).

Typically, the most effective modern materials are multicomponent systems. It means that they consist of two or more components that may have different chemical compositions and physical properties and may interact with each other to form new physical or chemical properties that differ from those of the original components (Cantor 2024). For example, these are high-entropy alloys (Buranich et al. 2020), multiphase composites (Christian et al. 2019), nanomaterials (Budnik et al. 2015), polymer blends (Górski et al. 2018), etc. Industry increasingly requires materials with clearly defined characteristics (Dzyubyk et al. 2024). It requires accurate prediction methods to determine material structure and properties at the calculation and modeling stage, including preventing material destruction (Usov et al. 2025). Given the composition and the possibility of microstructural interaction of multicomponent materials (Loskutova et al. 2025), classical approaches to predicting properties cannot always consider specific nuances affecting the quality of the prediction results. Traditional experimental methods, such as physicochemical and mechanical tests, require significant financial and time resources, which are always insufficient. In addition, the development of new materials can take years due to the need for numerous tests, which limit the speed of innovation (Stach et al. 2021).

In the era of Industry 4.0 and future Industry 5.0, with the rapid growth of computing power and its accessibility to every scientist, as well as the development of artificial intelligence, we have a chance to create new effective forecasting methods that surpass traditional empirical models (Raja Santhi and Muthuswamy 2023). Using numerical modeling (Krzywanski et al. 2024) and machine learning approaches (Stergiou et al. 2023) can allow accurate predictions without conducting expensive and time-consuming experiments (Kibrete et al. 2023).

Thus, combining modern computer technologies, artificial intelligence, and traditional methodology will significantly improve the accuracy of predicting the properties of multicomponent materials, reduce the time and costs of their development, and ensure an innovative breakthrough in materials science.

2 Fundamentals and Challenges

Predicting the properties of multicomponent materials is a central challenge in materials science within Industry 4.0 and future Industry 5.0, and is vital for developing new functional materials. Modern approaches include empirical methods based on experimental data and computational modeling, which utilize machine learning and quantum mechanical calculations.

2.1 Brief Overview of Computational Methods Applied in Materials Science

Computational methods are key to exploring and understanding materials in materials science. They allow researchers to model and predict material structure, properties, and behavior under diverse conditions. The classification of computational methods widely used in materials science is presented in Fig. 1. These modern approaches, including finite element methods, molecular dynamics, ab initio calculations, and machine learning, provide a multiscale view of materials (Bordonhos

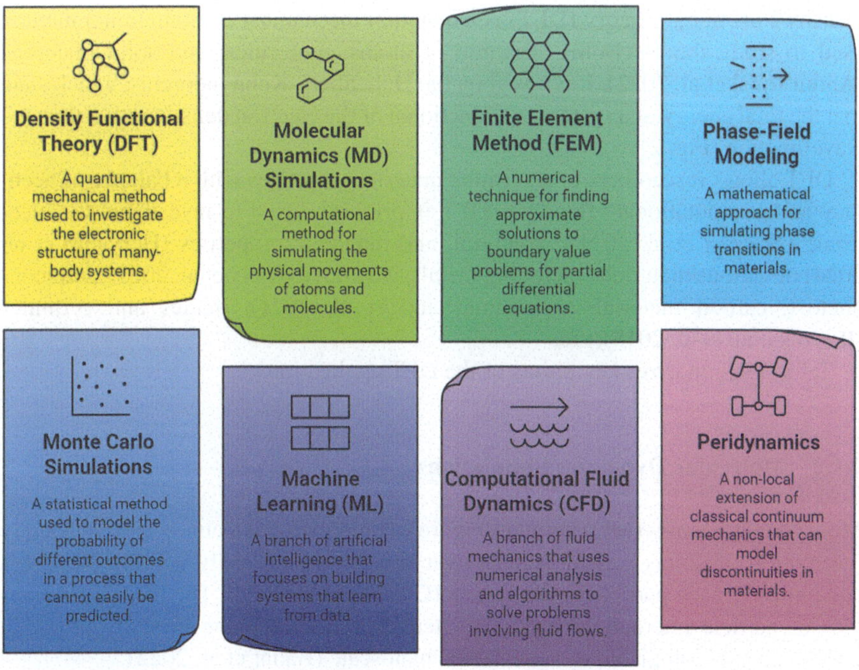

Fig. 1 Modern computational techniques for predicting the materials properties, behavior, and performance

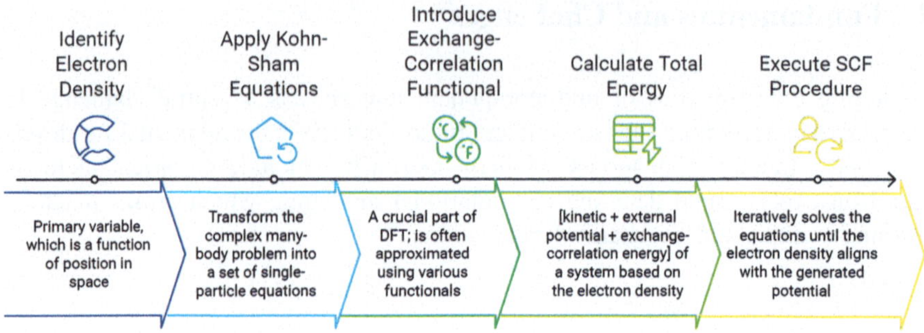

Fig. 2 Key steps of the Density Functional Theory algorithm

et al. 2023). These methods enable the investigation of processes that cannot be accessed through direct experimentation, thereby speeding up the development of new materials with specific properties. However, their implementation comes with several challenges, such as computational complexity, the necessity for careful model selection, and limited experimental validation (Singh et al. 2022).

2.1.1 Density Functional Theory

Density Functional Theory (DFT) is a quantum mechanical computational method used to study the electronic structure of atoms, molecules, and solid materials (Aamir Iqbal et al. 2022). It is based on the Hohenberg-Kohn theorems, which state that the total energy of a system is a functional of the electron density rather than the wavefunction (Fig. 2).

DFT allows researchers to determine material properties with a balance of accuracy and computational efficiency. DFT is primarily used to investigate the electronic (Marzari et al. 2021), structural, and magnetic properties (Heryanto et al. 2024) of various materials, including metals and alloys (Kiely et al. 2021), semiconductors, carbon materials (Qasemnazhand et al. 2021), oxides and ceramics (Rasukkannu et al. 2018), etc.

Table 1 summarizes key points of the DFT application.

2.1.2 Molecular Dynamics Simulations

Molecular Dynamics (MD) simulations are a computational technique used to study the motion of atoms and molecules over time by numerically solving Newton's equations for each particle in the system (Ciccotti et al. 2022). The forces and energies of particle interactions are calculated, which allows one to study physical, chemical, and biological processes at the nanoscale (Badar et al. 2022).

MD simulation algorithm involves initializing atomic positions and velocities, calculating forces based on interatomic potentials, updating atomic velocities and

Table 1 DFT applications and software

Brief description	Research objects	Software
Focuses on the electronic structure of many-body systems (atoms, molecules, solids), primarily examining their ground state properties Simplifies the complex many-body problem by using electron density as the central variable instead of the many-body wavefunction, significantly reducing computational complexity Enables the prediction of material properties and behaviors without empirical parameters, making it invaluable for designing new materials	Metals and alloys Semiconductors Carbon materials Oxides and ceramics Polymers and organic molecules Magnetic materials Superconductors	Open-source: Quantum ESPRESSO, ABINIT, SIESTA, GPAW, OpenMX Commercial: VASP, CASTEP, DMol^3, WIEN2k, ONETEP, ADF

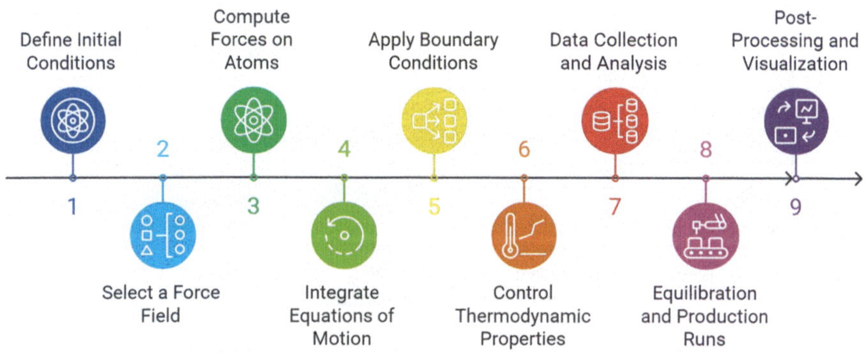

Fig. 3 Main steps of the Molecular Dynamics Simulation algorithm

positions using numerical integration, and iterating these steps over time to study the system's behavior (Fig. 3). MD applied in biochemistry, materials science, and pharmaceuticals to study protein structure, material properties, and drug development for different research objects (Garduño-Juárez et al. 2024; AlRawashdeh and Barakat 2024).

A brief summary of the MD approach is presented in Table 2.

2.1.3 Finite Element Modeling (FEM)

Finite Element Modeling (FEM) is a computational technique based on breaking down a complex geometric model into small, simple elements (finite elements) for which the equations of mechanics are solved numerically (Ashcroft and Mubashar 2018). The results for all elements are combined to obtain an approximate solution for the entire system, allowing analysis of stresses, deformations, heat transfer, etc. FEM methodology involves defining the geometry, discretizing the domain into elements, assigning material properties (Povstyanoy et al. 2024), applying boundary conditions and loads, formulating and assembling the element equations, solving for nodal displacements, and analyzing the results through postprocessing for

Table 2 MD Simulation applications and software

Brief description	Research objects	Software
Used to study the motion of atoms and molecules over time by numerically solving newton's equations for each particle in the system The forces and energies of particle interactions are calculated, which allows one to study physical, chemical, and biological processes at the nanoscale Applied in biochemistry, materials science, and pharmaceuticals to study protein structure, material properties, and drug development	Biomolecules Metals and alloys Polymers Ceramics and inorganic materials Carbon nanomaterials Semiconductors Liquids and solutions	Open-source: GROMACS, LAMMPS, AMBER, NAMD, OpenMM, HOOMD-blue, ESPResSo, GENESIS Commercial: Desmond (by Schrödinger), materials studio, MOE, CHARMM, accelerated MD, MedeA

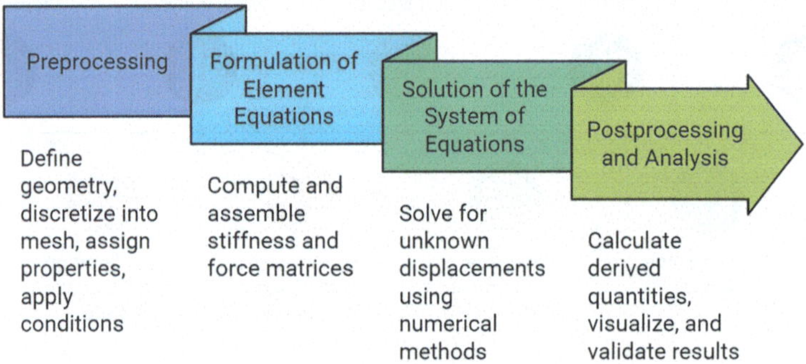

Fig. 4 Algorithm of Finite Element Modeling procedures

further interpretation and validation (Fig. 4). It is widely used in engineering (Le et al. 2025), materials science (Povstyanoy et al. 2021), and biomechanics (Atif et al. 2024) to model the behavior of structures and materials under external loads.

Table 3 summarizes key points of the FEM application.

2.1.4 Phase-Field Modeling

Phase-Field Modeling (PFM) is a numerical method for modeling the evolution of phase boundaries in complex systems without explicit interface tracking. It uses a continuous scalar field that smoothly changes between different phases, allowing processes such as crystallization, grain growth, and material failure to be described (Galfré et al. 2023). The methodology of PFM includes defining the system's initial phase distribution, formulating the governing equations for phase evolution, discretizing the domain, solving the equations using numerical methods, and postprocessing the results to analyze the dynamics of phase transitions and material behaviors (Fig. 5).

Table 3 FEM applications and software

Brief description	Research objects	Software
Based on breaking down a complex geometric model into small, simple elements (finite elements) for which the equations of mechanics are solved numerically The results for all elements are combined to obtain an approximate solution for the entire system, allowing analysis of stresses, deformations, heat transfer, etc. Widely used in engineering, materials science, and biomechanics to model the behavior of structures and materials under external loads	Metals and alloys Polymers Composites Ceramics and glass Concrete and building materials Biological materials and fabrics Porous and foam materials Carbon nanomaterials	Open-source: CalculiX, Code_Aster, Elmer FEM, OpenFOAM, FEniCS, FreeFEM, Kratos Multiphysics, Z88, FEBio, MOOSE Commercial: ANSYS, Abaqus, COMSOL Multiphysics, MSC Nastran & Patran, Altair HyperWorks, Siemens Simcenter, Autodesk fusion 360, Dassault Systèmes CATIA, LS-DYNA, Midas NFX & Midas GTS NX

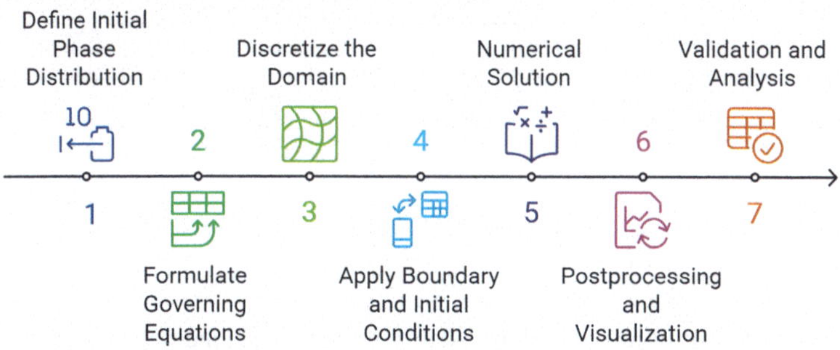

Fig. 5 The key steps of realizing the Phase-Field Modeling

Due to the efficient modeling of microstructural changes and phase boundary dynamics, PFM is an ideal tool for studying and optimizing materials science (Tourret et al. 2022) and engineering processes (Liu et al. 2024) of various materials. A brief summary of the PFM approach is presented in Table 4.

2.1.5 Monte Carlo Simulation

Monte Carlo Simulation is a computational technique used to model and analyze systems with inherent uncertainty using random sampling and statistical methods to simulate various possible outcomes (Hung 2024). The Monte Carlo Simulation algorithm defines the problem and input probability distributions, generates random samples, performs repeated simulations using a mathematical model, analyzes the results statistically, and interprets the outcomes to estimate probabilities and uncertainties (Fig. 6).

Table 4 PFM applications and software

Brief description	Research objects	Software
A numerical method for modeling the evolution of phase boundaries in complex systems without explicit interface tracking It uses a continuous scalar field that smoothly changes between different phases, allowing processes such as crystallization, grain growth, and material failure to be described Efficiently modeling microstructural changes and phase boundary dynamics, making it an ideal tool for studying and optimizing materials science and engineering processes	Metals and alloys Ceramics and glass Polymers Composites Biomaterials Geological materials Liquid and multi-component systems	Open-source: MOOSE, FiPy, PhaseField, MP-PM, OpenPhase, ALLOY Commercial: COMSOL Multiphysics, Abaqus, ANSYS, X-FEM, JMatPro, material studio

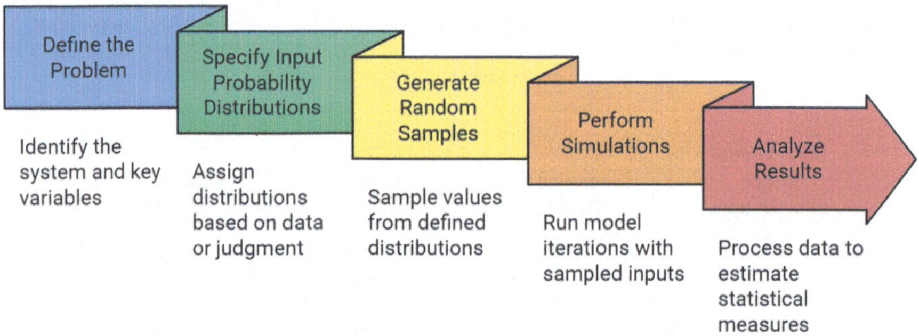

Fig. 6 Algorithm of Monte Carlo Simulation procedures

This type of simulation is widely used in materials science to model thermodynamic properties (Liu et al. 2021a), phase transitions, and atomic-scale lattice configurations (Ding et al. 2024), enabling the prediction of materials behavior under varying conditions with probabilistic accuracy. Table 5 summarizes key points of the Monte Carlo Simulation application.

2.1.6 Machine Learning

Machine learning (ML) is a pivotal subfield of artificial intelligence that enables systems to learn from data and improve their performance over time without being explicitly programmed (Taye 2023). ML algorithms can identify patterns, make predictions, and automate decision-making processes across various domains (Das et al. 2015).

A Machine Learning algorithm involves defining a problem, collecting and preprocessing data, selecting and training a model, evaluating its performance, tuning hyperparameters, deploying the model, and continuously improving it through monitoring and retraining (Fig. 7). It is usually used to predict properties such as

Table 5 Monte Carlo Simulation applications and software

Brief description	Research objects	Software
Using random numbers to model and analyze complex systems where an exact solution is impossible, making it suitable for study at both macro and micro levels Repeated random selection of input data (or parameters) and statistical evaluation of the results to obtain an approximate solution or distribution of possible outcomes Widely used in various fields, including physics and engineering, to assess uncertainty and predict probabilistic events	Metals and alloys Polymers Composites Ceramics and glass Liquids and gas mixtures Nanomaterials Biomaterials Building materials Geological and mineralogical materials	Open-source: PyMC3, SimPy, OpenMC, ROOT, MCSim, GAMBIT Commercial: MATLAB, crystal ball, @ RISK, Simul8, ToxEL, LHS, StatGraphics centurion, AnyLogic

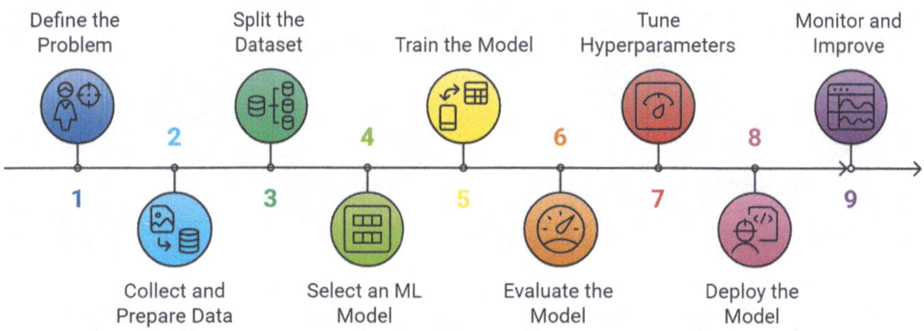

Fig. 7 Main steps of the Machine Learning algorithm

band gaps, mechanical strength (Fang et al. 2022), and phase transitions (Huang et al. 2023), accelerating the discovery (Yildirim and Ceyda Yelgel 2023) and design of novel materials with desired characteristics (Chen et al. 2020; Pai et al. 2025). A brief summary of the ML approach is presented in Table 6.

2.1.7 Computational Fluid Dynamics

Computational Fluid Dynamics (CFD) is a computational technique for analyzing and simulating fluid flow, heat transfer, and related physical phenomena (Soodmand et al. 2023). Its algorithm involves discretizing the fluid domain, formulating governing equations (such as the Navier-Stokes equations), applying boundary conditions, solving the equations numerically using iterative methods, and postprocessing the results to analyze fluid flow (Haider et al. 2024), heat transfer (Wijesooriya et al. 2023), and related phenomena (Fareez et al. 2024) (Fig. 8). Table 7 summarizes key points of the Computational Fluid Dynamics application.

Table 6 Machine learning applications and software

Brief description	Research objects	Software
Use algorithms to analyze large amounts of data and identify patterns without explicit programming instructions Build mathematical models that can predict outcomes, classify objects, or optimize processes by learning from training data Applied in various fields such as pattern recognition, trend prediction, diagnostics, and process automation	Metals and alloys Polymers Composites Nanomaterials Ceramics and glass Biomaterials Electronic and magnetic materials	Open-source: TensorFlow, PyTorch, Scikit-learn, Keras, XGBoost, LightGBM, H2O.Ai, apache spark MLlib, FastAI, Jupyter notebooks Commercial: MATLAB, IBM Watson studio, Google cloud AI platform, Microsoft azure machine learning, Amazon SageMaker, DataRobot, RapidMiner, Alteryx, SAS Viya, TIBCO data science

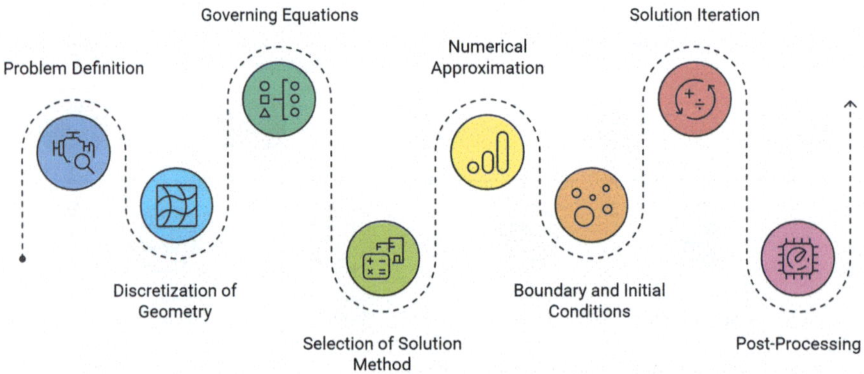

Fig. 8 Sequence of Computational Fluid Dynamics procedures

2.1.8 Peridynamics

Peridynamics is a non-local reformulation of classical continuum mechanics designed to model the behavior of materials with discontinuities, such as fractures and damage evolution (Dorduncu et al. 2024). This algorithm numerically solves material deformation and failure problems by discretizing the domain into material points, computing non-local interactions using integral equations, iterating force and displacement updates, and removing failed bonds to simulate crack propagation (Wang and Yin 2025; Liu et al. 2021b) (Fig. 9). A brief summary of the Peridynamics application is presented in Table 8.

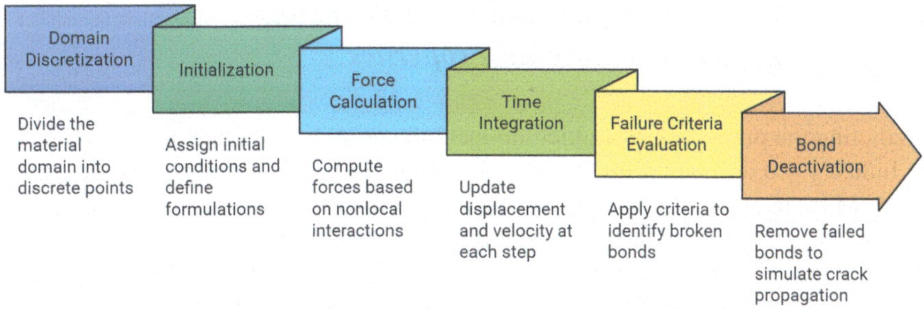

Fig. 9 Sequence of Peridynamics procedures

Table 7 CFD applications and software

Brief description	Research objects	Software
A method of solving equations describing the motion of liquids and gases using numerical methods and algorithms Dividing the flow area into small elements for which velocities, pressures, and other flow characteristics are calculated and then combined to obtain a global picture Used for modeling and analyzing complex physical processes such as aerodynamics, heat transfer, and chemical reactions in various engineering and scientific fields	Liquids and gases Solid-state materials in heat transfer Composites Powder materials and suspensions Nanomaterials Biomaterials Thermal insulation materials Magnetic and electrostatic materials	<u>Open-source:</u> OpenFOAM, SU2, Elmer, CFDLib, DualSPHysics, Gerris flow solver <u>Commercial:</u> ANSYS fluent, COMSOL Multiphysics, Siemens star-CCM+, Autodesk CFD, Bentley Systems, CFX (ANSYS), Altair HyperWorks, Flow3D, MSC software, SimScale

Table 8 Peridynamics applications and software

Brief description	Research objects	Software
A theory of mechanics that uses a "particle interactions at a distance" approach to model the deformation and failure of materials. Considers interactions between points that can be separated by a finite distance. Allows more accurate modeling of phenomena such as crack formation, plastic deformation, and other nonlinear processes in materials.	Metals and alloys Composite materials Polymers and elastomers Geomaterials (soils, concrete, rocks) Biomaterials Glass and ceramics	<u>Open-source:</u> Peridigm, MPM (material point method) and Peridynamics coupling, OpenPeridynamics <u>Commercial:</u> LS-DYNA, Altair HyperWorks, Abaqus (with Peridynamics module), Simulia, Autodesk simulation

2.2 Challenges and Opportunities in Predicting Multicomponent Material Properties

Predicting the properties of multicomponent materials is still challenging in the era of Industry 4.0 despite progress in the field (Pai et al. 2025). The papers (Pai et al. 2025; Muroga et al. 2023) discussed the integration of machine learning into polymer material design, leading to faster innovation cycles, improved predictive capabilities, and a more efficient exploration of material options. However, it also addresses significant challenges: the need for high-quality datasets, the black-box nature of many ML models that limit interpretability, and the necessity of balancing computational efficiency with data accuracy in feature engineering strategies (Pai et al. 2025). The focus on data quality, model interpretability, and the need for broader adoption of ML techniques among experimental researchers, emphase the importance of balancing traditional expertise with advanced computational methods for successful implementation (Muroga et al. 2023). Xue and Hong (2024) developed a materials properties prediction framework that utilizes a graph neural network approach to predict material properties based solely on chemical formulas. This method can solve the limitations of traditional techniques that require detailed crystal structures and specific descriptors. However, the framework faces challenges due to the limited quantity and quality of available data, affecting the model's performance and its inability to differentiate between polymorphs with the same chemical formulas. The machine learning model created in the paper (Zhang et al. 2024) showed a high level of accuracy in forecasting the mechanical properties of high-entropy transition metal carbides. It surpasses earlier models and successfully generalizes across the full range of compositions. However, the challenges include the vast compositional space of multicomponent materials, which complicate experimental exploration and fine-tuning, as well as the resource-intensive nature of ab initio calculations required to predict their mechanical properties effectively. The research (Zhang et al. 2022) was to integrate CALPHAD and density functional theory (DFT) calculations to develop a dataset for predicting the mechanical properties of refractory multicomponent alloys using ML. The main problem was accurately modeling and predicting the properties of alloys, as traditional mathematical models can significantly diverge from experimental data.

Thus, all challenges described above are primarily due to high-dimensional parameter spaces, nonlinear interactions, limited data, the complexity of multiphysics, and limitations in existing theories. The advancement of artificial intelligence and oversized data offers new opportunities but requires improvements in model interpretability and verification of their predictions. The challenge of developing universal approaches capable of considering both thermodynamic and kinetic aspects of material structure formation remains relevant.

3 Ideas and Solutions

3.1 *Finite Element Modeling of Structure-Property Relationships in Polymer Composites*

The study (Berladir et al. 2022a) examined PTFE-based polymer composites reinforced with chaotically distributed fillers to improve mechanical properties. It highlights the significance of optimizing filler concentrations and modeling composite behavior using FEM in ANSYS APDL. The numerical model was developed to predict the strength and load-bearing capacity of these materials, including damage accumulation analysis.

The research methodology followed a structured material preparation, processing, experimental testing, and numerical modeling workflow to ensure accurate property predictions (Fig. 10).

The study determined the optimal filler content for maximizing strength and wear resistance: 20 wt.% coke led to a 9–15 times reduction in wear intensity but caused a 27–44% drop in breaking strength compared to unfilled PTFE; 15 wt.% carbon fiber resulted in a 17–25 times reduction in wear intensity, along with a 12–39% decrease in breaking strength.

The FEM analysis showed that the model accurately predicts stress distribution for spherical inclusions, closely matching experimental values. Cylindrical inclusions can contain stress concentrators, which may require additional modeling refinements, such as cohesive zone modeling to simulate adhesion loss.

Although the predicted models provided a robust framework for designing reinforced composites, several limitations were noted:

- assumption of ideal filler-matrix adhesion—real-world imperfections and potential fiber pull-out effects were not fully modeled.
- frictional effects during pressing were neglected, which could influence residual stress.
- limited exploration of alternative filler distributions, such as fiber orientation control, which might enhance mechanical properties.

The study acknowledges challenges in accurately modeling cylindrical fillers due to stress concentration effects. The assumptions about perfect adhesion between the matrix and inclusions may not fully reflect real material behavior. Further refinements in computational techniques are needed to improve failure prediction accuracy.

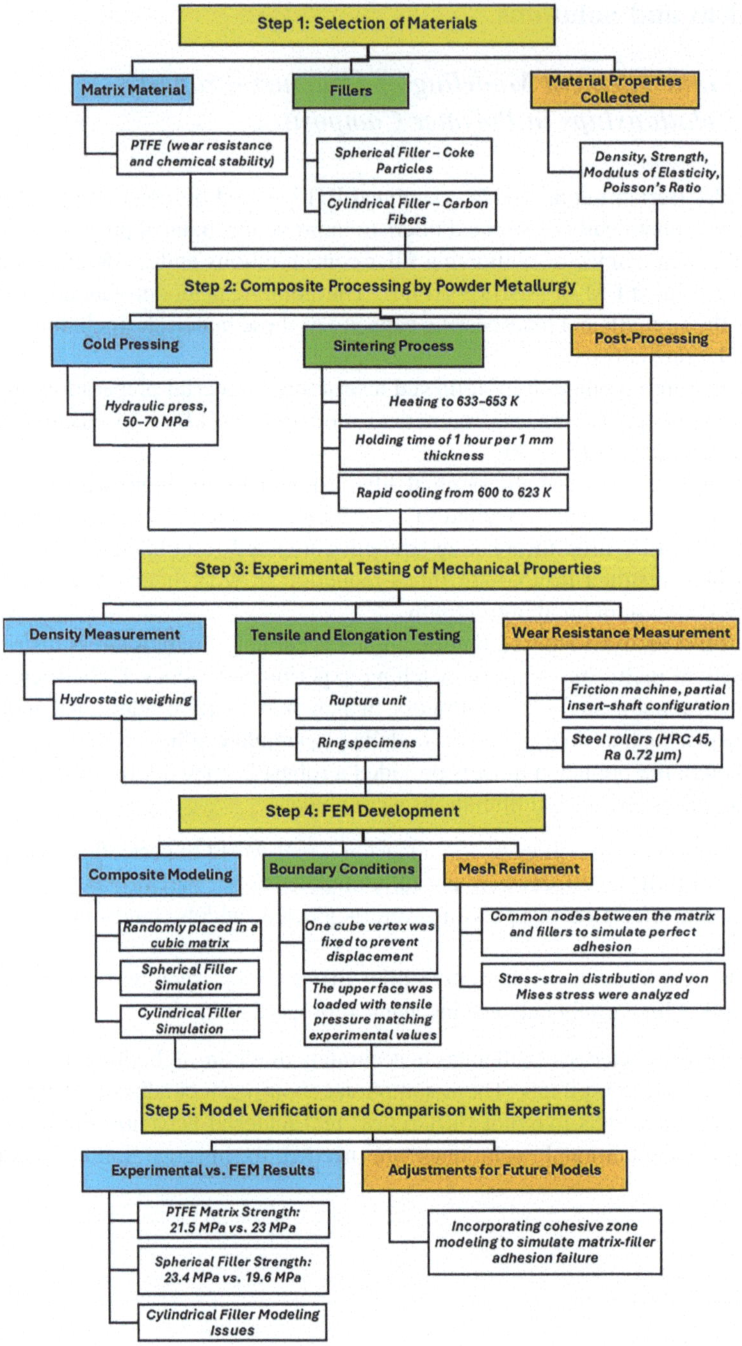

Fig. 10 Step-by-step methodology for modeling polymer composites with randomly distributed fillers

3.2 Simulation-Driven Prediction of Composite Properties: From Pressing to Sintering

The research (Berladir et al. 2022b) employs computer simulation techniques to investigate the mechanical behavior of polymer composite materials during pressing and sintering. The simulations were performed using the ANSYS Workbench software suite, which facilitates precise modeling of the pressing and sintering processes, offering insights into strain capacity and structural alterations in composite materials. The methodology in Fig. 11 provided a replicable framework for predicting composite material behavior, offering insights into optimizing processing conditions for enhanced mechanical performance. The authors used a micromechanical approach to simulate composite materials reinforced with spherical and short cylindrical fillers to construct the models. The finite element models were built and analyzed in ANSYS Meshing using the Patch Conforming algorithm, ensuring high-quality mesh generation, which is crucial for accurately predicting stress-strain relationships. The Weak Springs simulation technology was employed to stabilize structures and model interactions between fillers and the polymer matrix.

The experimental results were validated by comparing simulated stress-strain curves with mechanical tests. The study found that finite element simulations could accurately predict the ultimate strength, Poisson's ratio, and strain capacity of the tested polymer composites.

The first simulation stage focused on micromechanical modeling to determine material properties, while the second stage applied macromechanical simulations to assess the effect of mechanical processing before sintering. The obtained finite element simulations predicted the ultimate strength within a 6.02% error margin for PTFE + coke and 0.33% for PTFE + kaolin, demonstrating high reliability.

The general strain capacity of PTFE + coke increased by 1.3 times, while PTFE + kaolin showed a 15.4-fold improvement, suggesting significant polymerization effects post-sintering.

While the study successfully modeled composite behavior, some limitations include the ideal contact assumption, friction considerations, and alternative filler shapes. The simulation assumes perfect adhesion between the matrix and fillers, which may not reflect real-world conditions. Future studies should incorporate cohesive zone modeling to account for imperfect bonding. The model also does not include friction effects between the press mandrel and the composite material, which could affect material deformation and failure predictions. Future research should explore the impact of other filler geometries on material properties.

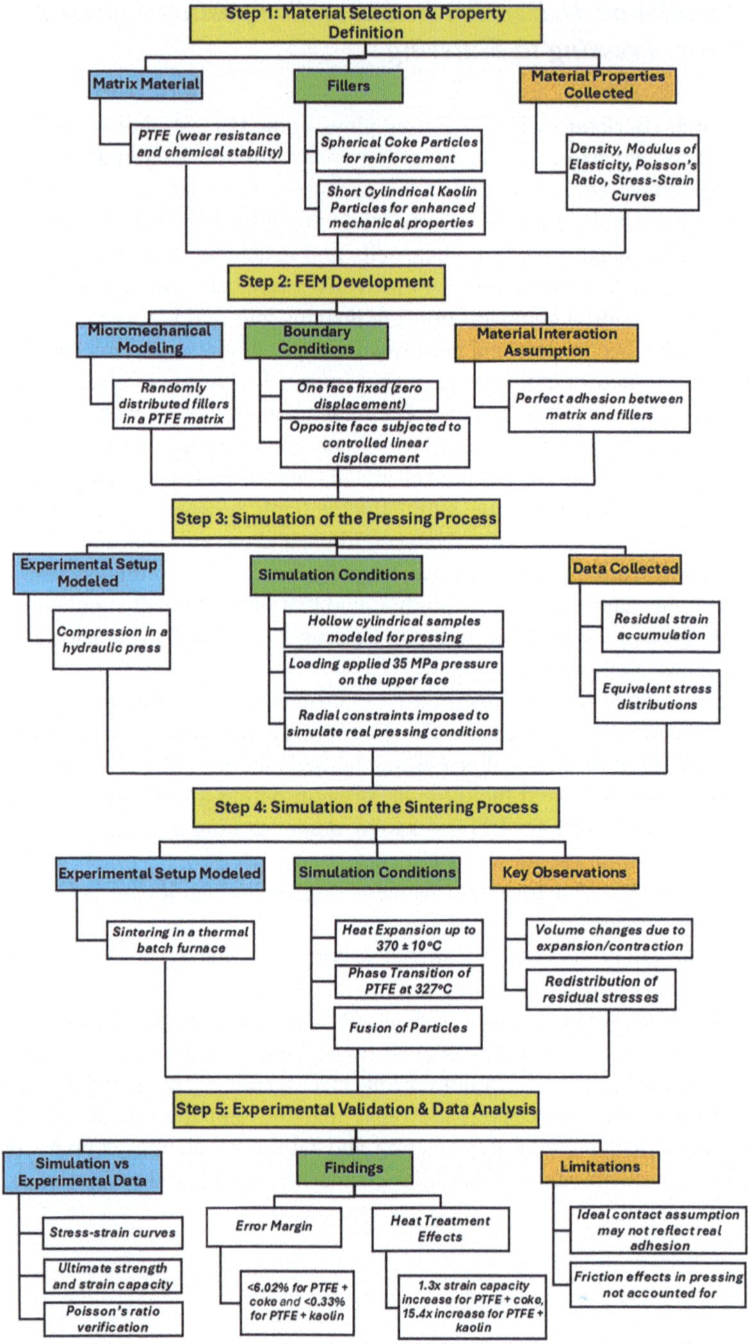

Fig. 11 Proposed methodology for researching and verifying a simulation model of the behavior of a two-component material

3.3 Challenges and Advances in Modeling the Chaotic Structure of Multicomponent Composites

The above-described studies were based on earlier experimental studies, including new computational methods to simulate realistic structure, deformation, and material behavior. In contrast to previous works in this field (Zhyhylii et al. 2025; Mezeix et al. 2023), which focused mainly on layered composites with clear, reproducible structures, these studies integrated micromechanical and macromechanical finite element models to predict material characteristics in terms of contact interactions between components and during processing. Modeling the chaotic structure of multicomponent composites with different fillers is a much more difficult task than laminates due to the random nature of the phase distribution, material heterogeneity, and the need to consider complex contact interactions between components (Table 9).

One of the main problems is the generation of a realistic microstructure, since fillers can have different geometries (spherical, fibrous, lamellar) and variable mechanical properties, which require random filling algorithms, boundary conditions, etc. In this case, particle intersection should be avoided, and a uniform phase distribution should be ensured, which complicates modeling and increases computational costs. Another critical aspect is modeling the contact interaction between phases since fillers may have weak adhesion to the matrix or even detach under load, which affects the actual mechanical behavior of the material. For accurate prediction, it is necessary to consider cohesive zones or use contact models with friction, increasing the calculations' complexity. Due to the chaotic distribution of fillers in the material, zones of local increased stress may arise, which affect the deformation and strength. Since the anisotropy of mechanical characteristics in such composites is not deterministic, its assessment requires significant computational resources and three-dimensional discrete modeling, significantly increasing the analysis's complexity. In contrast to multicomponent composites, laminates possess a clearly defined periodic structure, significantly simplifying their

Table 9 Simulation complexity comparison

Criterion	Multicomponent composites (chaotic structure)	Laminates (layered composites)
Geometric complexity	Complex, requires random placement of fillers	Simple, clearly defined layer boundaries
Contact interaction	Difficult to consider adhesion, sliding between particles is possible	Easily described in classical models
Mechanical heterogeneity	High, affects local stresses and strains	Moderate, depends on the orientation of the layers
Computational costs	High—3D models, large data volumes	Lower, 2D models, simplified plate approaches
Possibility of analytical calculation	Limited, mostly numerical methods	High, laminate theories are used

modeling. They consist of distinct layers with well-defined mechanical properties, allowing their behavior to be studied using analytical methods, such as classical laminate theory. In contrast to chaotic structures, laminates have clearly defined contact conditions between the layers, which facilitates predicting their behavior when subjected to load. In most cases, perfect adhesion between the layers is assumed, and cohesion models are used to account for possible defects. Since the structure of laminates is ordered, complex stochastic modeling is not required to predict their properties, so methods based on calculations of effective mechanical characteristics can be used. In the case of multicomponent chaotic composites, this approach is unsuitable due to the complex distribution of local stresses and the need to model each filler.

4 Conclusions

This study explored various computational approaches to predicting the properties of multicomponent materials, which are crucial for advancing materials science in the era of Industry 4.0 and the transition to Industry 5.0. The discussed methods offer powerful tools for predicting material properties, behavior, and performance, accelerating the design and discovery of novel materials. For example, DFT enables the investigation of electronic, structural, and magnetic properties, while MD simulations allow the study of atomic and molecular motion over time. FEM facilitates the analysis of stresses, deformations, and heat transfer in complex geometries. PFM is effective for modeling microstructural changes and phase boundary dynamics. Monte Carlo Simulation addresses systems with inherent uncertainty through statistical analysis. ML algorithms can identify patterns and predict material properties from large datasets. While each method offers unique advantages, they also present challenges such as computational complexity, careful model selection, and the necessity for experimental validation. Solving the challenges mentioned above is essential for unlocking the complete potential of computational materials science.

Practical experience in using the computational approach in materials science detailed was highlighted in the application of FEM in predicting the structure-property relationships within polymer composites based on PTFE, with a particular focus on materials containing randomly distributed fillers. FEM analysis accurately predicted stress distribution for spherical inclusions. However, stress concentrators around cylindrical inclusions necessitate further modeling refinements, potentially involving cohesive zone modeling to simulate adhesion loss. The study reported error margins of 6.02% for PTFE + coke and 0.33% for PTFE + kaolin when comparing simulated stress-strain curves with experimental data. The simulations also revealed significant changes in strain capacity post-sintering. The general strain capacity of PTFE + coke increased by 1.3 times, while PTFE + kaolin showed a more substantial 15.4-fold improvement, indicating significant polymerization effects. While the study successfully modeled composite behavior, some limitations include the ideal contact assumption, friction considerations, and alternative filler shapes. It is important to note that modeling multicomponent composites with chaotic structures presents significant challenges compared to layered structures. The random nature of phase distribution, material heterogeneity, and complex contact interactions require advanced computational resources and three-dimensional discrete modeling. Unlike laminates, analytical techniques are limited due to the complex distribution of local stresses. Thus, the FEM approach continues to evolve and improve with new technologies. Its prospects include improved model accuracy, integration with artificial intelligence, material composition optimization, and multi-physics approaches. It opens new possibilities for designing and predicting the behavior of multi-component composite materials, which is essential for high-tech industries in the context of Industry 4.0.

The prospect of further research is seen in the development and integration of hybrid approaches that combine different computational methods to create multi-scale models capable of capturing the behavior of materials at various spatial and temporal levels. Further improvement of machine learning models trained on complex experimental and computational data will contribute to increasing the accuracy and efficiency of predicting the properties of materials. The most significant potential for predicting and optimizing the properties of multicomponent materials is the combination of artificial intelligence and computational methods. It is the way to create new functional materials with improved characteristics to meet the requirements of modern industry and promote technological progress.

Acknowledgments The research was partially supported by the International Association for Technological Development and Innovations. The research was also supported by a grant from the Ministry of Education and Science of Ukraine (No. 0125U000447). This work was supported by the projects VEGA 1/0453/24, KEGA 012TUKE-4/2024 granted by the Ministry of Education, Research, Development and Youth of the Slovak Republic. The research was funded by the EU NextGenerationEU through the Recovery and Resilience Plan for Slovakia under projects no. 09I03-03-V01-00101. This work was supported by the Slovak Research and Development Agency under the Contract No. VV-MVP-24-0190.

References

Aamir Iqbal M, Ashraf N, Shahid W, Afzal D, Idrees F, Ahmad R (2022) Fundamentals of density functional theory: recent developments, challenges and future horizons. IntechOpen. https://doi.org/10.5772/intechopen.99019

Adel A (2022) Future of industry 5.0 in society: human-centric solutions, challenges and prospective research areas. J Cloud Comput 11:40. https://doi.org/10.1186/s13677-022-00314-5

AlRawashdeh, S., Barakat, K.H.: Applications of molecular dynamics simulations in drug discovery. In: Gore, M., Jagtap, U.B. (eds.) Computational drug discovery and design. Methods in molecular biology, vol 2714, pp. 127–141. Humana, New York (2024). https://doi.org/10.1007/978-1-0716-3441-7_7

Ashcroft IA, Mubashar A (2018) Numerical approach: finite element analysis. In: da Silva L, Öchsner A, Adams R (eds) Handbook of adhesion technology. Springer, Cham, pp 701–739. https://doi.org/10.1007/978-3-319-55411-2_25

Atif M, Tewari N, Reshikesh M, Chanda A, Mathur VP, Morankar R (2024) Methods and applications of finite element analysis in dental trauma research: a scoping review. Dent Traumatol 40:366–388. https://doi.org/10.1111/edt.12933
Badar MS, Shamsi S, Ahmed J, Alam MA (2022) Molecular dynamics simulations: concept, methods, and applications. In: Rezaei N (ed) Transdisciplinarity. Integrated science, vol 5. Springer, Cham, pp 131–151. https://doi.org/10.1007/978-3-030-94651-7_7
Berladir K, Zhyhylii D, Gaponova O, Krmela J, Krmelová V, Artyukhov A (2022a) Modeling of polymer composite materials chaotically reinforced with spherical and cylindrical inclusions. Polymers 14:2087. https://doi.org/10.3390/polym14102087
Berladir K, Zhyhylii D, Brejcha J, Pozovnyi O, Krmela J, Krmelová V, Artyukhov A (2022b) Computer simulation of composite materials behavior under pressing. Polymers 14:5288. https://doi.org/10.3390/polym14235288
Bordonhos M, Galvão TLP, Gomes JRB et al (2023) Multiscale computational approaches toward the understanding of materials. Adv Theory Simul 6:2200628. https://doi.org/10.1002/adts.202200628
Budnik AF, Rudenko PV, Berladir KV, Budnik OA (2015) Structured Nanoobjects of polytetrafluoroethylene composites. J Nano-Electron Phys 7(2):02022
Buranich V, Rogoz V, Postolnyi B, Pogrebnjak A (2020) Predicting the properties of the refractory high-entropy alloys for additive manufacturing-based fabrication and mechatronic applications. In: 2020 IEEE 10th international conference nanomaterials: applications & properties (NAP). IEEE, pp 1–5. https://doi.org/10.1109/NAP51477.2020.9309720
Cantor B (2024) Exploring multicomponent phase space to discover new materials. J Phase Equilib Diffus 45:188–218. https://doi.org/10.1007/s11669-024-01131-w
Chen A, Zhang X, Zhou Z (2020) Machine learning: accelerating materials development for energy storage and conversion. InfoMat 2:553–576. https://doi.org/10.1002/inf2.12094
Christian W, Dvurecenska K, Amjad K, Przybyla C, Patterson E (2019) Machine vision characterisation of the 3D microstructure of ceramic matrix composites. J Compos Mater 53(16):2285–2296. https://doi.org/10.1177/0021998319826355
Ciccotti G, Dellago C, Ferrario M et al (2022) Molecular simulations: past, present, and future (a topical issue in EPJB). Eur Phys J B 95:3. https://doi.org/10.1140/epjb/s10051-021-00249-x
Das S, Dey A, Pal A, Roy N (2015) Applications of artificial intelligence in machine learning: review and prospect. Int J Comput Appl 115:31–41. https://doi.org/10.5120/20182-2402
Ding, J., Su, J., Tang, H-K., Yu, W.Y. Efficient sampling using macrocanonical Monte Carlo and density of states mapping. Phys Rev Res 6, 043070 (2024). https://doi.org/10.1103/PhysRevResearch.6.043070
Dorduncu M, Ren H, Zhuang X, Silling S, Madenci E, Rabczuk T (2024) A review of peridynamic theory and nonlocal operators along with their computer implementations. Comput Struct 299:107395. https://doi.org/10.1016/j.compstruc.2024.107395
Dzyubyk A, Kusyi Y, Dzyubyk L, Nazar I, Ivanov V (2024) Influence of the welding cycle on the parameters of material damageability of the high-strength steel connection with an austenitic seam. Vibroeng Proc 55:201–207. https://doi.org/10.21595/vp.2024.24483
Fang J, Xie M, He X, Zhang J, Hu J, Chen Y, Yang Y, Jin Q (2022) Machine learning accelerates the materials discovery. Mater Today Commun 33:104900. https://doi.org/10.1016/j.mtcomm.2022.104900
Fareez UNM, Naqvi SAA, Mahmud M, Temirel M (2024) Computational fluid dynamics (CFD) analysis of bioprinting. Adv Healthc Mater 13:2400643. https://doi.org/10.1002/adhm.202400643
Galfré A, Huang X, Cogné C, Couenne F (2023) The phase field method-from fundamentals to practical applications in crystal growth. J Cryst Growth 620:127334. https://doi.org/10.1016/j.jcrysgro.2023.127334
Garduño-Juárez R, Tovar-Anaya DO, Perez-Aguilar JM, Lozano-Aguirre Beltran LF, Zubillaga RA, Alvarez-Perez MA, Villarreal-Ramirez E (2024) Molecular dynamic simulations for

biopolymers with biomedical applications. Polymers 16:1864. https://doi.org/10.3390/polym16131864
Górski F, Kuczko W, Wichniarek R, Hamrol A (2018) Mechanical properties of composite parts manufactured in FDM technology. Rapid Prototyp J 24(8):1281–1287. https://doi.org/10.1108/RPJ-11-2016-0197
Haider R, Li X, Shi W, Lin Z, Xiao Q, Zhao H (2024) Review of computational fluid dynamics in the design of floating offshore wind turbines. Energies 17:4269. https://doi.org/10.3390/en17174269
Heryanto H, Ardiansyah A, Rahmat R, Tahir D (2024) Science mapping analysis of density functional theory (DFT) for material design: a review. JOM 76:4629–4642. https://doi.org/10.1007/s11837-024-06644-w
Huang G, Guo Y, Chen Y, Nie Z (2023) Application of machine learning in material synthesis and property prediction. Materials 16:5977. https://doi.org/10.3390/ma16175977
Hung Y-C (2024) A review of Monte Carlo and quasi-Monte Carlo sampling techniques. WIREs Comput Stat 16(1):e1637. https://doi.org/10.1002/wics.1637
Jasiulewicz-Kaczmarek M, Antosz K, Zhang C, Ivanov V (2023) Industry 4.0 technologies for sustainable asset life cycle management. Sustainability 15:5833. https://doi.org/10.3390/su15075833
Kibrete F, Trzepieciński T, Gebremedhen HS, Woldemichael DE (2023) Artificial intelligence in predicting mechanical properties of composite materials. J Compos Sci 7:364. https://doi.org/10.3390/jcs7090364
Kiely E, Zwane R, Fox R, Reilly AM, Guerin S (2021) Density functional theory predictions of the mechanical properties of crystalline materials. CrystEngComm 23(34):99–110. https://doi.org/10.1039/d1ce00453k
Krzywanski J, Sosnowski M, Grabowska K, Zylka A, Lasek L, Kijo-Kleczkowska A (2024) Advanced computational methods for modeling, prediction and optimization – a review. Materials 17:3521. https://doi.org/10.3390/ma17143521
Le VL, Lee C-H, Alemayehu RW, Park MJ (2025) Finite element modeling of the behaviors of concrete-filled steel tube (CFT) columns at elevated temperatures. Buildings 15(11). https://doi.org/10.3390/buildings15010011
Liu X, Zhang J, Yin J, Bi S, Eisenbach M, Wang Y (2021a) Monte Carlo simulation of order-disorder transition in refractory high entropy alloys: a data-driven approach. Comput Mater Sci 187:110135. https://doi.org/10.1016/j.commatsci.2020.110135
Liu Z, Ye H, Qian D, Zhang H, Zheng Y (2021b) A time-discontinuous peridynamic method for transient problems involving crack propagation. Int J Numer Methods Eng 122:1824–1845. https://doi.org/10.1002/nme.6602
Liu X, Zhu J, He Y, Jia H, Li B, Fang G (2024) State-of-the-art review of the simulation of dynamic recrystallization. Metals 14:1230. https://doi.org/10.3390/met14111230
Loskutova T, Scheffler M, Ivanov V, Krueger M, Kharchenko N, Taran V, Hatala M, Karpets M, Stelmakh Y, Hasemann G (2025) Influence of diffusion siliconizing and boron–siliconizing in chlorine-containing environments on the structure, composition, and protective properties of TZM molybdenum alloy. Appl Surf Sci Adv 27:100769. https://doi.org/10.1016/j.apsadv.2025.100769
Marzari N, Ferretti A, Wolverton C (2021) Electronic-structure methods for materials design. Nat Mater 20:736–749. https://doi.org/10.1038/s41563-021-01013-3
Mezeix L, Rivas AS, Relandeau A, Bouvet C (2023) A new method to predict damage to composite structures using convolutional neural networks. Materials 16:7213. https://doi.org/10.3390/ma16227213
Muroga S, Miki Y, Hata K (2023) A comprehensive and versatile multimodal deep-learning approach for predicting diverse properties of advanced materials. Adv Sci 10(24):2302508. https://doi.org/10.1002/advs.202302508

Pai SM, Shah KA, Sunder S, Albuquerque RQ, Brütting C, Ruckdäschel H (2025) Machine learning applied to the design and optimization of polymeric materials: a review. Next Mater 7:100449. https://doi.org/10.1016/j.nxmate.2024.100449

Povstyanoy O, Mikhailov A, Imbirovich N, Dziubynska O, Herasymchuk H (2021) Simulation permeable porous materials of the complex shape during radial-isostatic compression. In: Tonkonogyi V et al (eds) Advanced manufacturing processes II. InterPartner 2020, Lecture notes in mechanical engineering. Springer, Cham, pp 339–348. https://doi.org/10.1007/978-3-030-68014-5_34

Povstyanoy O, Imbirovich N, Redko R, Redko O, Savaryn P (2024) Numerical evaluation of the properties of highly efficient titanium porous materials. In: Tonkonogyi V, Ivanov V, Trojanowska J, Oborskyi G, Pavlenko I (eds) Advanced manufacturing processes V. InterPartner 2023, Lecture notes in mechanical engineering. Springer, Cham, pp 307–317. https://doi.org/10.1007/978-3-031-42778-7_28

Qasemnazhand M, Khoeini F, Marsusi F (2021) Predicting the new carbon nanocages, fullerenes: a DFT study. Sci Rep 11:2511. https://doi.org/10.1038/s41598-021-82142-2

Raja Santhi A, Muthuswamy P (2023) Industry 5.0 or industry 4.0S? Introduction to industry 4.0 and a peek into the prospective industry 5.0 technologies. Int J Interact Des Manuf 17:947–979. https://doi.org/10.1007/s12008-023-01217-8

Rasukkannu M, Velauthapillai D, Bianchini F, Vajeeston P (2018) Properties of novel non-silicon materials for photovoltaic applications: a first-principle insight. Materials 11:2006. https://doi.org/10.3390/ma11102006

Singh V, Patra S, Murugan NA, Toncua D-C, Tiwari A (2022) Recent trends in computational tools and data-driven modeling for advanced materials. Mater Adv 3:4069–4087. https://doi.org/10.1039/d2ma00067a

Soodmand AM, Azimi B, Nejatbakhsh S et al (2023) A comprehensive review of computational fluid dynamics simulation studies in phase change materials: applications, materials, and geometries. J Therm Anal Calorim 148:10595–10644. https://doi.org/10.1007/s10973-023-12438-0

Stach E, DeCost B, Kusne AG (2021) Autonomous experimentation systems for materials development: a community perspective. Matter 4(9):2702–2726. https://doi.org/10.1016/j.matt.2021.06.036

Stergiou K, Ntakolia C, Varytis P, Koumoulos E, Karlsson A, Moustakidis S (2023) Enhancing property prediction and process optimization in building materials through machine learning: a review. Comput Mater Sci 220:112031. https://doi.org/10.1016/j.commatsci.2023.112031

Taye MM (2023) Understanding of machine learning with deep learning: architectures, workflow, applications and future directions. Computers 12:91. https://doi.org/10.3390/computers12050091

Tourret D, Liu H, LLorca J (2022) Phase-field modeling of microstructure evolution: recent applications, perspectives and challenges. Prog Mater Sci 123:100810. https://doi.org/10.1016/j.pmatsci.2021.100810

Usov A, Ivanov V, Kunitsyn M, Sikirash Y (2025) Influence of stochastically distributed defects on crack formation on grinding surfaces of materials prone to cracking. In: Tonkonogyi V, Ivanov V, Trojanowska J, Oborskyi G (eds) Advanced manufacturing processes VI. Interpartner 2024, Lecture notes in mechanical engineering. Springer, Cham, pp 651–662. https://doi.org/10.1007/978-3-031-82746-4_58

Wang L, Yin Z-Y (2025) Peridynamic modelling of time-dependent behaviour and creep damage in hyper-viscoelastic solids with pre-cracks. Comput Methods Appl Mech Eng 433(Part A):117512. https://doi.org/10.1016/j.cma.2024.117512

Wijesooriya K, Mohotti D, Lee C-K, Mendis P (2023) A technical review of computational fluid dynamics (CFD) applications on wind design of tall buildings and structures: past, present and future. J Build Eng 74:106828. https://doi.org/10.1016/j.jobe.2023.106828

Xue S-D, Hong Q-J (2024) Materials properties prediction (MAPP): empowering the prediction of material properties solely based on chemical formulas. Materials 17(17):4176. https://doi.org/10.3390/ma17174176

Yildirim E, Ceyda Yelgel Ö (2023) Using machine learning techniques to discover novel thermoelectric materials. IntechOpen. https://doi.org/10.5772/intechopen.1003210

Zhang C, Gu X, Bhandari U, Lei J, Guo S, Kourouma M, Karoui A, Yang S (2022) Machine learning prediction of the mechanical properties of refractory multicomponent alloys based on a dataset of phase and first principles simulation. Front Metals Alloys 1. https://doi.org/10.3389/ftmal.2022.1036656

Zhang J, He L, Xiong Y, Huang S, Xu B, Ma S, Xiang X, Fu H, Kai J, Wu Z, Zhao S (2024) Local-distortion-informed exceptional multicomponent transition-metal carbides uncovered by machine learning. NPJ Comput Mater 10(1). https://doi.org/10.1038/s41524-024-01351-1

Zhyhylii D, Pozovnyi O, Chai EY, Dvurecenska K, Christian WJ (2025) Quantified assessment of modelled impacts on asymmetric composite laminates. J Compos Mater 0(0). https://doi.org/10.1177/00219983251319085

The manufacturer's authorised representative in the EU is Springer Nature Customer Service Centre GmbH, Europaplatz 3, 69115 Heidelberg, Germany. If you have any concerns regarding our products, please contact ProductSafety@springernature.com

Printed and bound by CPI Group (UK) Ltd, Croydon, CR0 4YY

21/09/2026

02220012-0011